Metallothioneins in Biochemistry and Pathology

Metallothioneins in Biochemistry and Pathology

edited by

Paolo Zatta

University of Padova, Italy

World Scientific

NEW JERSEY · LONDON · SINGAPORE · BEIJING · SHANGHAI · HONG KONG · TAIPEI · CHENNAI

Published by

World Scientific Publishing Co. Pte. Ltd.
5 Toh Tuck Link, Singapore 596224
USA office: 27 Warren Street, Suite 401-402, Hackensack, NJ 07601
UK office: 57 Shelton Street, Covent Garden, London WC2H 9HE

British Library Cataloguing-in-Publication Data
A catalogue record for this book is available from the British Library.

METALLOTHIONEINS IN BIOCHEMISTRY AND PATHOLOGY

ISBN-13 978-981-277-893-2
ISBN-10 981-277-893-4

Typeset by Stallion Press
Email: enquiries@stallionpress.com

Printed by FuIsland Offset Printing (S) Pte Ltd, Singapore

CONTENTS

LIST OF CONTRIBUTORS

Michael Aschner
Department of Pediatrics
Vanderbilt University Medical Center
1162 21st Avenue South
B-3307 Medical Center North
Nashville, TN 37232-2495
USA

K. C. Balaji
University of Massachusetts Medical School
S4868, 55 Lake Avenue North
Worcester, MA 01655
USA

Boon-Huat Bay
Department of Anatomy
National University of Singapore
Singapore 117597
Singapore

Silvia Bolognin
Department of Pharmacological Sciences
University of Padova, Padova
Italy

Lu Cai
Departments of Medicine and Radiation Oncology
University of Louisville School of Medicine
511 South Floyd Street, MDR 533
Louisville, KY 40202
USA

Roger S. Chung
School of Medicine and Menzies Research Institute
University of Tasmania
Private Bag 58, Hobart
Tasmania 7001
Australia

Renata D'Incà
Department of Surgical and Gastroenterological Sciences
University of Padova
Via Giustiniani 2
35100 Padova
Italy

Samantha J. Fung
Menzies Research Institute
University of Tasmania
Private Bag 58, Hobart
Tasmania 7001
Australia

Robertina Giacconi
Immunology Center
Section Nutrigenomic and Immunoscience
INRCA Research Department, Ancona
Italy

Juan Hidalgo
Department of Cellular Biology, Physiology and Immunology
Animal Physiology Unit
Faculty of Sciences
Autonomous University of Barcelona
Bellaterra, Barcelona
Spain

Srinivasan Dinesh Kumar
Department of Anatomy
Yong Loo Lin School of Medicine
National University of Singapore
4 Medical Drive, MD10
Singapore 117597
Singapore

Yiyang Lai
Department of Anatomy
National University of Singapore
Singapore 117597
Singapore

Marco Malavolta
Immunology Center
Section Nutrigenomic and Immunoscience
INRCA Research Department, Ancona
Italy

Wolfgang Maret
Department of Preventive Medicine and Community Health
The University of Texas Medical Branch
Ewing Hall, 700 Harborside Drive
Galveston, TX 77555-1109
USA

Valentina Medici
Division of Gastroenterology and Hepatology
Department of Internal Medicine
University of California, Davis
4150 V Street, Suite 3500
Sacramento, CA 95817
USA

Gabriele Meloni
Institute of Biochemistry
University of Zürich
Winterthurerstrasse 190
CH-8057 Zürich
Switzerland

Eugenio Mocchegiani
Immunology Center
Section Nutrigenomic and Immunoscience
INRCA Research Department, Ancona
Italy

Narassa Narayani
University of Massachusetts Medical School
S4868, 55 Lake Avenue North
Worcester, MA 01655
USA

Lydia Oliva
Department of Surgical and Gastroenterological Sciences
University of Padova
Via Giustiniani 2
35100 Padova
Italy

Sarah E. Owens
Department of Pediatrics
Vanderbilt University Medical Center
1162 21st Avenue South
B-3307 Medical Center North
Nashville, TN 37232-2495
USA

Daisuke Shirasaka
Department of Clinical Molecular Medicine
Kobe University Graduate School of Medicine, Kobe
Japan

Giacomo Carlo Sturniolo
Department of Surgical and Gastroenterological Sciences
University of Padova
Via Giustiniani 2
35100 Padova
Italy

Marshall L. Summar
Department of Pediatrics
Vanderbilt University Medical Center
1162 21st Avenue South
B-3307 Medical Center North
Nashville, TN 37232-2495
USA

Puay-Hoon Tan
Department of Pathology
Singapore General Hospital
Outram Road, Singapore 169608
Singapore

Stamatios E. Theocharis
Department of Forensic Medicine and Toxicology
School of Medicine
National and Kapodistrian University of Athens
75 Mikras Asias Street, Goudi
Athens 11527
Greece

Milan Vašák
Institute of Biochemistry
University of Zürich
Winterthurerstrasse 190
CH-8057 Zürich
Switzerland

Georg Weinlich
Clinical Department of Dermatology and Venerology
Medical University Innsbruck
Anichstr. 35
A-6020 Innsbruck
Austria

Adrian West
School of Medicine and Menzies Research Institute
University of Tasmania
Private Bag 58, Hobart
Tasmania 7001
Australia

George Wai-Cheong Yip
Department of Anatomy
Yong Loo Lin School of Medicine
National University of Singapore
4 Medical Drive, MD10
Singapore 117597
Singapore

Paolo Zatta
CNR–Institute for Biomedical Technologies
Metalloproteins Unit
Department of Biology
University of Padova, Padova
Italy

Bernhard Zelger
Clinical Department of Dermatology and Venerology
Medical University Innsbruck
Anichstr. 35
A-6020 Innsbruck
Austria

PART I

METALLOTHIONEIN IN NEUROLOGICAL DISORDERS

Chapter 1

METALLOTHIONEIN STRUCTURE AND REACTIVITY

Milan Vašák and Gabriele Meloni

The structure and chemistry of mammalian metallothioneins (MTs) with divalent (Zn^{II}, Cd^{II}) and monovalent (Cu^{I}) metal ions pertinent to their role in biological systems are discussed. In human, four MT isoforms designated MT-1 through MT-4 are found. The characteristic feature of these cysteine- and metal-rich proteins is the presence of two metal-thiolate clusters located in independent protein domains. The structure of these clusters is highly dynamic, allowing a fast metal exchange and metal transfer to modulate the activity and function of zinc-binding proteins. Despite the fact that the protein thiolates are involved in metal binding, they show a high reactivity toward electrophiles and free radicals, leading to cysteine oxidation and/or modification and metal release. The unusual structural properties of MT-3 are responsible for its neuronal growth inhibitory activity, involvement in trafficking of zinc vesicles in the central nervous system (CNS), and protection against copper-mediated toxicity in Alzheimer's disease. MT-1/MT-2 also play a role in cellular resistance against a number of metal-based drugs.

Keywords: Metallothionein; three-dimensional structure; structure dynamics; metal-thiolate cluster; metal-cluster reactivity; zinc; copper.

1. Introduction

Metallothioneins (MTs) are a superfamily of low-molecular-mass cysteine- and metal-rich proteins or polypeptides conserved through evolution and present in all eukaryotes and certain prokaryotes. They were discovered initially in the search for the tissue constituent responsible for the natural accumulation of cadmium in equine and human kidney in 1957 by Margoshes and Vallee (Margoshes and

Vallee, 1957). This protein was subsequently named "metallothionein" to reflect the extremely high thiolate sulfur and metal content, both of the order of 10% (w/w) (Kägi and Vallee, 1960). Although MTs are still the only macromolecules in which cadmium accumulates naturally, this metal is only one of several optional metallic components of this protein, the others being usually zinc and copper. In fact, in subsequent studies, zinc was found to be the most abundant and often sole metallic component in mammalian tissues under normal physiological conditions. Besides Zn^{II}, Cu^{I}, and Cd^{II}, MTs bind *in vitro* and in certain cases also *in vivo* (e.g. Pt^{II} in cancer chemotherapy) a variety of other metal ions such as Co^{II}, Ni^{II}, Fe^{II}, Hg^{II}, Pt^{II}, Au^{I}, Ag^{I}, In^{III}, Sb^{III}, Bi^{III}, As^{III}, and TcO^{III} (Vašák and Romero-Isart, 2005). At present, it is becoming increasingly clear that MTs fulfill different functions in a number of biological processes. In mammalian cells, these include homeostasis and transport of physiologically essential metals (Zn, Cu), metal detoxification (Cd, Hg), protection against oxidative stress, maintenance of intracellular redox balance, regulation of cell proliferation and apoptosis, protection against neuronal injury and degeneration, and regulation of neuronal outgrowth (reviewed in Palmiter, 1998; Maret, 2000; Miles *et al.*, 2000; Hidalgo *et al.*, 2001). The binding of copper to mammalian MTs plays mainly a role in copper sequestration in copper-related diseases such as Menkes and Wilson's diseases (Tapiero *et al.*, 2003). MTs also play a critical role in the chemotherapy of certain cancers, both in the development of tolerance to chemotherapeutics and as an adjunct to reduce toxic side-effects (Cherian *et al.*, 2003; Theocharis *et al.*, 2004).

Differential expression of mammalian MT isoforms is tightly regulated during development and in pathological situations (Miles *et al.*, 2000; Theocharis *et al.*, 2004). In mammals, four distinct MT isoforms exist designated MT-1 through MT-4. They are acidic 6–7-kDa proteins possessing a novel type of metal-thiolate cluster. MT-1 and MT-2 are expressed in almost all tissues. Their biosynthesis is inducible by a variety of stress conditions and compounds including glucocorticoids, cytokines, reactive oxygen species, and metal ions (Miles *et al.*, 2000); whereas MT-3 and MT-4 are relatively unresponsive to these inducers. Although MT-1/MT-2 are cytosolic proteins, during development they have also been detected in the nucleus (Nartey *et al.*, 1987). MT-3, also known as the growth inhibitory factor (GIF), has been discovered as a factor deficient in the brain of Alzheimer's disease patients. The protein occurs intracellularly and extracellularly, and exhibits neuronal growth inhibitory activity in neuronal cell cultures (Uchida *et al.*, 1991). MT-3 expression is primarily confined to the central nervous system (CNS), where it represents a major component of the intracellular Zn^{II} pool in zinc-enriched neurons (ZENs) (Masters *et al.*, 1994). Lower expression levels of MT-3 have also been reported in pancreas, kidney, stomach, heart, salivary glands, organs of the reproductive system, and maternal deciduum (Sogawa *et al.*, 2001; Irie *et al.*, 2004). The expression of the lastly identified mammalian metallothionein isoform, MT-4, has been found restricted to cornified and stratified squamous epithelium (Quaife *et al.*, 1994). The developmentally regulated MT-4 expression in maternal deciduum together with the expression of the entire *MT* gene locus have been reported in mouse (Liang *et al.*, 1996).

2. Gene Structure and Regulation

Based on the classification by Binz and Kägi (1999), the mammalian *MT* gene family belongs to the vertebrate family (family 1), which is characterized by the consensus sequence K-Xaa(1,2)-C-C-Xaa-C-C-P-Xaa(2)-C, where Xaa stands for other amino acids. Within this family, mammalian metallothioneins constitute four different subfamilies designated m1 (MT-1) through m4 (MT-4). In *Homo sapiens*, a significant genetic polymorphism exists. The human *MT* genes are located on chromosome 16q13 and are encoded by a multigene cluster of tightly linked genes. There are at least seven functional *MT-1* genes (*MT-1A, MT-B, MT-E, MT-F, MT-G, MT-H,* and *MT-X*) and a single gene encoding the other MT isoforms MT-2 (*MT-2A*), MT-3, and MT-4 (Cherian *et al.*, 2003). A number of other *MT* or *MT*-like genes and pseudogenes with a significant homology to functional *MT* genes exist within the human genome, but their functionality is so far unknown. The *MT* genes encode for two-domain proteins. The three exons composing the human *MT* genes sequentially encode for the N-terminal region of the β-domain (exon 1), the rest of the β-domain (from residue 11 or 12; exon 2), and all of the α-domain (from residue 31 or 32; exon 3). Exons 2 and 3 are spliced at the junction of codons for the Lys/Lys or Arg/Lys residues in the interdomain region.

The promoter regions of the *MT-1* and *MT-2* genes contain several metal-responsive elements (MREs) and glucocorticoid-responsive elements (GREs) as well as elements involved in basal level transcription (BLEs). MT expression is also regulated by oxidative stress by antioxidant responsive elements (AREs) or by MREs which are also responsive to oxidants. The metal-regulatory transcription factor (MTF-1), which is essential for basal expression and induction by zinc, binds to proximal promoter MREs. MTF-1 binds to MREs through its six C2H2 zinc fingers (Heuchel *et al.*, 1994). The DNA sequences responsible for cell-specific regulation of *MT-3* and *MT-4* genes are currently unknown. The regulation of tissue-specific *MT-3* gene expression does not appear to involve a repressor; thus, other mechanisms such as chromatin organization and epigenetic modifications may account for the presence or absence of *MT-3* transcription.

3. Structure of Mammalian Metallothioneins

3.1. *Metallothionein-1 and Metallothionein-2*

The metal-free protein, also named apo-metallothionein or thionein, possesses a predominantly disordered structure, which renders it vulnerable to proteolysis (Winge and Miklossy, 1982). However, a well-defined structure develops upon metal binding. The structural features of mammalian metallothioneins were forthcoming almost exclusively from spectroscopic studies. The first direct evidence for the existence of metal-thiolate clusters in MTs came from the ^{113}Cd nuclear magnetic resonance (NMR) studies of the reconstituted $^{113}Cd_7$MT-2 (Otvos and Armitage, 1980). The studies revealed that 20 cysteine residues and 7 divalent metal ions are partitioned between two metal-thiolate clusters, a cyclohexane-like three-metal cluster ($M_3^{II}(Cys)_9$) in the N-terminal β-domain (residues 1–30) and an adamantane-related four-metal cluster ($M_4^{II}(Cys)_{11}$) in the C-terminal α-domain (residues 31–61). In these clusters, the metal ions are coordinated by both terminal and μ_2-bridging thiolate ligands (Otvos and Armitage, 1980) in a tetrahedral-type symmetry (Vašák, 1980).

A great deal of knowledge on the chemical features of MT-1/MT-2 pertinent to their function has arisen from the determination of their three-dimensional (3D) structures. The 3D structures of MT-1/MT-2 from various species were obtained mainly by NMR spectroscopy (Arseniev *et al.*, 1988; Schultze *et al.*, 1988; Messerle *et al.*, 1990b; Zangger *et al.*, 1999), but also by X-ray crystallography (Robbins *et al.*, 1991; Braun *et al.*, 1992) (Figs. 1 and 2). The reported 3D structures reveal a similar monomeric two-domain protein of a

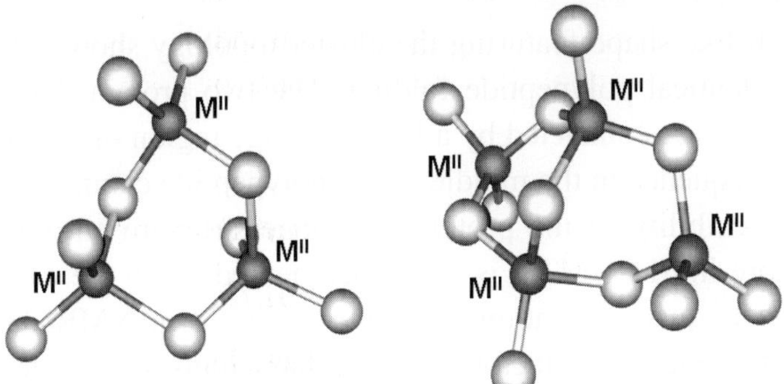

Fig. 1 Schematic drawing of the two metal-thiolate clusters in mammalian Zn_2Cd_5MT-2 (Robbins *et al.*, 1991). The metal atoms (M^{II}) are shown as shaded spheres connected to sulfur atoms. The models were generated with the program PyMOL v0.99 (http://www.delanoscientific.com/) using the Protein Data Bank (PDB) coordinate 4mt2.

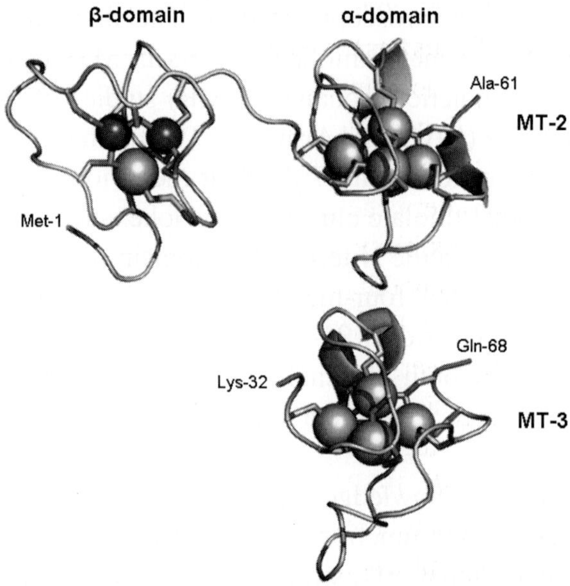

Fig. 2 The three-dimensional crystal structure of rat Zn_2Cd_5MT-2 (top) (Robbins *et al.*, 1991) and the NMR solution structure of the α-domain of human $^{113}Cd_7$MT-3 (bottom) (Wang *et al.*, 2006). The Cd^{2+} and Zn^{2+} ions are shown as shaded spheres connected to the protein backbone by cysteine thiolate ligands. The models were generated with the program PyMOL v0.99 (http://www.delanoscientific.com/) using the PDB coordinates 4mt2 and 2f5h.

dumbbell-like shape featuring the cluster topology shown in Fig. 1 and an identical polypeptide folding. The two protein domains in MT-1/MT-2 are connected by a flexible hinge region of a conserved Lys-Lys sequence in the middle of the polypeptide chain.

The flexibility of the protein backbone structure enfolding the metal core in MT-1/MT-2 is well documented. Both the calculated root mean square deviation (RMSD) values from NMR data and the crystallographic B-factors indicate that a considerable degree of dynamic structural disorder exists (Schultze *et al.*, 1988). Direct evidence for the nonrigid MT structure came from the ^{1}H NMR studies of ^{1}H–^{2}H amide exchange in Cd_7-MT-1/Cd_7-MT-2 (Messerle *et al.*, 1990a; Zangger *et al.*, 1999). In these studies, the enhanced flexibility of the less structurally constrained β-domain compared to the α-domain in both isoforms has been demonstrated. Molecular dynamics simulations of the β-domain of rat liver MT-2 in aqueous solution also show that the polypeptide loops between cysteine ligands exhibit an extraordinary flexibility without disrupting the geometry of the three-metal cluster (Berweger *et al.*, 2000).

Because of the flexibility of the polypeptide loops between cysteine ligands, it was generally assumed that the protein could accommodate a wide range of divalent metal ions of different sizes without any selectivity. However, in inorganic polynuclear adamantane-like cages, both the metal size and the varying length of the metal-thiolate bonds give rise to widely differing cluster sizes and thus cluster volumes (Hagen *et al.*, 1982). As a consequence of this effect, the binding of different metal ions to MT should lead to an expansion of the cluster core, thus influencing the energetics of the folding process. The metal selectivity of MT-1/MT-2 structures has been studied by offering seven equivalents of two divalent metal ions — i.e. Co^{II}/Cd^{II}, Zn^{II}/Cd^{II}, Co^{II}/Zn^{II}, and Fe^{II}/Cd^{II} — in a 3:4 ratio to the apoprotein, followed by the determination of the respective metal distributions within and between the clusters by electronic absorption, magnetic circular dichroism (MCD), ^{113}Cd NMR, ^{57}Fe Mössbauer spectroscopy, and electron paramagnetic resonance (EPR) spectroscopy, as required (Good *et al.*, 1991; Pountney and Vašák, 1992). As a result, offering Co^{II}/Cd^{II} ions resulted in $Co_3^{II}Cd_4MT$-2 in which two homometallic clusters, i.e. the Co_3^{II}-thiolate cluster in the β-domain and the Cd_4-thiolate cluster in the α-domain, were present. Conversely, in the case of Fe^{II}/Cd^{II}, the homometallic species $Fe_7^{II}MT$-1 and Cd_7MT-1 in the ratio of added metal ions were formed. Although Co^{II}/Zn^{II} and Zn^{II}/Cd^{II} gave rise to heterometallic clusters in $Co_3^{II}Zn_4MT$-1 and Zn_3Cd_4MT-2, they showed a distinct metal distribution. Thus, an appreciable selectivity in metal partitioning among and within the clusters is indeed observed in this protein. These results are opposite to those obtained in the studies of inorganic adamantane-like cages with monodentate thiolate ligands of the general formula

$[M_4(SPh)_{10}]^{2-}$ ($M = Cd^{II}, Zn^{II}, Co^{II}, Fe^{II}$) where, despite the differences in the metal-thiolate affinities and in the homometallic cluster volumes, a simple mixing of two homometallic cages always produced heterometallic cage complexes with an almost statistical distribution of both metal ions (Hagen *et al.*, 1982). Since the same metal ions were employed in the studies of MT-1/MT-2, properties of the MT structure are responsible for this effect. It has been concluded that interplay between both the chemistry of metal ions and the steric requirements of the protein structure determines the diversity of metal-thiolate cluster structures in MT-1/MT-2. In this context, it should be noted that small structural differences between Zn_7-MT-2 and Cd_7-MT-2, due to differences in the cluster volumes (approximately 20%), have been observed (Braun *et al.*, 1992).

The coordination of the metal ions is the major determinant in the folding of the polypeptide chain around the two clusters of MT-1/MT-2. The pathway of cluster formation has been determined for Cd_7MT-2 and Co_7MT-2 by ^{113}Cd NMR and EPR/paramagnetic ^{1}H NMR, respectively. Because of the similar size and chemistry of Co^{II} and Zn^{II}, the isostructural substitution of Zn^{II} by Co^{II} is a widely used method to probe the spectroscopically silent zinc sites in proteins. The studies revealed that the formation of both clusters in Cd_7MT-2 is cooperative and sequential, with the four-metal cluster in the α-domain being formed first (Good *et al.*, 1988). However, the formation of Co^{II} clusters in Co_7MT-2, and by inference also Zn^{II} clusters, proceeds by a different pathway; in this case, prior to the cluster formation in the α-domain, the first four Co^{II} are bound in independent Co^{II} sites in tetrathiolate coordination (Vašák and Kägi, 1981; Bertini *et al.*, 1989).

At neutral pH, closely similar average apparent stability constants have been determined for MT-1/MT-2 by different methods, with values of the order of 10^{11} M^{-1} and 10^{14} M^{-1} for Zn^{II} and Cd^{II}, respectively (reviewed in Romero-Isart and Vašák, 2002). The recent development of fluorescent metal chelators with different metal affinities allowed the dissection of average apparent stability constants. Although each of the seven Zn^{II} ions in Zn_7MT-2 is bound in a tetrathiolate coordination environment, analysis of Zn^{II} binding to thionein revealed at least three classes of binding sites with affinities that differ by four orders of magnitude (Krezel and Maret, 2007). One Zn^{II} ion in Zn_7MT-2 is relatively weakly bound ($\log K = 7.7$), making MT a zinc donor. Moreover, it has been suggested that physiological ligands, including thionein, would further modulate the metal binding and the redox reactivity of thiols in the cellular environment. These chemical characteristics suggest how the molecular structures and redox chemistries of fully and partially metallated MT and thionein may contribute to the variety of different functions that MT may serve in the cell.

Copper binds readily to MT *in vivo* and *in vitro* (Kägi and Schäffer, 1988). Both Wilson's and Menkes diseases in human are inborn disorders of copper metabolism in which excess copper accumulates intracellularly in MT. In all instances examined to date, copper is bound to MT as Cu^I. The only available 3D crystal structure of Cu^I-thiolate clusters in MT is that of yeast Cu_8MT. The structure shows the largest known oligonuclear Cu^I_8-thiolate cluster in biomolecules, consisting of six trigonally and two diagonally coordinated Cu^I ions (Calderone *et al.*, 2005). However, the 3D structure of Cu^I containing mammalian MT-1/MT-2 is so far unknown. The structural features of Cu^I-thiolate clusters in mammalian MTs have been studied by various spectroscopic techniques (Stillman, 1995). The current knowledge is limited to the fact that fully Cu^I-loaded MT-1/MT-2 bind 12 Cu^I ions into two metal-thiolate clusters, where, in contrast to divalent metal ions, the monovalent copper ions are coordinated by two or three cysteine ligands forming two independent Cu^I_6-thiolate clusters (Nielson *et al.*, 1985; Stillman, 1995). In these studies, the Zn_7MT-1 form was titrated with Cu^I ions. However, the binding studies of Cu^I to thionein revealed that, at a lower Cu^I/protein stoichiometry, two distinct Cu^I_4-thiolate clusters are formed in both protein domains (Pountney *et al.*, 1994). There is evidence to suggest that Cu^I binds preferably to the less structurally constrained β-domain (Nielson and

Winge, 1984), and Cd^{II} and Zn^{II} to the α-domain (Nielson and Winge, 1983). A recent NMR investigation of both synthetic domains of MT-1 filled with Cu^{I} or Zn^{II} ions revealed significant differences in their respective polypeptide folds. This observation signifies the different coordination geometries required for the binding of monovalent and divalent metal ions (Dolderer *et al.*, 2007).

3.2. *Metallothionein-3*

Although MT-1/MT-2 and MT-3 share a conserved array of 20 cysteine residues, evolutionarily conserved changes in the primary structure of MT-3 exist. These are a T_5 insert followed by a unique C_6-P-C-P_9 sequence within the β-domain, and an acidic hexapeptide insert within the α-domain (Fig. 3).

The structural studies on recombinant Zn_7MT-3 and Cd_7MT-3 established the presence of two mutually interacting protein domains, resembling those reported for MT-1 and MT-2, with each domain encompassing a metal-thiolate cluster (Faller and Vašák, 1997; Bogumil *et al.*, 1998; Hasler *et al.*, 1998; Faller *et al.*, 1999). The presence of a three-metal cluster in the N-terminal β-domain (residues 1–31) and a four-metal cluster in the C-terminal α-domain (residues 32-68) of Zn^{II} and Cd^{II} containing M_7^{II}MT-3 was inferred from the studies of separate α- and β-domains. The metal-binding affinity of MT-3 is weaker than that of MT-1/MT-2, and the metal binding to MT-3 is noncooperative (Hasler *et al.*, 2000; Palumaa *et al.*, 2002). When compared to Cd_7MT-1/Cd_7MT-2, a markedly increased structure flexibility and cluster dynamics exist in Cd_7MT-3. This information was forthcoming from ^{113}Cd NMR studies of

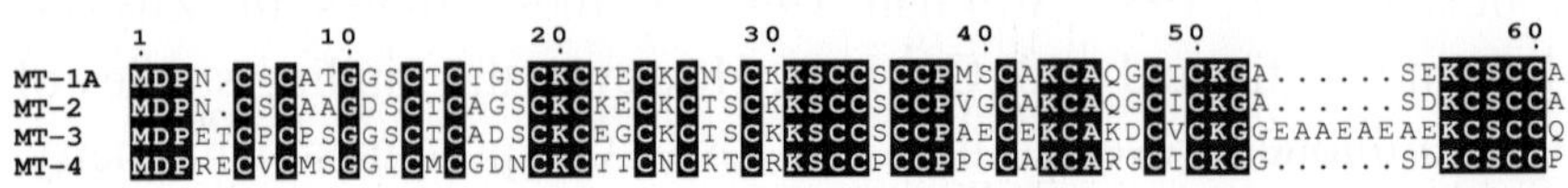

Fig. 3 KALIGN amino acid sequence alignment of four human metallothionein isoforms, with the conserved residues highlighted. The figure was generated with the program ESPript version 2.2.

Cd_7MT-3, in which a significant broadening of all NMR resonances and a very low and temperature-independent intensity of the Cd_3Cys_9 cluster resonances have been interpreted in terms of dynamic processes acting on two different NMR time scales: (1) fast exchange between conformational cluster substates and (2) additional, very slow exchange processes between configurational cluster substates in the β-domain (Faller *et al.*, 1999). The changes in conformational substates may be visualized as minor dynamic fluctuations of the metal coordination environment, and those of the configurational substates as major structural alterations brought about by temporary breaking and reforming of the metal-thiolate bonds. Therefore, only the 3D structure of the α-domain of mouse and human $^{113}Cd_7$MT-3 could be determined by NMR (Fig. 2, bottom). The structure reveals a peptide fold and a cluster organization very similar to those found in Cd_7MT-1/Cd_7MT-2, with the exception of an extended flexible loop encompassing the acidic hexapeptide insert (Öz *et al.*, 2001; Wang *et al.*, 2006).

Since the mutation of conserved proline residues in the T_5CPCP_9 motif of MT-3 to Ala and Ser — the amino acids present in MT-2 — abolished the neuroinhibitory activity and cluster dynamics, it has been suggested that dynamic events centered at the three-metal cluster of MT-3 would include a partial unfolding of the β-domain and that the *cis*/*trans* interconversion of Cys-Pro amide bonds would govern the kinetics of this process (Hasler *et al.*, 2000). Additional insights into this process were provided by a molecular dynamics (MD) simulation of the β-domain of Cd_7MT-3 (Ni *et al.*, 2007). The studies revealed that, due to the structural constraints introduced by the T_5CPCP_9 motif, an unusual conformation of the N-terminal fragment (aa 1–13) is formed when compared with Cd_7MT-2, and that the formation of the *trans*/*trans* isomer is energetically more favorable. Further simulation of the partial unfolding supported the proposed role for *cis*/*trans* interconversion of Cys-Pro amide bonds in the folding/unfolding process of the β-domain. Other studies using the chimeric MT forms, generated by swapping the MT domains of

MT-3 and MT-1, showed that the α-domain — via domain–domain interactions — modulates the bioactivity of the β-domain of MT-3 and that, besides T_5, P_7, and P_9 mutations (Hasler *et al.*, 2000), the E23K mutation also abolishes the growth inhibitory activity of MT-3 (Ding *et al.*, 2006; Ding *et al.*, 2007). Although the mechanisms underlining these effects remain to be elucidated, the results obtained so far suggest that the structure of the β-domain of MT-3 is subjected to a fine tuning. Taken together, the changes in the primary structure of MT-3 in comparison with MT-1/MT-2 result in extraordinary and unprecedented structure dynamics and thus in the alteration of surface topology important for its bioactivity, playing a putative role in protein–protein or protein–receptor interactions.

The specific binding of intracellularly occurring neuronal Zn_7MT-3 to the small GTPase Rab3A in its GDP-bound state has been demonstrated, and the nature of the complex characterized. Rab3A is critically involved in the exo-endocytotic cycle of synaptic vesicles, including neuronal zinc vesicles. This finding indicates that Zn_7MT-3 actively participates in the synaptic vesicle trafficking upstream of vesicle fusion, playing a chaperone, sensor, and/or effector role in this process (Knipp *et al.*, 2005). The identification of MT-3 as a component of a brain multiprotein complex with heat shock protein 84 (HSP84) and creatine kinase (CK) has been linked to the peculiar structural properties of the protein described above (El Ghazi *et al.*, 2006).

As isolated from the brain, MT-3 contains both Cu^I and Zn^{II} ions in Cu_4,Zn_{3-4}MT-3. The extended X-ray absorption fine structure (EXAFS) studies of this species revealed the presence of two homometallic clusters: a Zn_{3-4}-thiolate cluster and a Cu^I_4-thiolate cluster with Zn^{II} and Cu^I ions tetrahedrally and trigonally coordinated, respectively (Bogumil *et al.*, 1998). A striking feature of the Cu^I_4-thiolate cluster, which by immunochemical methods was found to be localized in the β-domain (Roschitzki and Vašák, 2002), is its remarkable stability against air oxidation. By contrast, the studies on a well-defined Cu_4Zn_4MT-3 form prepared by a selective metal reconstitution of both domains *in vitro* revealed the presence of a redox-labile zinc site in the Zn_4-thiolate cluster located in the α-domain. While under anaerobic or reducing conditions the Zn_4-thiolate cluster is stable, a partial oxidation of specific thiolate ligands in air results in Zn^{II} release, giving rise to a Zn_3-thiolate cluster in the α-domain (Roschitzki and Vašák, 2003). It may be noted that the neuroinhibitory activity of MT-3 in cell cultures was found for both Cu_4Zn_{3-4}MT-3 and Zn_7MT-3 metalloforms (Uchida *et al.*, 1991; Erickson *et al.*, 1994). However, comparative biological studies on well-defined metalloforms are currently lacking.

In metal-linked neurodegenerative disorders like Alzheimer's disease, a dysregulated copper homeostasis and related neurotoxicity are linked to the production of reactive oxygen species (ROS). A protective role of extracellularly occurring Zn_7MT-3 against copper-mediated toxicity has been suggested based on its reactivity with free Cu^{II} ions *in vitro*. The studies showed that Zn_7MT-3, through Cu^{II} reduction to Cu^{I} by thiolate ligands and binding to the protein forming an air-stable $Cu(I)_4Zn_4$MT-3 species, can efficiently scavenge and redox-silence the toxic free Cu^{II} ions (Meloni *et al.*, 2007).

3.3. *Metallothionein-4*

The last identified member of the mammalian MT family is MT-4. This isoform consists of 62 amino acids, showing an insert of Glu in position 5 relative to MT-1/MT-2 proteins (Fig. 3). MT-4 appears to be present exclusively in cornified and stratified squamous epithelium. Much of the information on MT-4 deals with gene regulation molecular biology and expression profiles in mammalian maternal deciduum (Liang *et al.*, 1996), and during epithelium development and physiology (Quaife *et al.*, 1994; Schlake and Boehm, 2001). All of these studies revealed that the *MT-4* gene is subject to a strict developmental regulation. However, the question of whether MT-4 is involved in copper or zinc metabolism in epithelia is still debated.

Insights into the function of a protein have been inferred from its structural properties. From the studies of metal-binding abilities of MT-4 using heterologously expressed MT-4 with zinc, cadmium, and copper in combination with the *in silico* protein sequence analyses, the copper-binding nature has been suggested (Tio *et al.*, 2004). However, from biological studies, a special role of MT-4 in the regulation of zinc-dependent processes in keratinocytes has also been proposed (Quaife *et al.*, 1994). This function is supported by the structural studies on Cd_7MT-4 in which similar average apparent metal-binding affinities of two metal-thiolate clusters in Cd_7MT-4 and Cd_7MT-1/Cd_7MT-2, overall similarities in cluster topologies to that determined for Cd_7MT-1/Cd_7MT-2 (Fig. 2), and the pathway of cluster assembly have been found (Meloni *et al.*, 2006). Thus, MT-4 may be involved in both zinc and copper metabolism in keratinocytes.

4. Reactivity of Metal-Thiolate Clusters

Although the thiol groups in MT are masked through their interaction with metal ions, they retain a substantial degree of the nucleophilicity seen with the metal-free protein. This property is reflected by the extremely high reactivity of the coordinated cysteine ligands with alkylating and oxidizing agents such as iodoacetamide or 5,5′-dithiobis-(2-nitrobenzoic acid) (DTNB), respectively (Shaw *et al.*, 1991). Different sulfur reactivities in both domains have also been reported. Kinetic, mass spectrometric, and NMR studies have shown that the kinetically preferred reaction of mammalian MT with electrophiles may be localized in either the α- or the β-domain, depending on the specific attacking reagent. For instance, whereas iodoacetamide and *p*-(hydroxymercuri)benzoate have been found to react preferentially with the β-domain, DTNB, aurothiomalate, melphalan, and chlorambucil have been shown to react preferentially with the α-domain (Yu *et al.*, 1995; Zaia *et al.*, 1996; Muñoz *et al.*, 1999). Cysteine residues of the zinc-thiolate clusters in Zn_7MT-1/Zn_7MT-2 can also be oxidized by mild cellular oxidants like oxidized glutathione (GSSG), releasing bound metal ions in this process (Jacob *et al.*, 1998). Experiments in the presence of the GSH/GSSG redox pair provided evidence for an oxidoreductive mechanism modulating the zinc affinity of the cysteine thiolate ligands in Zn_7MT *in vitro* (Jiang *et al.*, 1998; Maret and Vallee, 1998). The importance of redox-active cysteine ligands in zinc proteins, especially Zn_7MTs, in converting redox signals to zinc signals in a cellular environment has also been discussed (Krezel *et al.*, 2007).

Another interesting aspect of MT reactivity is its ability to react with radical species. Thus, it has been shown that mammalian M^{II}_7MT ($M = Zn^{II}$ and/or Cd^{II}) are efficient scavengers of the ROS such as hydroxyl ($^{\bullet}OH$) and superoxide ($^{\bullet}O_2^-$) radicals or the reactive nitrogen species (RNS) such as nitric oxide ($^{\bullet}NO$) (Thornalley and Vašák, 1985; Kröncke *et al.*, 1994). In all instances, the free radical attack occurs at the metal-bound thiolates, leading to the protein oxidation and/or modification and subsequent metal release. Interestingly, in many instances, these effects could be reversed under reductive conditions and in the presence of the appropriate metal ion. The protective role of MTs against ROS damage in biological systems is well documented (reviewed in Fabisiak *et al.*, 2002). Although the reactivity of MT-1/MT-2/MT-3 with ROS is comparable, MT-3 can scavenge free NO and NO from *S*-nitrosothiols more efficiently than MT-1 and MT-2. In this process, the zinc-thiolate bonds are targets for both a direct attack by free NO and *S*-nitrosothiols, leading to the Cys*S*-NO formation and metal release. Further reaction of these Cys*S*-NO groups results in the formation of intramolecular disulfide bonds in the MT structure. ^1H and ^{113}Cd NMR studies of Cd_7MT-1 revealed that, whereas NO selectively releases the metals from the N-terminal β-domain, the C-terminal α-domain is less sensitive to the NO attack (Zangger *et al.*, 2001); in contrast, *S*-nitrosothiols efficiently release zinc from both the α- and β-domains of Zn_7MT-3. The *S*-nitrosylation of MT-3 by *S*-nitrosothiols occurs via a transnitrosation reaction. The increased reactivity of MT-3 with free NO and

S-nitrosothiols led to the proposal that Zn_7MT-3 may specifically convert NO signals into zinc signals (Chen *et al.*, 2002).

Studies on ligand substitution reactivity have revealed that small multidentate ligands (e.g. polyamines, polyaminocarboxylates, bis(thiosemicarbazones)) are effective competitors for zinc bound to MT in reactions which are much faster than the dissociation rate constant for zinc in Zn_7MT-1/Zn_7MT-2, implying a direct competition mechanism (Petering *et al.*, 1992). Biphasic kinetics and differential reactivity of the two metal clusters, α- and β-domains, in mammalian M^{II}_7MT have been measured with ethylenediaminetetraacetic acid (EDTA) showing $\beta > \alpha$ (Gan *et al.*, 1995), and in opposite order $\alpha > \beta$ with nitrilotriacetic acid (NTA) (Li and Otvos, 1998); on the other hand, bidentate ligands such as ethylenediaminediacetic acid and triethylenetetramine were found to be ineffective, even at thermodynamically competent concentrations. From these studies, it has been suggested that a tripod configuration of chelating ligands is required in thiolate substitution and that only specific regions of the protein domains may provide an easy access to the metal clusters. In this context, it may be noted that, in the crystal structure of Zn_2Cd_5MT-2, each domain contains a solvent-exposed cleft containing three accessible cysteine sulfurs. In summary, the reactivity of the MT structure, which is pertinent to its function, is dominated by the chemistry of the nucleophilic thiolate groups.

5. Structure Dynamics

Despite the high thermodynamic stability of the metal-thiolate clusters in MT, and as a consequence of the structure dynamics, they are kinetically very labile, i.e. the thiolate ligands allow a rapid metallation and demetallation. The rate of metal exchange in MT-1/MT-2 follows the order $Hg^{II} > Cd^{II} > Zn^{II}$. A comparison with similar studies on inorganic complexes affords the most plausible explanation for the kinetics of metal exchange (Martell *et al.*, 1994). In the latter studies, a correlation between the level of ligand preorganization and complex lability revealed that a decreasing rigidity of the ligand structure results in an increasing lability of the metal complexes. Based on the already-discussed properties of metal-free and metal-containing MTs, the apoprotein with multidentate cysteine thiolate ligands resembles chelating inorganic ligands with long bridges, for which a low level of ligand preorganization and hence a high kinetic lability has been shown. Evidence for the kinetic lability of metals in Cd_7MT was provided by ^{113}Cd NMR saturation transfer experiments, which established the presence of intermolecular and/or intramolecular metal exchange within the three-metal cluster of the β-domain with a half-life of the order of 0.5 seconds. The occurrence of similar processes taking place within the four-metal cluster, but with a half-life of about 16 minutes, was afforded by metal exchange studies using the radioactive ^{109}Cd isotope (Otvos *et al.*, 1993). The importance of protein dynamics for metal exchange has been recognized in the NMR studies of Cd_7MT-1 in which the enhanced backbone flexibility, when compared to Cd_7MT-2, resulted in a much faster intersite cadmium exchange in the three-metal cluster (Zangger *et al.*, 1999). In this context, it may be noted that in zinc enzymes such as alkaline phosphatase and carboxypeptidase, where a well-defined catalytic site is present in a rather rigid protein structure, the exchange half-life of zinc is in the order of hours and days. Thus, the actual exchange rates of metal ions in proteins are not an intrinsic attribute of their binding properties, but rather are determined by the energetics and kinetics of protein folding.

Intermolecular zinc transfer between zinc proteins and Zn_7MT *in vitro* has also been studied, leading to the apoenzyme activation and the modulation of zinc-dependent transcription factors (Jacob *et al.*, 1998; Roesijadi *et al.*, 1998). Moreover, Zn_7MT has been shown to be able to transfer one zinc ion to apoenzymes possessing a lower affinity compared to the apparent average binding constant for zinc in Zn_7MT. However, the presence of one weakly bound metal ion in the structure of Zn_7MT has recently been demonstrated (Krezel and Maret, 2007).

6. Metallothioneins in Inorganic Pharmacology

Metallothionein, owing to its affinity to a wide range of metal ions, can through a direct interaction with inorganic drugs influence their efficacy. Among others, the MT-1/MT-2 reactivity with anticancer platinum-based drugs, bismuth drugs, and gold-containing drugs used in rheumatoid arthritis has attracted particular interest. Platinum drugs are effective chemotherapeutic agents for the treatment of testicular cancer and are used in combination therapy for a variety of other tumors. However, the occurrence of intrinsic resistance in some tumors, and that acquired after initial treatment, are the major drawbacks of these chemotherapeutics. Among other responses leading to the resistance, the high reactivity of platinum compounds with the major intracellular thiols glutathione (GSH) and MT confer resistance to these drugs. However, the cysteine ligands in Zn_7MT react with cisplatin much faster than the thiol group in GSH. As a consequence, in an anticancer treatment with platinum-based electrophylic antineoplastic drugs including cisplatin, an overexpression of MTs in cancer cells is observed. A direct interaction of these drugs with MTs is believed to be the primary cause of tumor cell resistance (Kelley *et al.*, 1988). A detailed study on the interaction of Zn_7MT-2 with a number of *cis*- and *trans*-Pt^{II} compounds, including a new-generation Pt^{II} drug with potential use in the clinic, showed that all ligands in *cis*-Pt^{II} compounds, including cisplatin, are replaced by cysteine thiolates of the protein, thereby fully deactivating the drug (Knipp *et al.*, 2007). In contrast, the protein-bound *trans*-Pt^{II} compounds retained their N-donor ligands, thus remaining in a potentially active form; however, no *trans*-Pt^{II} transfer from MT-2 to plasmid DNA was observed. During the binding process of Pt^{II}, the corresponding amount of Zn^{II} is released from MT-2. The binding of released zinc to the metal-responsive transcription factor (MTF-1), which activates MT-2 expression, represents an unrecognized aspect contributing to the cellular resistance against platinum-based anticancer drugs.

The preinduction of MT-1/MT-2 by bismuth compounds, which induce these proteins almost exclusively in the kidney, can reduce the renal toxicity of cisplatin. In addition, bismuth complexes are also used as antiulcer drugs. Bi^{III} binds strongly to MTs, forming $Bi_7^{III}MT$. EXAFS studies of $Bi_7^{III}MT$ revealed that the metal coordination sphere is composed of three to four sulfur atoms with some additional oxygen atoms. This suggests that a widely different MT structure with Bi^{III} ions exists when compared with those reported for Zn^{II}- and Cd^{II}-containing MTs.

It has been suggested that MT, besides playing a possible role in the detoxification of gold drugs in kidney, may also contribute to the retention and localization of gold in the tissues during rheumatoid arthritis treatment with Au^I compounds. Moreover, a high content of MT in cells is apparently responsible for their resistance to the growth inhibitory effects of Au^I-auranofin (a rheumatoid arthritis drug). The acquired resistance to the gold drug aurothiomalate may be due in part to MT induction by a mechanism described above for Pt^{II} compounds. The structural studies of gold-containing MT, upon the reactions of $M_7^{II}MT$ with gold thiomalate (AuSTm), revealed that an excess of Zn_7MT results in the formation of an $Au^I(SCys)_2$ complex in which the thiomalate ligand is replaced by the thiolate ligand. However, an excess of AuSTm gave rise to a monodentate Au^ISCys coordination and the retention of the thiomalate ligand, i.e. CysS-Au-STm (Laib *et al.*, 1985). Furthermore, mass spectrometry measurements revealed that a gold-binding mechanism to MT is similar to that reported for the formation of gold nanoclusters (Mercogliano and DeRosier, 2006).

Structural information on other metal derivatives of MT, such as Ni^{II}, Fe^{II}, Hg^{II}, Ag^{I}, In^{III}, Sb^{III}, As^{III}, and TcO^{III}, is also available (Vašák and Romero-Isart, 2005). In view of an increasing number of new metal-based drugs for cancer chemotherapy and treatment of other pathologies, with some drugs already in preclinical or clinical testing, further studies of MT reactivity with the metal-based drugs are required to shed light on the chemical and biological rationale behind the mode of action and efficacy of these compounds.

References

Arseniev A, Schultze P, Wörgötter E, *et al.* Three-dimensional structure of rabbit liver Cd_7metallothionein-2a in aqueous solution determined by nuclear magnetic resonance. *J Mol Biol* 1988; **201**:637–657.

Bertini I, Luchinat C, Messori L, Vašák M. Proton NMR studies of the cobalt(II)-metallothionein system. *J Am Chem Soc* 1989; **111**:7296–7300.

Berweger CD, Thiel W, van Gunsteren WF. Molecular-dynamics simulation of the beta domain of metallothionein with a semi-empirical treatment of the metal core. *Proteins* 2000; **41**:299–315.

Binz PA, Kägi JHR. Metallothionein: Molecular evolution and classification. In: Klaassen C (ed.), *Metallothionein IV*, Birkhäuser Verlag, Basel, 1999, pp. 7–13.

Bogumil R, Faller P, Binz PA, *et al.* Structural characterization of Cu(I) and Zn(II) sites in neuronal-growth-inhibitory factor by extended X-ray absorption fine structure (EXAFS). *Eur J Biochem* 1998; **255**:172–177.

Braun W, Vašák M, Robbins AH, *et al.* Comparison of the NMR solution structure and the X-ray crystal structure of rat metallothionein-2. *Proc Natl Acad Sci USA* 1992; **89**:10124–10128.

Calderone V, Dolderer B, Hartmann HJ, *et al.* The crystal structure of yeast copper thionein: The solution of a long-lasting enigma. *Proc Natl Acad Sci USA* 2005; **102**:51–56.

Chen Y, Irie Y, Keung WM, Maret WS. Nitrosothiols react preferentially with zinc thiolate clusters of metallothionein III through transnitrosation. *Biochemistry* 2002; **41**:8360–8367.

Cherian G, Jayasurya A, Bay BH. Metallothioneins in human tumors and potential roles in carcinogenesis. *Mutat Res* 2003; **533**:201–209.

Ding ZC, Teng XC, Cai B, *et al.* Mutation at Glu23 eliminates the neuron growth inhibitory activity of human metallothionein-3. *Biochem Biophys Res Commun* 2006; **349**:674–682.

Ding ZC, Zheng Q, Cai B, *et al.* Effect of alpha-domain substitution on the structure, property and function of human neuronal growth inhibitory factor. *J Biol Inorg Chem* 2007; **12**:1173–1179.

Dolderer B, Echner H, Beck A, *et al.* Coordination of three and four Cu(I) to the α- and β-domain of vertebrate Zn-metallothionein-1, respectively, induces significant structural changes. *FEBS J* 2007; **274**:2349–2362.

El Ghazi I, Martin BL, Armitage IM. Metallothionein-3 is a component of a multiprotein complex in the mouse brain. *Exp Biol Med (Maywood)* 2006; **231**:1500–1506.

Erickson JC, Sewell AK, Jensen LT, *et al.* Enhanced neurotrophic activity in Alzheimer's disease cortex is not associated with down-regulation of metallothionein-III (GIF). *Brain Res* 1994; **649**:297–304.

Fabisiak JP, Borisenko GG, Liu SX, *et al.* Redox sensor function of metallothioneins. *Methods Enzymol* 2002; **353**:268–281.

Faller P, Hasler DW, Zerbe O, *et al*. Evidence for a dynamic structure of human neuronal growth inhibitory factor and for major rearrangements of its metal-thiolate clusters. *Biochemistry* 1999; **38**:10158–10167.

Faller P, Vašák M. Distinct metal-thiolate clusters in the N-terminal domain of neuronal growth inhibitory factor. *Biochemistry* 1997; **36**:13341–13348.

Gan T, Munoz A, Shaw CF, Petering DH. Reaction of $^{111}Cd_7$-metallothionein with EDTA. A reappraisal. *J Biol Chem* 1995; **270**:5339–5345.

Good M, Hollenstein R, Sadler PJ, Vašák M. ^{113}Cd NMR studies on metal-thiolate cluster formation in rabbit Cd(II)-metallothionein: Evidence for a pH dependence. *Biochemistry* 1988; **27**:7163–7166.

Good M, Hollenstein R, Vašák M. Metal selectivity of clusters in rabbit liver metallothionein. *Eur J Biochem* 1991; **197**:655–659.

Hagen KS, Stephan DW, Holm RH. Metal-ion exchange-reactions in cage molecules — The systems $[M_{4\text{-Eta}}M'_{\text{eta}}(SC_6H_5)^{10}]^{2-}$ (M, M′ = Fe(II), Co(II), Zn(II), Cd(II)) with adamantane-like stereochemistry and the structure of $[Fe_4(SC_6H_5)^{10}]^{2-}$. *Inorg Chem* 1982; **21**:3928–3936.

Hasler DW, Faller P, Vašák M. Metal-thiolate clusters in the C-terminal domain of human neuronal growth inhibitory factor (GIF). *Biochemistry* 1998; **37**:14966–14973.

Hasler DW, Jensen LT, Zerbe O, *et al*. Effect of the two conserved prolines of human growth inhibitory factor (metallothionein-3) on its biological activity and structure fluctuation: Comparison with a mutant protein. *Biochemistry* 2000; **39**:14567–14575.

Heuchel R, Radtke F, Georgiev O, *et al*. The transcription factor MTF-1 is essential for basal and heavy metal-induced metallothionein gene expression. *EMBO J* 1994; **13**:2870–2875.

Hidalgo J, Aschner M, Zatta P, Vašák M. Roles of the metallothionein family of proteins in the central nervous system. *Brain Res Bull* 2001; **55**:133–145.

Irie Y, Mori F, Keung WM, *et al*. Expression of neuronal growth inhibitory factor (metallothionein-III) in the salivary gland. *Physiol Res* 2004; **53**:719–723.

Jacob C, Maret W, Vallee BL. Control of zinc transfer between thionein, metallothionein, and zinc proteins. *Proc Natl Acad Sci USA* 1998; **95**:3489–3494.

Jiang LJ, Maret W, Vallee BL. The glutathione redox couple modulates zinc transfer from metallothionein to zinc-depleted sorbitol dehydrogenase. *Proc Natl Acad Sci USA* 1998; **95**:3483–3488.

Kägi JHR, Schäffer A. Biochemistry of metallothionein. *Biochemistry* 1988; **27**: 8509–8515.

Kägi JHR, Vallee B. Metallothionein: A cadmium- and zinc-containing protein from equine renal cortex. *J Biol Chem* 1960; **235**:3460–3465.

Kelley SL, Basu A, Teicher BA, *et al*. Overexpression of metallothionein confers resistance to anticancer drugs. *Science* 1988; **241**:1813–1815.

Knipp M, Karotki AV, Chesnov S, *et al*. Reaction of Zn_7metallothionein with *cis*- and *trans*-$[Pt(N\text{-donor})_2Cl_2]$ anticancer complexes: *trans*-Pt(II) complexes retain their N-donor ligands. *J Med Chem* 2007; **50**:4075–4086.

Knipp M, Meloni G, Roschitzki B, Vasak M. Zn_7metallothionein-3 and the synaptic vesicle cycle: Interaction of metallothionein-3 with the small GTPase Rab3A. *Biochemistry* 2005; **44**:3159–3165.

Krezel A, Hao Q, Maret W. The zinc/thiolate redox biochemistry of metallothionein and the control of zinc ion fluctuations in cell signaling. *Arch Biochem Biophys* 2007; **463**:188–200.

Krezel A, Maret W. Dual nanomolar and picomolar Zn(II) binding properties of metallothionein. *J Am Chem Soc* 2007; **129**:10911–10921.

Kröncke KD, Fehsel K, Schmidt T, *et al.* Nitric oxide destroys zinc-sulfur clusters inducing zinc release from metallothionein and inhibition of the zinc finger-type yeast transcription activator LAC9. *Biochem Biophys Res Commun* 1994; **200**:1105–1110.

Laib JE, Shaw CF, Petering DH, *et al.* Formation and characterization of aurothioneins: Au,Zn,Cd-thionein, Au,Cd-thionein, and (thiomalato-Au)chi-thionein. *Biochemistry* 1985; **24**:1977–1986.

Li H, Otvos JD. Biphasic kinetics of Zn^{2+} removal from Zn metallothionein by nitrilotriacetate are associated with differential reactivity of the two metal clusters. *J Inorg Biochem* 1998; **70**:187–194.

Liang L, Fu K, Lee DK, *et al.* Activation of the complete mouse metallothionein gene locus in the maternal deciduum. *Mol Reprod Dev* 1996; **43**:25–37.

Maret W. The function of zinc metallothionein: A link between cellular zinc and redox state. *J Nutr* 2000; **130**:1455S-1458S.

Maret W, Vallee BL. Thiolate ligands in metallothionein confer redox activity on zinc clusters. *Proc Natl Acad Sci USA* 1998; **95**:3478–3482.

Margoshes M, Vallee BL. A cadmium protein from equine kidney cortex. *J Am Chem Soc* 1957; **79**:4813–4814.

Martell AE, Hancock RD, Motekaitis RJ. Factors affecting stabilities of chelate, macrocyclic and macrobicyclic complexes in solution. *Coord Chem Rev* 1994; **133**:39–65.

Masters BA, Quaife J, Erickson JC, *et al.* Metallothionein III is expressed in neurons that sequester zinc in synaptic vesicles. *J Neurosci* 1994; **14**:5844–5857.

Meloni G, Faller P, Vašák M. Redox silencing of copper in metal-linked neurodegenerative disorders: Reaction of Zn_7metallothionein-3 with Cu^{2+} ions. *J Biol Chem* 2007; **282**:16068–16078.

Meloni G, Zovo K, Kazantseva J, *et al.* Organization and assembly of metal-thiolate clusters in epithelium-specific metallothionein-4. *J Biol Chem* 2006; **281**:14588–14595.

Mercogliano CP, DeRosier DJ. Gold nanocluster formation using metallothionein: Mass spectrometry and electron microscopy. *J Mol Biol* 2006; **355**:211–223.

Messerle BA, Bos M, Schäffer A, *et al.* Amide proton exchange in human metallothionein-2 measured by nuclear magnetic resonance spectroscopy. *J Mol Biol* 1990a; **214**: 781–786.

Messerle BA, Schäffer A, Vašák M, *et al.* Three-dimensional structure of human [$^{113}Cd_7$]metallothionein-2 in solution determined by nuclear magnetic resonance spectroscopy. *J Mol Biol* 1990b; **214**:765–779.

Miles AT, Hawksworth GM, Beattie JH, Rodilla V. Induction, regulation, degradation, and biological significance of mammalian metallothioneins. *Crit Rev Biochem Mol Biol* 2000; **35**:35–70.

Muñoz A, Petering DH, Shaw CF. Reactions of electrophilic reagents that target the thiolate groups of metallothionein clusters: Preferential reaction of the alpha-domain with 5,5′-dithio-bis(2-nitrobenzoate) (DTNB) and aurothiomalate (AuSTm). *Inorg Chem* 1999; **38**:5655–5659.

Nartey NO, Banerjee D, Cherian MG. Immunohistochemical localization of metallothionein in cell nucleus and cytoplasm of fetal human liver and kidney and its changes during development. *Pathology* 1987; **19**:233–238.

Ni FY, Cai B, Ding ZC, *et al.* Structural prediction of the beta-domain of metallothionein-3 by molecular dynamics simulation. *Proteins* 2007; **68**:255–266.

Nielson KB, Atkin CL, Winge DR. Distinct metal-binding configurations in metallothionein. *J Biol Chem* 1985; **260**:5342–5350.

Nielson KB, Winge DR. Order of metal binding in metallothionein. *J Biol Chem* 1983; **258**:13063–13069.

Nielson KB, Winge DR. Preferential binding of copper to the beta domain of metallothionein. *J Biol Chem* 1984; **259**:4941–4946.

Otvos JD, Armitage IM. Structure of the metal clusters in rabbit liver metallothionein. *Proc Natl Acad Sci USA* 1980; **77**:7094–7098.

Otvos JD, Liu X, Li H, *et al.* Dynamic aspects of metallothionein structure. In: Suzuki KT, Imura N, Kimura M (eds.), *Metallothionein III*, Birkhäuser Verlag, Basel, 1993, pp. 57–74.

Öz G, Zangger K, Armitage IM. Three-dimensional structure and dynamics of a brain specific growth inhibitory factor: Metallothionein-3. *Biochemistry* 2001; **40**:11433–11441.

Palmiter RD. The elusive function of metallothioneins. *Proc Natl Acad Sci USA* 1998; **95**:8428–8430.

Palumaa P, Eriste E, Njunkova O, *et al.* Brain-specific metallothionein-3 has higher metal-binding capacity than ubiquitous metallothioneins and binds metals noncooperatively. *Biochemistry* 2002; **41**:6158–6163.

Petering DH, Krezoski S, Chen P, *et al.* Kinetic reactivity of metallothioneins. In: Stillman MJ, Shaw III CF, Suzuki KT (eds.), *Metallothioneins — Synthesis, Structure and Properties of Metallothioneins, Phytochelatins and Metal-Thiolate Complexes*, VCH Publishers, New York, 1992, pp. 164–183.

Pountney DL, Schauwecker I, Zarn J, Vašák M. Formation of mammalian Cu_8-metallothionein *in vitro*: Evidence for the existence of two $Cu(I)_4$-thiolate clusters. *Biochemistry* 1994; **33**:9699–9705.

Pountney DL, Vašák M. Spectroscopic studies on metal distribution in Co(II)/Zn(II) mixed-metal clusters in rabbit liver metallothionein 2. *Eur J Biochem* 1992; **209**:335–341.

Quaife CJ, Findley SD, Erickson JC, *et al.* Induction of a new metallothionein isoform (MT-IV) occurs during differentiation of stratified squamous epithelia. *Biochemistry* 1994; **33**:7250–7259.

Robbins AH, McRee DE, Williamson M, *et al.* Refined crystal structure of Cd, Zn metallothionein at 2.0 Å resolution. *J Mol Biol* 1991; **221**:1269–1293.

Roesijadi G, Bogumil R, Vašák M, Kägi JH. Modulation of DNA binding of a Tramtrack zinc finger peptide by the metallothionein–thionein conjugate pair. *J Biol Chem* 1998; **273**:17425–17432.

Romero-Isart N, Vašák M. Advances in the structure and chemistry of metallothioneins. *J Inorg Biochem* 2002; **88**:388–396.

Roschitzki B, Vašák M. A distinct Cu_4-thiolate cluster of human metallothionein-3 is located in the N-terminal domain. *J Biol Inorg Chem* 2002; **7**:611–616.

Roschitzki B, Vašák M. Redox labile site in a Zn_4 cluster of Cu_4,Zn_4-metallothionein-3. *Biochemistry* 2003; **42**:9822–9828.

Schlake T, Boehm T. Expression domains in the skin of genes affected by the nude mutation and identified by gene expression profiling. *Mech Dev* 2001; **109**:419–422.

Schultze P, Wörgötter E, Braun W, *et al.* Conformation of Cd_7-metallothionein-2 from rat liver in aqueous solution determined by nuclear magnetic resonance spectroscopy. *J Mol Biol* 1988; **203**:251–268.

Shaw III CF, Savas MM, Petering DH. Ligand substitution and sulfhydryl reactivity of metallothionein. *Methods Enzymol* 1991; **205**:401–414.

Sogawa CA, Asanuma M, Sogawa N, *et al.* Localization, regulation, and function of metallothionein-III/growth inhibitory factor in the brain. *Acta Med Okayama* 2001; **55**:1–9.

Stillman MJ. Metallothioneins. *Coord Chem Rev* 1995; **144**:461–511.

Tapiero H, Townsend DM, Tew KD. Trace elements in human physiology and pathology. Copper. *Biomed Pharmacother* 2003; **57**:386–398.

Theocharis SE, Margeli AP, Klijanienko JT, Kouraklis GP. Metallothionein expression in human neoplasia. *Histopathology* 2004; **45**:103–118.

Thornalley PJ, Vašák M. Possible role for metallothionein in protection against radiation-induced oxidative stress. Kinetics and mechanism of its reaction with superoxide and hydroxyl radicals. *Biochim Biophys Acta* 1985; **827**:36–44.

Tio L, Villarreal L, Atrian S, Capdevila M. Functional differentiation in the mammalian metallothionein gene family: Metal binding features of mouse MT4 and comparison with its paralog MT1. *J Biol Chem* 2004; **279**:24403–24413.

Uchida Y, Takio K, Titani K, *et al.* The growth inhibitory factor that is deficient in the Alzheimer's disease brain is a 68 amino acid metallothionein-like protein. *Neuron* 1991; **7**:337–347.

Vašák M. Spectroscopic studies on cobalt(II) metallothionein: Evidence for pseudotetrahedral coordination. *J Am Chem Soc* 1980; **102**:3953–3955.

Vašák M, Kägi JH. Metal thiolate clusters in cobalt(II)-metallothionein. *Proc Natl Acad Sci USA* 1981; **78**:6709–6713.

Vašák M, Romero-Isart N. Metallothioneins. In: King RB (ed.), *Encyclopedia of Inorganic Chemistry*, 2nd ed., John Wiley & Sons, New York, 2005, pp. 3208–3221.

Wang H, Zhang Q, Cai B, *et al.* Solution structure and dynamics of human metallothionein-3 (MT-3). *FEBS Lett* 2006; **580**:795–800.

Winge DR, Miklossy KA. Domain nature of metallothionein. *J Biol Chem* 1982; **257**: 3471–3476.

Yu X, Wu Z, Fenselau C. Covalent sequestration of melphalan by metallothionein and selective alkylation of cysteines. *Biochemistry* 1995; **34**:3377–3385.

Zaia J, Jiang LC, Han MS, *et al.* A binding site for chlorambucil on metallothionein. *Biochemistry* 1996; **35**:2830–2835.

Zangger K, Öz G, Haslinger E, *et al.* Nitric oxide selectively releases metals from the amino-terminal domain of metallothioneins: Potential role at inflammatory sites. *FASEB J* 2001; **15**:1303–1305.

Zangger K, Öz G, Otvos JD, Armitage IM. Three-dimensional solution structure of mouse Cd_7-metallothionein-1 by homonuclear and heteronuclear NMR spectroscopy. *Protein Sci* 1999; **8**:2630–2638.

Chapter 2

THIOL REACTIVITY AS A CENTRAL ASPECT OF METALLOTHIONEIN'S MECHANISM OF ACTION

Wolfgang Maret

Studies involving metallothioneins (MTs) are often performed without reference to a primary biological function or a chemical mechanism of the protein, despite the fact that the critical cysteinyl residues in MT, and their neighboring amino acids, have unique characteristics and determine specific functions. Fluorescence methods have provided new insights into how the redox state of the sulfur donors and the interactions with zinc ions are controlled in mammalian MTs. Two zinc/thiolate clusters modulate zinc affinities over several orders of magnitude, thus allowing fine-tuned control over the cellular availability of free zinc ions. *In vivo*, not all of the sulfur donors of MT are reduced, and less than seven zinc ions are bound. Thus, MT exists in different states depending on zinc availability and redox poise. Dynamic changes in the sulfur redox state and in zinc binding demonstrate a function of mammalian MT in the interconversion of zinc and redox signals. In the cellular signaling network, MT is an integrated part of the following pathway: signal → reactive species → MT → Zn^{2+} → target. MT is translocated within cells, secreted from cells, and taken up by cells. Its function is important for regulation of the physiological processes that depend on zinc availability, and the pathological processes of cell injury induced by oxidative stress or by thiol-reactive compounds.

Keywords: Metallothionein; sulfur chemistry; zinc metabolism; redox signaling; oxidative stress.

1. Introduction to Metallothionein

The discovery of metallothionein (MT) was reported 50 years ago (Margoshes and Vallee, 1957). The name "thionein" reflects the relatively high sulfur (cysteine) content of the protein (Kägi and Vallee, 1960), which is the single most important structural and, as it turns out, functional feature of MT. The prefix "metallo-" is a reminder of the fact that the protein isolated from its original source, i.e. horse kidney, contains cadmium and trace amounts of other metals in addition to zinc (Kägi and Vallee, 1960). Structural studies of MT turned out to be challenging. From the time of its discovery, it took 20 years to obtain the amino acid sequence of one of the two major MT isoforms isolated from horse kidney (MT-1b) (Kojima *et al.*, 1976). The structure is comprised of a 61-amino-acid, single-chain protein with characteristic CC, CxC, and CxxC motifs for its 20 cysteines. Four major isoforms are recognized in mammals: MT-1 to MT-4. MT-3 and MT-4 have a more restricted tissue-specific expression than MT-1 and MT-2 (Uchida *et al.*, 1991; Quaife *et al.*, 1994). The gene for MT-1 experienced considerable duplication in humans. This article will focus on mammalian zinc MTs because, in most tissues, MTs are mainly zinc proteins with a major role in zinc metabolism (Vallee, 1995; Cousins *et al.*, 2006), although, under specific conditions, MT can bind copper. Cadmium and other toxic metals accumulate in MT and can interfere with its primary functions when an organism is exposed to such metals.

Isolated MTs are not homogeneous with regard to their metal content and/or the redox state of their cysteines. Therefore, the isolated MTs are treated with thiol reductants and reconstituted with a given metal ion to obtain defined species for structural analysis and further experimentation (Vašák, 1991). The stoichiometry is seven for divalent transition metal ions (Zn, Cd), but it is higher for monovalent transition metal ions (Cu). For technical reasons, the solution structure was determined on rabbit MT-2a with seven cadmium ions (Arseniev *et al.*, 1988). The crystal structure of MT was solved from a species with the metal composition Cd_5Zn_2–MT-2, which was isolated from the livers of rats after induction of the protein with cadmium (Robbins *et al.*, 1991). The zinc ions in MT are bound exclusively to the sulfur donors of the 20 cysteines in Zn_3S_9 and Zn_4S_{11} clusters that have characteristic sulfur (thiolate) ligand bridges between the ions. The clusters are located in the two domains of MT. Most of the work in the MT field is interpreted with reference to these structures. However, these earlier studies did not resolve the issue of exactly what the structure of the protein is *in vivo*. Clearly, such knowledge is critically important in order to understand the *in vivo* function of the protein.

It now turns out that MT with 20 reduced cysteinyl residues and seven bound zinc ions is not the physiologically important state. Therefore, the name "metallothionein" with reference to its three-dimensional (3D) structure is ambiguous when applied to the state of the protein in tissues, and inferences about its function from its 3D structure can be erroneous. Studies aimed at defining the *in vivo* state of MT employed chemical modification with a thiol-specific fluorescent probe under conditions where three states of the cysteinyl residues can be distinguished: reduced, oxidized, and metal-bound (Yang *et al.*, 2001; Haase and Maret, 2004; Krężel and Maret, 2007a). In this way, it was demonstrated that the protein occurs in tissues and cells in the holoform as "metallothionein", in the apoform as "thionein" (Yang *et al.*, 2001), and in oxidized forms as "thionin" (Krężel and Maret, 2007a). Thus, under normal physiological conditions, MT is neither fully loaded with zinc ions nor are its cysteines fully reduced. For example, in rat liver, 20% of MT is in the apoform and 7% is oxidized; however, the distribution of these forms is dynamic and changes when cells are incubated with zinc ions or when the cellular thiol/disulfide redox balance is perturbed (Krężel and Maret, 2007a). Overexpression of MT in cells leads to a much larger fraction of thionein (St. Croix *et al.*, 2002). Also, disulfides in the protein have been detected when it is overexpressed in the livers of mice; the number of disulfides increases when these mice are oxidatively stressed (Feng *et al.*, 2006). Thus, the metal load and the redox state of MT are variable. Three species can be prepared from the isolated protein: fully metal-loaded metallothionein (MT or Zn_7T), metal-depleted and fully reduced thionein (T_R), and metal-depleted and fully oxidized thionin (T_O). However, *in vivo* it is likely that MT is one molecule with different states of the cysteinyl residues, reflecting both the availability of metal ions and the redox state of the cell. These states describe a protein that is either partially zinc-depleted and fully reduced ($Zn_{7-x}T_R$) or partially zinc-depleted and partially oxidized ($Zn_{7-x}T_O$).

2. The Sulfur Chemistry of Metallothionein

Examination of zinc binding to T_R and dissociation of zinc from MT using fluorescent zinc-chelating agents supports the concept that, *in vivo,* MT neither binds a full complement of metal ions nor has fully reduced cysteines. Despite the binding of all seven zinc ions in tetrathiolate (Cys_4) coordination environments in the clusters, the affinities of the sites differ by four orders of magnitude (Krężel and Maret, 2007b). Four zinc ions are bound tightly ($\log K = 11.8$), two zinc ions are bound with intermediate strength ($\log K \sim 10$), and one zinc ion is bound relatively weakly ($\log K = 7.7$). This finding has significant implications for the functions of MT. It suggests that MT does not bind seven zinc ions under normal physiological conditions because, as will be explained later, there are not enough free zinc ions available in the cell to saturate its binding site with $\log K = 7.7$. Such a description of MT with Zn_4T, Zn_5T, Zn_6T, and Zn_7T species is fundamentally different from the one given previously, where all seven zinc ions were thought to bind tightly with similar or equal affinities, and accordingly, only subpicomolar concentrations of free zinc ions would be available (Kägi, 1993). In contrast, different zinc affinities make picomolar to nanomolar concentrations of free zinc ions available from MT, making it possible for MT to participate actively in zinc metabolism rather than trapping any zinc ions in a thermodynamically very stable complex.

In addition to providing coordination environments with fine-tuned affinities for zinc, the sulfur donors from the cysteine ligands have another important chemical activity: they permit reversible redox reactions with concomitant release and binding of zinc (Fig. 1) (Maret and Vallee, 1998). It is most remarkable that MT is a redox

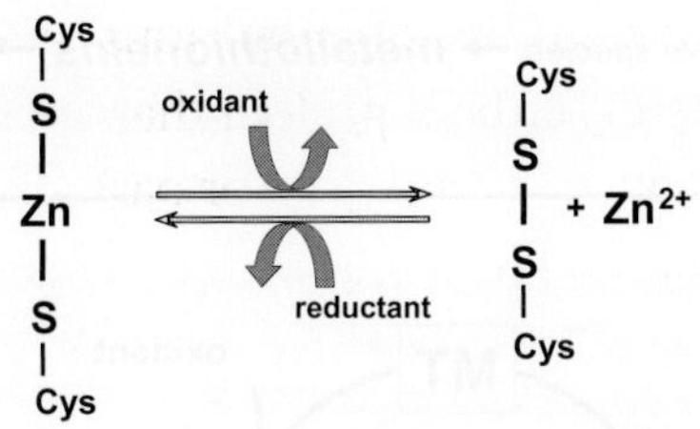

Fig. 1 Chemical mechanism of redox-dependent zinc binding and release in metallothionein. Oxidation and reduction of the sulfur donors is coupled with the binding and release of free zinc ions.

protein because zinc proteins were generally not considered to be redox-active (Maret, 2006). In this way, the availability of zinc ions, which are redox-inert in biology, can be controlled by redox metabolism, thus mobilizing them from their tight binding sites under oxidizing conditions and binding them under reducing conditions (Maret, 2000; Maret, 2003). The biological significance of this relationship is that the unique coordination environments of the clusters in MT, in essence, convert redox signals into zinc signals in a signaling pathway (Fig. 2) (Maret, 2004; Maret, 2006).

While discussions in the literature address almost exclusively the direction of the signaling pathway shown in Fig. 2, the redox cycle of MT (Chen and Maret, 2001) produces reversibility for the functions of zinc ions at their target sites. Reducing conditions recycle the protein, to which zinc can then bind, thereby abrogating any functions of zinc at its target sites (Fig. 2).

3. Metallothionein and Cellular Zinc Regulation

Before continuing the discussion of the biology of MT in this signaling pathway, it is necessary to provide some background on one particular aspect of cellular zinc homeostasis that is highly relevant for understanding the functions of MT in zinc metabolism. A major question in zinc biology is how zinc is redistributed inside the cell, specifically whether it is transferred between proteins

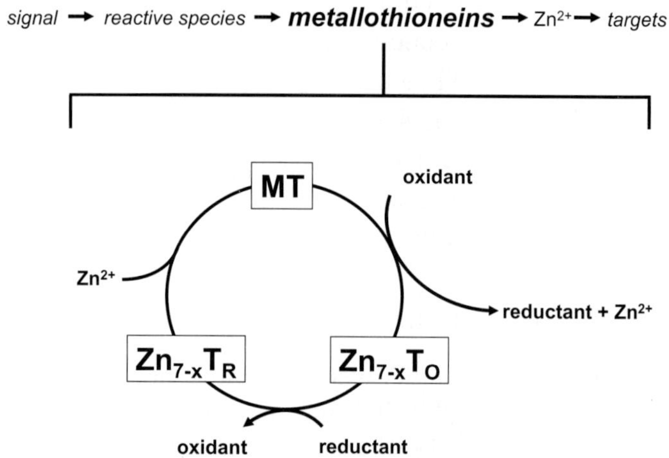

Fig. 2 Metallothionein redox cycle in a cellular signaling pathway. One redox couple controls zinc release from MT, the zinc-loaded state of the protein, to a partially oxidized and partially zinc-depleted state ($Zn_{7-x}T_O$). Another redox couple controls the equilibrium between this state and a fully reduced and partially zinc-depleted state ($Zn_{7-x}T_R$). Zinc binding to this state completes the cycle. In the presence of catalytic selenium compounds, the redox cycle of metallothionein couples with the glutathione/glutathione disulfide redox pair (Chen and Maret, 2001).

by protein/protein interactions without ever being free or whether it dissociates first and then reassociates with another protein. To resolve this issue, it is crucial to determine cellular free zinc ion concentrations.

Most of the zinc is tightly bound to proteins with picomolar affinities. Accordingly, free zinc ion concentrations must be very low, although total cellular zinc concentrations are a few hundred micromolar and thus relatively high. Estimates of the concentrations of free zinc ions were made by various methods as early as 1971. They are in the range of hundreds of picomolars in rabbit skeletal muscle (Peck and Ray, 1971), 24 pM in erythrocytes (Simons, 1991), and 500 pM in neuroblastoma cells (Adebodun and Post, 1995; Benters *et al.*, 1997). More recent estimates for eukaryotic cells employing highly sensitive fluorescence methods are 5–10 pM for pheochromocytoma (PC12) cells (Bozym *et al.*, 2006) and 784 pM for proliferating human colon cancer (HT29) cells (Krężel and Maret, 2006). These concentrations depend on the state of the cell. Thus, growth-arrested HT29 cells have lower free zinc ion concentrations, and differentiating and apoptotic cells have higher free zinc ion concentrations (Krężel *et al.*, 2007). Moreover, the free zinc ion concentrations correlate with the cellular glutathione/glutathione disulfide redox state and the amount of thionin (Krężel and Maret, 2006).

Different signals elicit free zinc ion fluctuations in the range of picomolar to nanomolar concentrations (Atar *et al.*, 1995; Turan *et al.*, 1997; Choi and Koh, 1988; Aizenman *et al.*, 2000; St. Croix *et al.*, 2002; Smith *et al.*, 2002; Sensi *et al.*, 2003; Spahl *et al.*, 2003). Such fluctuations prompt the question: what are the targets of zinc? One target of increased free zinc ion concentrations is metal response element-binding transcription factor-1 (MTF-1), which is essential for basal and zinc-induced expression of MT (Heuchel *et al.*, 1994; Giedroc *et al.*, 2001; Laity and Andrews, 2007). Its function as a cellular zinc sensor is based on the coordination chemistry of at least two of its six zinc fingers with a Cys_2His_2 coordination motif. The affinities of the fingers for zinc are thought to vary about 10–50-fold (Potter *et al.*, 2005); this variation is despite the fact that the donor atoms are all the same in the fingers. The affinities are in the nanomolar range, which is lower than in "classic" zinc fingers that typically have picomolar affinities (Laity and Andrews, 2007), but are in the range where zinc sensing is expected on the basis of the above quantitative considerations. The relationship between MTF-1 and MT is such that MT and T regulate the availability of zinc for MTF-1 activation (Zhang *et al.*, 2003). Zinc release from MT, by displacement with cadmium or thiol oxidation by hydrogen peroxide, activates MTF-1. The concentrations of fluctuating zinc ions also match nanomolar or lower zinc affinities of some proteins that are generally not recognized as zinc proteins (Maret *et al.*, 1999). Because of such strong interactions, free zinc ions are very potent cellular effectors and may have regulatory functions at protein sites.

Low nanomolar concentrations of zinc inhibit enzymes of intermediary and energy metabolism, signaling, and mitochondrial function (Maret *et al.*, 1999; Ye *et al.*, 2001). IC_{50} values are 17 nM for protein tyrosine phosphatase 1B (PTP-1B) (Haase and Maret, 2003) and <10 nM for caspase-3 (Maret *et al.*, 1999). When calculations are based on free instead of total zinc concentrations, zinc inhibition of Ca^{2+}-ATPase occurs at approximately 80 pM (Hogstrand *et al.*, 1999). Tonic inhibition of mitochondrial respiration, phosphorylation signaling, and membrane proteins such as the *N*-methyl-D-aspartate (NMDA) receptor is further evidence that zinc is a physiological modulator of protein function (Paoletti *et al.*, 1997; Ye *et al.*, 2001; Haase and Maret, 2003).

Fluctuations must occur with tight control to avoid unspecific reactions and cytotoxic effects of zinc ions. MT is the ideal candidate to exert such control. It is a zinc donor for activating the apoform of sorbitol dehydrogenase (SDH), which has a log K of 11.2 for zinc, and for inhibiting PTP-1B, which has a log K of 7.8 for zinc (Krężel and Maret, 2008). Whether or not these cytosolic enzymes with zinc affinities that differ by at least three orders of magnitude receive zinc from MT depends on the redox poise and the T/(MT+T) molar ratio. At ratios between 0.08 and 0.31 prevailing in tissues and cells, picomolar concentrations of free zinc ions are available from MT for reconstituting apoenzymes with zinc. Under conditions of decreased ratios, such as are found during oxidative stress, nanomolar concentrations of free zinc ions become available from MT/T and affect enzymes that are not zinc metalloenzymes. The close match between zinc availability from MT and the zinc affinity of these proteins suggests a function of MT in controlling cellular free zinc ions. It is also clear from this discussion that the concentrations of free zinc ions are too low to saturate the weak binding site in MT, making it possible for some fraction of the protein to exist in the apoform (thionein), as discussed above.

4. From Chemical Activities to Biological Functions of Metallothionein

Many chemically different compounds can react with the sulfur donors in MT and release zinc (Maret and Vallee, 1998). Among those, major classes are reducible sulfur and selenium compounds (Jacob *et al.*, 1999). Without being comprehensive, several examples may serve to illustrate under which conditions oxidants and electrophiles react with MT. It is known that MT can be nitrosylated (Kröncke *et al.*, 1994). MT-3 reacts much faster with *S*-nitrosothiols than MT-1 or MT-2 owing to specific consensus motifs for nitrosylation, such as ^{21}KCE, in its sequence (Chen *et al.*, 2002). Agents that elevate cytoplasmic calcium and activate calmodulin-dependent nitric oxide (NO) synthase also release zinc from MT (Pearce *et al.*, 2000). The released zinc then activates zinc-dependent MTF-1, which controls the expression of cytoprotective genes (Stitt *et al.*, 2006). This Ca^{2+}-NO-Zn^{2+} pathway is cytoprotective in the lung (St. Croix *et al.*, 2005). However, it can be cytotoxic in the brain because calcium influx into neurons activates syntheses of NO and superoxide, which combine to form peroxynitrite and release cytotoxic levels of zinc (Bossy-Wetzel *et al.*, 2004). Furthermore, MT can be thionylated (Maret, 1994; Maret, 1995). Zinc release during homocysteinylation activates the zinc finger transcription factor Egr-1 (Barbato *et al.*, 2007). Homocysteinylated MT no longer scavenges superoxide ions. Such an impairment of zinc and redox homeostasis has been proposed as a mechanism of how hyperhomocysteinemia, a major risk factor for heart disease and stroke, leads to increased production of reactive species, chronic inflammation, and atherothrombotic disease.

Reactive carbonyls are another group of compounds that react with MT. Aldehydes modify the cysteine ligands in MT, which results in concomitant zinc release (Hao and Maret, 2006). Aldehydes are formed during a variety of conditions such as oxidative stress, lipid peroxidation, hyperglycemia-induced glycations, and ethanol metabolism, or they affect an organism through environmental exposures. The potential for aldehydes to release zinc or to perturb zinc homeostasis provides a new mechanism for their actions. Dopamine quinone is another keto compound that reacts with MT, thus affording protection of dopaminergic neurons (Miyazaki *et al.*, 2007). The threshold of free zinc ion concentrations that separates pathology from physiology in the above processes has not been defined, and this issue requires attention in future studies.

Modulation of zinc availability by sulfur reactivity is only one aspect of the zinc–sulfur interaction, which can also be viewed as modulation of the reducing capacity of sulfur by zinc. MT, therefore, complements the other cellular thiol/disulfide-based redox systems as a zinc-dependent redox system. Given that there are micromolar concentrations of MT in some tissues and that MT is a highly inducible protein, the reducing capacity of MT is certainly not insignificant. Specific redox pairs that couple with T_R and T_O are thioredoxin and methionine sulfoxide reductase (Sagher *et al.*, 2006). A function of MT as an antioxidant and scavenger of reactive species is widely acknowledged (Sato and Bremner, 1993; Kang 1999). However, when such functions are discussed, attention is rarely given to the consequences of zinc release in these reactions.

The capacity of zinc to protect the cell against oxidative damage has been recognized for many years and has led to the notion that zinc itself is an antioxidant (Bray and Bettger, 1990; Powell, 2000). Because zinc is redox-inert in biology, this function must be indirect, i.e. it can only be "pro-antioxidant" (Hao and Maret, 2005). One basis for this function is that part of the cellular zinc-buffering capacity depends on thiols (Krężel *et al.*, 2007). These thiols are protected against oxidation when they bind zinc, but they remain unprotected when there is not enough zinc, e.g. under zinc deficiency which is a pro-oxidant condition (Hennig *et al.*, 1999; Oteiza *et al.*, 2000; Ho *et al.*, 2003). The pro-antioxidant functions also involve the zinc activation of MTF-1. This activation induces thionein (T_R), which has high reducing capacity owing to its 20 cysteines. However, if the cellular zinc-buffering capacity is exhausted, concentrations of free zinc ions will increase, and zinc will bind to other targets and become a pro-oxidant, which results in depletion of cellular energy and increased mitochondrial production of reactive oxygen species (Dineley *et al.*, 2003). Thus, it is a matter of the concentrations of zinc ions and the cellular zinc and redox buffering capacity as to whether zinc is cytotoxic or cytoprotective (Maret and Krężel, 2007; Maret, 2008). MT exerts control over these pro-oxidant and pro-antioxidant functions of zinc.

In summary, the discussion of biological functions of MT requires a speciation analysis *in vivo* because its state determines whether it is a zinc donor, a zinc acceptor, a reductant, or an oxidant. Antioxidant functions and scavenging of reactive species are necessarily linked to perturbation of cellular free zinc ion concentrations.

5. Integrative Biology of Metallothionein: Regulation of Gene Expression and Cellular Translocations

Control of the expression of the T genes is superimposed on the control of the availability of free zinc ions by MT biochemistry (Fig. 3). MT is highly dynamic in terms of the total amount available through isoform- and tissue-specific expression. There is an over 400-fold variation of expression in different cell lines (Woo *et al.*, 1997). T is induced by various forms of stress and by the acute phase response and inflammation (Kägi, 1993), interferons, cytokines, growth factors, and agents that act on nuclear hormone receptors or through the adenylate cyclase, phospholipase C, and Jak/Stat pathways (Andrews, 1990; Ghoshal and Jacob, 2001). A comprehensive treatment of the transcriptional regulation and signaling pathways leading to expression of T is beyond the scope of this chapter; suffice it to say that different levels of control establish that the zinc and redox functions of MT and T are an integral part of cellular biology. Given the low cellular free zinc ion concentrations discussed above, any T synthesized will equilibrate with any MT already

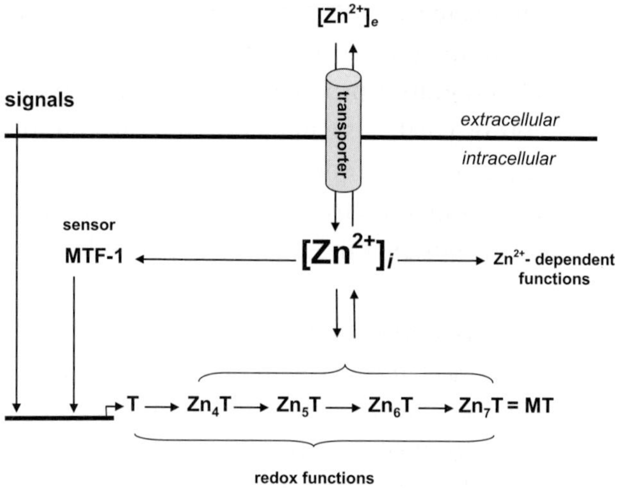

Fig. 3 Integrative cell biology of metallothionein. Membrane transporters (importers and exporters) located at the plasma membrane and at membranes of cellular organelles participate in the control of cellular zinc. Metal response element-binding transcription factor-1 (MTF-1) is a sensor for cellular free zinc ion concentrations, $[Zn^{2+}]_i$. MTF-1 controls basal and zinc-induced expression of thionein (T) genes. Additional zinc-containing transcription factors, such as Sp1 and nuclear hormone receptors, and other signaling pathways also participate in the induction of T. T functions either as a reducing agent or can bind zinc to form MT. The different zinc-loaded forms of MT provide control over free zinc ion concentrations that are potent effectors of cellular proteins.

existing and zinc can only be acquired if it is available. Therefore, the functions of these inducers could relate primarily to the properties of T as a zinc acceptor and reductant.

A major issue is how to link the sulfur and zinc chemistries of MTs to their biological activities. Important developments leading toward this goal are studies showing that MTs occur both extracellularly and intracellularly, that they are redistributed to different compartments in the cell, and that they do not function alone but in concert with other proteins. For these interactions, MTs employ specific structural features.

MT-3 has been isolated from brain tissue (Uchida *et al.*, 1991). Extracellular MT-3 is a growth inhibitory factor (GIF) for neurons (Uchida *et al.*, 1991). This activity of MT is specific for the MT-3 isoform and has been mapped to an N-terminal β-domain ^{6}CPCP motif, which is characteristic of this isoform (Sewell *et al.*, 1995). The molecular basis for the growth inhibitory activity of MT-3 is unknown. MT-3 is released from cultured astrocytes (Uchida *et al.*, 2002) and functions in a pathway that loads neuronal synaptic vesicles with zinc ions (Cole *et al.*, 2000). These synaptic zinc-containing vesicles colocalize with some glutamate-containing vesicles and release their zinc ions into the synaptic cleft when the nerve is stimulated (Frederickson *et al.*, 2005). In this process, MT-3 interacts with the small GTPase Rab3a that interacts with docking and trafficking proteins in the exo-endocytotic cycle of synaptic vesicles (Knipp *et al.*, 2005).

Several cell lines secrete MT (Moltedo *et al.*, 2000; Trayhurn *et al.*, 2000). Cells also take up MT from the extracellular space. This was first shown for kidney cell lines as an endocytotic process that involves the scavenger receptors megalin/cubilin and receptor-associated protein RAP (Erfurt *et al.*, 2003; Klassen *et al.*, 2004; Wolff *et al.*, 2006). When uptake was studied in hepatocarcinoma (HepG2) cells, it was found to be cholesterol-dependent, and that the metals, but not the protein, are translocated to the cytoplasm (Hao *et al.*, 2007). MT-1 and MT-2 bind to megalin and the lipoprotein receptor-related protein-1 (LRP) on cerebellar granule neurons, suggesting that extracellular MTs act in this way on neurons when they promote neurite outgrowth and survival (Fitzgerald *et al.*, 2007; Ambjørn *et al.*, 2008). These authors identified the ^{19}CKCK sequence in MT-2 to be important in this process. Thus, in this experimental system, the biological activity of MT-2 is opposite to that of MT-3, which prevents neurite outgrowth (Uchida *et al.*, 2002).

Another biological activity of MT is its effect on mitochondrial respiration (Ye *et al.*, 2001). MT inhibits mitochondrial respiration in a zinc-dependent manner, while T activates respiration by removing zinc from the inhibitory site. MT is imported into the intermembrane space of liver mitochondria. MT does not have a signal peptide for mitochondrial import. Apparently, additional factors direct trafficking. Proteins, such as MT, are imported into the intermembrane space of mitochondria through a reaction cycle of the protein Mia40 (Tim 40) and the sulfhydryl oxidase, Erv1. Following oxidation of Mia40 by Erv1, a transient disulfide between Mia40 and the protein to be imported is formed. Thiol/disulfide interchange then generates the oxidized imported protein and reduced Mia40 (Mesecke *et al.*, 2005). The observation that only the N-terminal β-domain of MT is active in inhibiting the respiratory chain (Ye *et al.*, 2001) would seem to indicate that this domain is responsible for formation of the transient disulfide and that the isolated α-domain cannot be imported, perhaps because it fails to form the transient disulfide. The lysines of MT are also involved in this process because their modification prevents translocation of MT into the intermembrane space.

There is also nucleocytoplasmic shuttling of MT. This process requires energy (ATP), and it depends on signaling kinases, the state of the cell, and the oxidation of a cytosolic factor (Tsujukawa *et al.*, 1991; Schmidt and Beyersmann, 1999; Apostolova *et al.*, 2000; Woo *et al.*, 2000; Takahashi *et al.*, 2005).

Extensive trafficking of MT between cells and within cells is stimulating further research into the question of whether the receptor-mediated cellular uptake observed in renal, hepatic, and neuronal cells is a general feature of all cells; and is also stimulating investigations to determine the amounts, redox state, and metal load of MT in different subcellular compartments.

Acknowledgments

This work was supported in part by a grant GM 065388 from the National Institutes of Health, a grant from the John Sealy Memorial Endowment Fund, and a sponsored research agreement with Neurobiotex, Inc. (Galveston, TX).

References

Adebodun F, Post JF. Role of intracellular free Ca(II) and Zn(II) in dexamethasone-induced apoptosis and dexamethasone resistance in human leukemic CEM cell lines. *J Cell Physiol* 1995; **163**:80–86.

Aizenman E, Stout AK, Hartnett KA, *et al.* Induction of neuronal apoptosis by thiol oxidation: Putative role of intracellular zinc release. *J Neurochem* 2000; **75**:1878–1888.

Ambjørn M, Asmussen JW, Lindstam M, *et al.* Metallothionein and a peptide modeled after metallothionein, EmtinB, induce neuronal differentiation and survival through binding to receptors of the low-density lipoprotein receptor family. *J Neurochem* 2008; **104**:21–37.

Andrews GK. Regulation of metallothionein gene expression. *Prog Food Nutr Sci* 1990; **14**:193–258.

Apostolova MD, Ivanova IA, Cherian MG. Signal transduction pathways, and nuclear translocation of zinc and metallothionein during differentiation of myoblasts. *Biochem Cell Biol* 2000; **78**:27–37.

Arseniev A, Schultze P, Wörgötter E, *et al.* Three-dimensional structure of rabbit liver [Cd7]metallothionein-2a in aqueous solution determined by nuclear magnetic resonance. *J Mol Biol* 1988; **201**:637–657.

Atar D, Backx PH, Appel MM, *et al.* Excitation-transcription coupling mediated by zinc influx through voltage-dependent calcium channels. *J Biol Chem* 1995; **270**: 2473–2477.

Barbato JC, Catanescu O, Murray K, *et al.* Targeting of metallothionein by L-homocysteine. A novel mechanism for disruption of zinc and redox homeostasis. *Arterioscler Thromb Vasc Biol* 2007; **27**:49–54.

Benters J, Flögel U, Schäfer T, *et al.* Study of the interactions of cadmium and zinc ions with cellular calcium homeostasis using ^{19}F-NMR spectroscopy. *Biochem J* 1997; **322**:793–799.

Bossy-Wetzel E, Talantova MV, Lee WD, *et al.* Crosstalk between nitric oxide and zinc pathways to neuronal cell death involving mitochondrial dysfunction and p38-activated K^+ channels. *Neuron* 2004; **41**:351–365.

Bozym RA, Thompson RB, Stoddard AK, Fierke CA. Measuring picomolar intracellular exchangeable zinc in PC-12 cells using a ratiometric fluorescence biosensor. *ACS Chem Biol* 2006; **1**:103–111.

Bray TM, Bettger WJ. The physiological role of zinc as an antioxidant. *Free Radic Biol Med* 1990; **8**:281–291.

Chen Y, Irie Y, Keung WM, Maret W. *S*-nitrosothiols react preferentially with zinc thiolate clusters of metallothionein III through transnitrosation. *Biochemistry* 2002; **41**: 8360–8367.

Chen Y, Maret W. Catalytic selenols couple the redox cycles of metallothionein and glutathione. *Eur J Biochem* 2001; **268**:3346–3353.

Choi DW, Koh JY. Zinc and brain injury. *Annu Rev Neurosci* 1988; **21**:347–375.

Cole TB, Robbins CA, Wenzel HJ, *et al.* Seizures and neuronal damage in mice lacking vesicular zinc. *Epilepsy Res* 2000; **39**:153–169.

Cousins RJ, Liuzzi JP, Lichten LA. Mammalian zinc transport, trafficking, and signals. *J Biol Chem* 2006; **281**:24085–24089.

Dineley KE, Votyakova TV, Reynolds IJ. Zinc inhibition of cellular energy production: Implications for mitochondria and neurodegeneration. *J Neurochem* 2003; **85**:563–570.

Erfurt C, Roussa E, Thévenod F. Apoptosis by Cd^{2+} or CdMT in proximal tubule cells: Different uptake routes and permissive role of endo/lysosomal CdMT uptake. *Am J Physiol Cell Physiol* 2003; **285**:C1367–C1376.

Feng W, Benz FW, Cai J, *et al.* Metallothionein disulfides are present in metallothionein-overexpressing transgenic mouse heart and increase under conditions of oxidative stress. *J Biol Chem* 2006; **281**:681–687.

Fitzgerald M, Nairn P, Bartlett CA, *et al.* Metallothionein-IIA promotes growth via the megalin receptor. *Exp Brain Res* 2007; **183**:171–180.

Frederickson CJ, Koh J-Y, Bush AI. The neurobiology of zinc in health and disease. *Nat Rev Neurosci* 2005; **6**:449–462.

Ghoshal K, Jacob ST. Regulation of metallothionein gene expression. *Prog Nucleic Acid Res Mol Biol* 2001; **66**:357–384.

Giedroc DP, Chen X, Apuy JL. Metal response element (MRE)-binding transcription factor-1 (MTF-1): Structure, function, and regulation. *Antioxid Redox Signal* 2001; **3**:577–596.

Haase H, Maret W. Intracellular zinc fluctuations modulate protein tyrosine phosphatase activity in insulin/insulin-like growth factor-1 signaling. *Exp Cell Res* 2003; **291**: 289–298.

Haase H, Maret W. A differential assay for the reduced and oxidized states of metallothionein and thionein. *Anal Biochem* 2004; **333**:19–26.

Hao Q, Hong S-H, Maret W. Lipid raft-dependent endocytosis of metallothionein in HepG2 cells. *J Cell Physiol* 2007; **210**:428–435.

Hao Q, Maret W. Imbalance between pro-oxidant and pro-antioxidant functions of zinc in disease. *J Alzheimers Dis* 2005; **8**:161–170.

Hao Q, Maret W. Aldehydes release zinc from proteins. A pathway from oxidative stress/lipid peroxidation to cellular functions of zinc. *FEBS J* 2006; **273**:4300–4310.

Hennig B, Meerarani P, Toborek M, McClain CJ. Antioxidant-like properties of zinc in activated endothelial cells. *J Am Coll Nutr* 1999; **18**:152–158.

Heuchel R, Radtke F, Georgiev O, *et al.* The transcription factor MTF-1 is essential for basal and heavy-metal-induced metallothionein gene expression. *EMBO J* 1994; **13**: 2870–2875.

Ho E, Courtemanche C, Ames BN. Zinc deficiency induces oxidative DNA damage and increases p53 expression in human lung fibroblasts. *J Nutr* 2003; **133**:2543–2548.

Hogstrand C, Verbost PM, Wendelaar Bonga SE. Inhibition of human erythrocyte Ca^{2+}-ATPase by Zn^{2+}. *Toxicology* 1999; **133**:139–145.

Jacob C, Maret W, Vallee BL. Selenium redox biochemistry of zinc-sulfur coordination sites in proteins and enzymes. *Proc Natl Acad Sci USA* 1999; **96**:1910–1914.

Kägi JHR. Evolution, structure and chemical activity of class I metallothioneins: An overview. In: Suzuki KT, Imura N, Kimura M (eds.), *Metallothionein III*, Birkhäuser, Basel, 1993, pp. 29–55.

Kägi JHR, Vallee BL. Metallothionein: A cadmium- and zinc-containing protein from equine renal cortex. *J Biol Chem* 1960; **235**:3460–3465.

Kang YJ. The antioxidant function of metallothionein in the heart. *Proc Soc Exp Biol Med* 1999; **222**:263–273.

Klassen RB, Crenshaw K, Kozyraki R, *et al.* Megalin mediates renal uptake of heavy metal metallothionein complexes. *Am J Physiol Renal Physiol* 2004; **287**:F393–F403.

Knipp M, Meloni G, Roschitzki B, Vašák M. Zn_7metallothionein-3 and the synaptic vesicle cycle: Interaction of metallothionein-3 with the small GTPase Rab3A. *Biochemistry* 2005; **44**:3159–3165.

Kojima Y, Berger C, Vallee BL, Kägi JHR. Amino-acid sequence of equine renal metallothionein-1B. *Proc Natl Acad Sci USA* 1976; **73**:3413–3417.

Krężel A, Hao Q, Maret W. The zinc/thiolate redox biochemistry of metallothionein and the control of zinc ion fluctuations in cell signaling. *Arch Biochem Biophys* 2007; **463**: 188–200.

Krężel A, Maret W. Zinc buffering capacity of a eukaryotic cell at physiological pZn. *J Biol Inorg Chem* 2006; **11**:1049–1062.

Krężel A, Maret W. Different redox states of metallothionein/thionein in biological tissue. *Biochem J* 2007a; **402**:551–558.

Krężel A, Maret W. The dual nanomolar and picomolar Zn(II) binding properties of metallothionein. *J Am Chem Soc* 2007b; **129**:10911–10921.

Krężel A, Maret W. Thionein/metallothionein control Zn(II) availability and the activity of enzymes. *J Biol Inorg Chem* 2008; **13**:401–409.

Kröncke K-D, Fehsel K, Schmid T, *et al.* Nitric oxide destroys zinc-sulfur clusters inducing zinc release from metallothionein and inhibition of the zinc finger-type yeast transcription factor LAC9. *Biochem Biophys Res Commun* 1994; **200**:1105–1110.

Laity JH, Andrews GK. Understanding the mechanism of zinc-sensing by metal-response element binding transcription factor-1 (MTF-1). *Arch Biochem Biophys* 2007; **463**:201–210.

Maret W. Oxidative metal release from metallothionein via zinc-thiol/disulfide interchange. *Proc Natl Acad Sci USA* 1994; **91**:237–241.

Maret W. Metallothionein/disulfide interactions, oxidative stress, and the mobilization of cellular zinc. *Neurochem Int* 1995; **27**:111–117.

Maret W. The function of zinc metallothionein: A link between cellular zinc and redox state. *J Nutr* 2000; **130**:1455S–1458S.

Maret W. The cellular zinc and redox states converge in the metallothionein/thionein pair. *J Nutr* 2003; **133**:1460S–1462S.

Maret W. Zinc and sulfur: A critical biological partnership. *Biochemistry* 2004; **43**: 3301–3309.

Maret W. Zinc coordination environments in proteins as redox sensors and signal transducers. *Antioxid Redox Signal* 2006; **8**:1419–1441.

Maret W. Metallothionein redox biology in the cytoprotective and cytotoxic functions of zinc. *Exp Gerontol* 2008; **43**:363–369.

Maret W, Jacob C, Vallee BL, Fischer EH. Inhibitory sites in enzymes: Zinc removal and reactivation by thionein. *Proc Natl Acad Sci USA* 1999; **96**:1936–1940.

Maret W, Krężel A. Cellular zinc and redox buffering capacity of metallothionein/thionein in health and disease. *Mol Med* 2007; **13**:371–375.

Maret W, Vallee BL. Thiolate ligands in metallothionein confer redox activity on zinc clusters. *Proc Natl Acad Sci USA* 1998; **95**:3478–3482.

Margoshes M, Vallee BL. A cadmium protein from equine kidney cortex. *J Am Chem Soc* 1957; **79**:4813.

Mesecke N, Terziyska N, Kozany C, *et al.* A disulfide relay system in the intermembrane space of mitochondria that mediates protein import. *Cell* 2005; **121**:1059–1069.

Miyazaki I, Asanuma M, Hozumi H, *et al.* Protective effects of metallothionein against dopamine-induced dopaminergic neurotoxicity. *FEBS Lett* 2007; **581**:5003–5008.

Moltedo O, Verde C, Capasso A, *et al.* Zinc transport and metallothionein secretion in the intestinal cell line Caco-2. *J Biol Chem* 2000; **275**:31819–31825.

Oteiza PI, Clegg MS, Zago MP, Keen CL. Zinc deficiency induces oxidative stress and AP-1 activation in 3T3 cells. *Free Radic Biol Med* 2000; **28**:1091–1099.

Paoletti P, Ascher P, Neyton J. High-affinity zinc inhibition of NMDA NR1-NR2A receptors. *J Neurosci* 1997; **17**:5711–5725.

Pearce LL, Gandley RE, Han W, *et al.* Role of metallothionein in nitric oxide signaling as revealed by a green fluorescent fusion protein. *Proc Natl Acad Sci USA* 2000; **97**: 477–482.

Peck EJ Jr, Ray WJ Jr. Metal complexes of phosphoglucomutase *in vivo*. *J Biol Chem* 1971; **246**:1160–1167.

Potter BM, Feng LS, Parasuram P, *et al.* The six fingers of metal-responsive element binding transcription factor-1 form stable and quasi-ordered structures with relatively small differences in zinc affinities. *J Biol Chem* 2005; **280**:28529–28540.

Powell SR. The antioxidant properties of zinc. *J Nutr* 2000; **130**:1447S–1454S.

Quaife CJ, Findley SD, Erickson JC, *et al.* Induction of a new metallothionein isoform (MT-IV) occurs during differentiation of stratified squamous epithelia. *Biochemistry* 1994; **33**:7250–7259.

Robbins AH, McRee DE, Williamson M, *et al.* Refined crystal structure of Cd, Zn metallothionein at 2.0 Å resolution. *J Mol Biol* 1991; **221**:1269–1293.

Sagher D, Brunell D, Hejtmancik JF, *et al.* Thionein can serve as a reducing agent for the methionine sulfoxide reductases. *Proc Natl Acad Sci USA* 2006; **103**:8656–8661.

Sato M, Bremner I. Oxygen free radicals and metallothionein. *Free Radic Biol Med* 1993; **14**:325–337.

Schmidt C, Beyersmann D. Transient peaks in zinc and metallothionein levels during differentiation of 3T3L1 cells. *Arch Biochem Biophys* 1999; **364**:91–98.

Sensi SL, Ton-That D, Sullivan PG, *et al.* Modulation of mitochondrial function by endogenous Zn^{2+} pools. *Proc Natl Acad Sci USA* 2003; **100**:6157–6162.

Sewell AK, Jensen LT, Erickson JC, *et al.* The bioactivity of metallothionein-3 correlates with its novel beta-domain sequence rather than metal binding properties. *Biochemistry* 1995; **34**:4740–4747.

Simons TJB. Intracellular free zinc and zinc buffering in human red blood cells. *J Membr Biol* 1991; **123**:63–71.

Smith PJ, Wiltshire M, Davies S, *et al.* DNA damage-induced [Zn(2+)](i) transients: Correlation with cell cycle arrest and apoptosis in lymphoma cells. *Am J Physiol Cell Physiol* 2002; **283**:C609–C622.

Spahl DU, Berendji-Grün D, Suschek CV, *et al.* Regulation of zinc homeostasis by inducible NO synthase-derived NO: Nuclear metallothionein translocation and intranuclear Zn^{2+} release. *Proc Natl Acad Sci USA* 2003; **100**:13952–13957.

St. Croix CM, Leelavaninchkul K, Watkins SC, *et al.* Nitric oxide and zinc homeostasis in acute lung injury. *Proc Am Thorac Soc* 2005; **2**:236–242.

St. Croix CM, Wasserloos KJ, Dineley KE, *et al.* Nitric oxide-induced changes in intracellular zinc homeostasis are mediated by metallothionein/thionein. *Am J Physiol Lung Cell Mol Physiol* 2002; **282**:L185–L192.

Stitt MS, Wasserloos KJ, Tang X, *et al.* Nitric oxide-induced nuclear translocation of the metal responsive transcription factor, MTF-1 is mediated by zinc release from metallothionein. *Vasc Pharmacol* 2006; **44**:149–155.

Takahashi Y, Ogra Y, Suzuki KT. Nuclear trafficking of metallothionein requires oxidation of a cytosolic partner. *J Cell Physiol* 2005; **202**:563–569.

Trayhurn P, Duncan JS, Wood AM, Beattie JH. Regulation of metallothionein gene expression and secretion in rat adipocytes differentiated from preadipocytes in primary culture. *Horm Metab Res* 2000; **32**:542–547.

Tsujikawa K, Imai K, Katutani M, *et al.* Localization of metallothionein in nuclei of growing primary cultured adult rat hepatocytes. *FEBS Lett* 1991; **283**:239–242.

Turan B, Fliss H, Désilets M. Oxidants increase intracellular free Zn^{2+} concentration in rabbit ventricular myocytes. *Am J Physiol Heart Circ Physiol* 1997; **272**: H2095–H2106.

Uchida Y, Gomi F, Masumizu T, Miura Y. Growth inhibitory factor prevents neurite extension and the death of cortical neurons caused by high oxygen exposure through hydroxyl radical scavenging. *J Biol Chem* 2002; **277**:32353–32359.

Uchida Y, Takio K, Titani K, *et al.* The growth inhibitory factor that is deficient in the Alzheimer's disease brain is a 68 amino acid metallothionein-like protein. *Neuron* 1991; **7**:337–347.

Vallee BL. The function of metallothionein. *Neurochem Int* 1995; **27**:23–33.

Vašák M. Metal removal and substitution in vertebrate and invertebrate metallothioneins. *Methods Enzymol* 1991; **205**:452–458.

Wolff NA, Abouhamed M, Verroust PJ, Thévenod F. Megalin-dependent internalization of cadmium-metallothionein and cytotoxicity in cultured renal proximal tubule cells. *J Pharmacol Exp Ther* 2006; **318**:782–791.

Woo ES, Dellapiazza D, Wang AS, Lazo JS. Energy-dependent nuclear binding dictates metallothionein localization. *J Cell Physiol* 2000; **182**:69–76.

Woo ES, Monks A, Watkins SC, *et al.* Diversity of metallothionein content and subcellular localization in the National Cancer Institute tumor panel. *Cancer Chemother Pharmacol* 1997; **41**:61–68.

Yang Y, Maret W, Vallee BL. Differential fluorescence labeling of cysteinyl clusters uncovers high tissue levels of thionein. *Proc Natl Acad Sci USA* 2001; **98**:5556–5559.

Ye B, Maret W, Vallee BL. Zinc metallothionein imported into liver mitochondria modulates respiration. *Proc Natl Acad Sci USA* 2001; **98**:2317–2322.

Zhang B, Georgiev O, Hagmann M, *et al.* Activity of metal-responsive transcription factor 1 by toxic heavy metals and H_2O_2 *in vitro* is modulated by metallothionein. *Mol Cell Biol* 2003; **23**:8471–8485.

Chapter 3

METALLOTHIONEINS AND NEURODEGENERATIVE DISEASES

Silvia Bolognin and Paolo Zatta

Although metallothioneins (MTs) were discovered nearly 40 years ago, their functional role has still not been completely clarified. The role of MTs in the central nervous system has in particular become an intense focus of scientific research. Many papers have confirmed the active and peculiar role played by these proteins in neurodegenerative disorders, even if contrasting results are still present. The involvement of MTs in various neurodegenerative diseases (Alzheimer's disease, frontotemporal dementia, Binswanger's disease, Parkinson's disease, amyotrophic lateral sclerosis, and prion protein disease) is herein reported.

Keywords: Neurodegeneration; aging; oxidative stress; metal ions.

1. Introduction

The precise role of MTs in the central nervous system (CNS) is yet to be fully established (reviewed in Hidalgo *et al.*, 2001), despite the wide number of papers published. The presence in the brain of the three isoforms MT-I, MT-II, and MT-III has been well documented, but their function is far from being completely understood. It has been demonstrated in rats (Fig. 1) and bovines (Zatta *et al.*, 2006) that aging *per se* is able to bring an increase in MT levels in the brain. Moreover, recent studies revealed that MTs tend to be highly expressed in both astrocytes and hippocampal neurons as well as in the aging human brain (Mocchegiani *et al.*, 2004). This increased expression might represent a physiological defense against stress

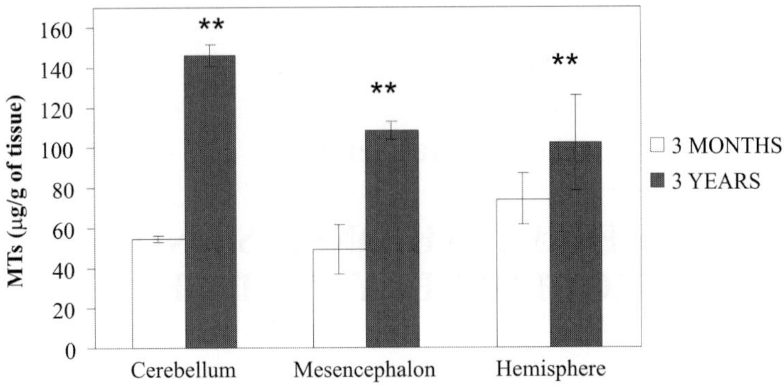

Fig. 1 Determination of MT content in three brain regions of young (3 months) and old (3 years) rats by silver-saturation method. The data represented are mean ± standard deviation (SD). ** $P < 0.01$.

factors in order to protect neurons. This natural response seems to be altered or enhanced in pathological conditions related to aging, such as neurodegenerative diseases.

In this chapter, we review the relationship between this family of proteins and some of the most common age-related disorders.

2. Alzheimer's Disease

Alzheimer's disease (AD) is the most common form of dementia, and its incidence is increasing together with the growth of the elderly population. It is characterized by the deterioration of selective cognitive domains, mainly related to memory (Mattson, 2004). The AD brain is marked by a severe loss of synapses and neurons and a huge imbalance of neurotrophic factors in the hippocampus and cerebral cortex. The neuropathological hallmarks of AD include the extracellular deposition of the β-amyloid peptide in senile plaques, and the intraneuronal aggregates of paired helical filaments of the hyperphosphorylated tau protein in neurofibrillary tangles in characteristic brain regions.

Several observations indicate that metal homeostasis is altered in AD (Liu *et al.*, 2006; Lovell *et al.*, 1998; Miu and Benga, 2006), and

MTs are a group of proteins with an unusually high metal-binding capacity (see Chapter 1) and with proposed functions in metal detoxification and storage as well as in prevention of free-radical cellular damage. Moreover, the hippocampus, which is greatly affected in AD, is the only region in the cortex where MT-I-/MT-II and MT-III are coexpressed (Aschner *et al.*, 1997). These considerations may support the link between some pathological features of the disease and MT expression in the brains of AD patients.

A distinction must be made between the three isoforms of MTs present in the brain, since there is compelling evidence that MT-I and MT-II are upregulated in Alzheimer's disease (Duguid *et al.*, 1989; Adlard *et al.*, 1998; Zambenedetti *et al.*, 1998; Chuah and Getchell, 1999), but contrasting results are reported for MT-III.

Immunohistochemical study of the cerebrum of AD patients revealed that the expression of MT-I/MT-II occurs in astrocytes and microcapillaries, both in the cortex and in the white matter [Fig. 2(B)] (Zambenedetti *et al.*, 1998). They are especially prominent in layers II–VI of the frontal cortex and in the hippocampus (Jorgensen and Balazs, 1993). On the contrary, in the same topographical areas from control patients, a much lower positivity in MT expression was detected [Fig. 2(A)]. The astrocytes in AD brain samples seem to be of the protoplasmatic type instead of fibrous, and have a swollen cytoplasm (Zambenedetti *et al.*, 1998). Among many hypotheses trying to explain the neurodegenerative mechanisms underlying Alzheimer's disease, it has been proposed that activated microglia may release proinflammatory cytokines and other inflammatory mediators, as well as oxidative species, that may account for the neuronal damage in patients with AD (Carrasco *et al.*, 1999; Barnham *et al.*, 2004; Streit, 2004). Understanding the role played by astrocytes, and the reason for their abundant activation reported in AD, may elucidate one important etiopathogenetic aspect of the disease. In contrast, some authors reported that the increased MT-I/MT-II expression correlated with the increase in glial fibrillary acidic protein (GFAP) mRNA, and so they postulated that MT-I/MT-II overexpression may

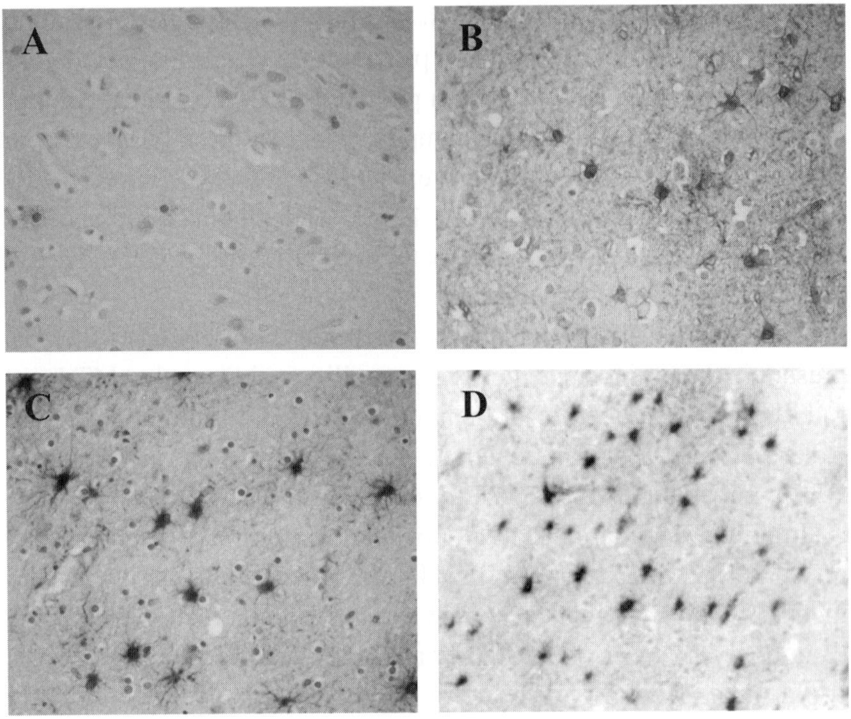

Fig. 2 Examples of immunohistochemical staining for MT-I/MT-II in (A) control, (B) Alzheimer's disease, (C) Pick's disease, and (D) Binswanger's disease cases. Calibration bar: 50 μm for all images.

follow glial cell proliferation (Nakajima and Suzuki, 1995). In this view, this increase may be more closely related to a general aspect of the pathology of AD rather than a specific response to pathological hallmark features such as metal ion dishomeostasis or β-amyloid plaque formation (Adlard *et al.*, 1998). All of these observations appear to be important in shedding some light on the mechanisms underlying the activation of astrocytes, and on the effects of this event on the development of AD. The relevant expression of MTs in astrocytes and microcapillaries may be the consequence of many possible synergistic dismetabolic phenomena observed in AD.

Even more uncertain is the information about the role of MT-III in AD. Many reports indicate that mRNA for MT-III is downregulated in tissue from Alzheimer's disease patients (Uchida *et al.*, 1991; Tsuji *et al.*, 1992; Naruse *et al.*, 1994; Carrasco *et al.*, 1999; Yu *et al.*, 2001), and that this may therefore increase the aberrant neuronal sprouting associated with the disease (Vickers, 1997). However, other studies did not confirm these data (Erickson *et al.*, 1994; Amoureux *et al.*, 1997), and instead showed immunohistochemically that MT-III remains essentially unaltered in AD and control patients (Hidalgo *et al.*, 2001; Hidalgo *et al.*, 2006). Since MT-III is able to inhibit neuronal growth (Uchida *et al.*, 1991), it has been speculated that decreased MT-III expression may be an endogenous protective response to facilitate neurite extension (Yuguchi *et al.*, 1995). An alternative explanation is that the decreasing MT-III induced by β-amyloid plaque formation has a neuroprotective function and prevents the development of AD (El Ghazi *et al.*, 2006). The peculiar role played by MT-III is confirmed by the fact that transcriptional regulation of MT-III differs from that of MT-I/MT-II (Palmiter, 1995; Belloso *et al.*, 1996; Hidalgo *et al.*, 1997). These data suggest that MT-III is not involved in metal ion homeostasis as MT-I/MT-II seem to be. Moreover, MT-III, but not MT-I/MT-II, was reported to modulate both the neurotropic and neurotoxic effects of β-amyloid (Irie and Keung, 2001). According to Miyazaki *et al.* (2002), MT-III in aged animals may be closely related to neurodegeneration. In addition, Martin *et al.* (2006) showed that in brains from the Tg2576 transgenic mouse model of AD, the protein level of MT-III was reduced compared with same-age, transgenic-negative control mice. The decrease in MT-III was attributed to its degradation in response to the pathological changes that develop in this mouse model of AD.

Altogether, these data suggest that, despite the similarity among MT isomers, their involvement in AD is different. Further studies are necessary to shed some light on this important issue, considering also that the distribution of MT isoforms in the brain correlates with the neuroanatomical distribution of senile plaques (Masters *et al.*, 1994). Clarification of the roles of MTs in AD will lead to a clearer understanding of the physiological role of MTs and the molecular mechanism of AD.

3. Frontotemporal Dementia

Frontotemporal dementia (FTD) is a general term coined for a group of presenile, slowly progressive neurodegenerative dementias, which have a distinctive pathological substrate but common clinical manifestations based on neuronal degeneration in the frontotemporal lobe (Lund and Manchester Groups, 1994). Diseases classified as types of FTD include Pick's disease (PiD), corticobasal degeneration (CBD), FTD with motor neuron disease (FTD+MND), frontal lobe degeneration (FLD), dementia-parkinsonism-amyotrophy complex (DPAC), familial nonspecific dementia mapping to chromosome 3, non-Alzheimer degenerative dementia, and a number of other infrequent syndromes with dementia and focal neurological signs (Cooper *et al.*, 1995; Jackson and Lowe, 1996; Lowe, 1997; Lowe and Spillantini, 1998).

We investigated immunohistochemically the regional expression of MT-I/MT-II in human brains affected both by classical Pick's disease, with typical intracytoplasmic inclusions (i.e. Pick bodies), and by FLD without Pick bodies (Zatta *et al.*, 2005b) [Fig. 2(c)]. MT expression was found highly increased in the frontal, temporal, and allocortical regions, and less significantly in the occipital region in all subtypes of FTD. A similar positivity was clearly observed in the hippocampal area for the FTD case with respect to the control. Important differences were observed among the three layers. FTD and PiD cases displayed an elevated MT-I/MT-II expression in the deep layer compared to controls, while this difference was less marked in the molecular layer and in the white matter. MT expression was also rather relatively high in the subcortical white matter, which is typically involved in the atrophy of FTD. It was not possible to register any relevant difference in MT expression in the two subtypes of FTD, probably because of the relatively small sample size available (13 cases).

The explanation for the increment of MTs may lie in both metal ions and free radical scavenging which is acted on by MTs. As a matter of fact, oxidative damage can favor the probability of age-dependent neurodegenerative diseases. A consequence of oxidative stress was considered to be abnormal phosphorylation of proteins, which occurs in many different diseases such as AD, Down syndrome, and PiD (Foster *et al.*, 1997). It has also been speculated that overaccumulation of some metal ions, especially in the hippocampus, could be linked to the pathology; and with this view, a chelation therapy using ethylenediaminetetraacetic acid (EDTA) was proposed (Tissot *et al.*, 1975). MT overexpression in the FTD, as in other neurodegenerative diseases, may reflect an endogenous response to persistent oxidative stress and/or to free metal ion toxicity.

4. Binswanger's Disease

Binswanger's disease (BD) is a subacute form of hypertensive encephalopathy characterized by a marked arteriosclerotic vascular impairment with involvement of small subcortical arteries (Furuta *et al.*, 1991; Tanoi *et al.*, 2000) and atrophy of the central hemispheric white matter (Esiri *et al.* 1997; Jellinger and Neumayer, 1964).

The cause of BD is still unclear. There is a general agreement to consider cortical and/or white matter astrocytosis and fibrillary gliosis as typical events occurring in BD (Esiri *et al.*, 1997). Immunohistochemical analysis of the brains of patients affected by Binswanger's disease revealed that MT-I/MT-II expression is higher compared with control patients [Fig. 2(D)]. This increase was evident in both white and gray matter (Fig. 3) (Zambenedetti *et al.*, 2002).

According to our data, a significant astrocytosis with marked expression of MT-I/MT-II is present in BD. MT-I/MT-II high positivity may coincide with the demonstration of a high, statistically significant oxidation of α-tocopherol-quinone in BD (Tohgi *et al.*, 1990), providing compelling evidence for the occurrence of oxidative stress phenomena in this disorder. It has also been proposed

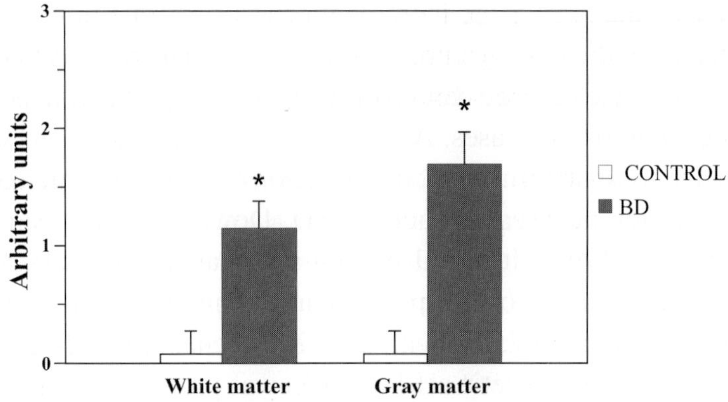

Fig. 3 Immunohistochemical analysis of the brains of patients affected by Binswanger's disease and the controls. The histogram represents the positivity of astrocytes to anti–MT-I/MT-II antibodies in both the white and gray matter. Arbitrary units for the grading system were established to grade the extent of MT-I/MT-II–positive astrocytes per optic field ($\times 50$) observed ($n = 10$), ranging from 0 to 3 elements (a score of 0); from 4 to 5 elements (a score of 0.5); from 7 to 10 elements (a score of 1); from 10 to 14 elements (a score of 2); and for more than 15 elements (a score of 3). The data represented are mean $\pm$ SD. $^*P < 0.02$.

that, in the pathogenesis of BD, alteration of the blood-brain barrier (BBB) may contribute to produce diffuse white matter degeneration (Takahashi *et al.*, 1996) with enhanced expression of MTs. Moreover, numerous reactive astrocytes in the pathology were positive for endothelin-1, which is a strong vasoconstrictor peptide that has been detected in the plasma of BD patients (Chen *et al.*, 1996). As with many cytokines and interleukins, endothelin-1 was able to induce the expression of MTs (Kaji *et al.*, 1993), suggesting that the significance of MTs may likely be an attempt to protect the brain by reactive astrocytes.

5. Parkinson's Disease

Parkinson's disease (PD) is characterized by a progressive loss of dopaminergic neurons in the substantia nigra zona compacta, resulting in muscular rigidity, postural irregularity, and body tremors.

The cause of the wide cell death which occurs in neurons of PD patients is not known. Dysregulated copper homeostasis and copper-mediated oxidative stress have been suggested to play a role in PD (Rasia *et al.*, 2005).

Another important contributing factor to the neuronal damage seems to be the elevated activation of nitric oxide synthase (NOS) and nitric oxide (NO) in the substantia nigra of patients with PD. Significant literature is available confirming the involvement of oxidative stress and nitrative stress in the etiopathogenesis of PD (Beal, 1998; Dawson and Dawson, 1998; Bringold *et al.*, 2000; Hirsch and Hunot, 2002), and therefore the involvement of MT proteins in the disease has been hypothesized. This hypothesis finds support in experiments performed in transgenic mice which showed that regional induction of MTs in the brain might be helpful in the prevention of PD in the aging brain (Ebadi *et al.*, 2005a). Free radicals produced by damaged neurons could activate lipid peroxidation of membranes and the peroxidation of protein and DNA (Jenner, 1998). It is believed that MTs, with their thiol groups, could act as a free-radical scavenger (Hussain *et al.*, 1996). The responsiveness of MTs to oxidative stress and their attributed role against reactive oxygen species (ROS) may explain the link with this family protein to the pathology.

Even if the exact mechanism of MT neuroprotection against reactive species is yet to be fully established, Ebadi and Sharma (2006) proposed that MTs could provide neuroprotection from $ONOO^-$ by increasing the rate of coenzyme Q_{10} (CoQ_{10}) synthesis. It has been proved that MT gene overexpression in the brain of MT transgenic mice inhibits nitration of α-synuclein and preserves mitochondrial CoQ_{10} levels to afford neuroprotection against oxidative and nitrative stresses of the aging brain (Cai *et al.*, 2001; Ebadi *et al.*, 2005b; Ebadi and Sharma, 2006; Sharma and Ebadi, 2003). Moreover, it has been reported that MT overexpression in neurons can reduce ROS synthesis, caspase-3 activation, and apoptosis (Ebadi and Sharma, 2006).

Another important aspect is that the neuronal death which occurs in PD can be correlated with inflammatory reactions characterized by the activation of microglial cells (McGeer *et al.*, 1988). This is confirmed by the elevated levels of cytokines found in the cerebrospinal fluid of PD patients (Mogi *et al.*, 1996). Synthesis of MT-I/MT-II was induced by cytokines such as interleukin-1 (IL-1), IL-6, and interferon-α (Aschner, 1998). According to Mirza *et al.* (2000), MT-I/MT-II were not upregulated in astrocytes of the substantia nigra of PD patients; on the contrary, microglial cells appeared to be consistently activated (Banati *et al.*, 1998). The lack of astrocytosis, which occurs in other neurodegenerative diseases such as AD, suggests that the inflammatory process in PD is different and slowly progressive. The fact that MT-I/MT-II in astrocytes were not upregulated in PD may also indirectly suggest that the blood-brain barrier (BBB) is left unaltered (Mirza *et al.*, 2000). In addition, release of metals capable of inducing MT-I/MT-II, such as zinc and copper, during the inflammatory process seems to be unable to induce MT-I/MT-II gene expression in astrocytes (Mirza *et al.*, 2000).

Other authors proposed that MTs may reduce the dangerous effects of free radicals and ROS by releasing zinc to the neuronal membrane (Shiraga *et al.*, 1993), postulating an antioxidant role for both zinc and MTs (Bray and Bettger, 1990; Sato and Bremner, 1993). Zinc may prevent excessive NO production (Persechini *et al.*, 1995), and one possible source of zinc is its release from MTs (Cuajungco and Lees, 1997). A shift in the glutathione redox balance, as a consequence of oxidative stress events, can accelerate the release of zinc from MTs. Dyshomeostasis of zinc ions has been actually found in PD, albeit with contrasting results (Forsleff *et al.*, 1999; Jimenéz-Jimenéz *et al.*, 1998; Dexter *et al.*, 1991). It has also been proposed that brain areas such as striatum, which contains a high level of iron and low amounts of MTs, may be particularly susceptible to oxidative stress (Rojas *et al.*, 2001).

Taken together, all of these findings suggest that oxidative stress is part of the pathological events leading to PD, and that the increase of

MT-I/MT-II in the PD brain could be a useful tool to prevent damage (Ebadi *et al.*, 1998). The contribution of MT-III to the pathology is less clear. Sogawa *et al.* (2001) reported a downregulation of MT-III; it has also been suggested that MT-III could be the source of toxic zinc amounts in neurons, confirming that MTs may not always have a protective role (Lee *et al.*, 2003). Clarification of this issue will probably lead to the use of tools capable of promoting an antioxidant strategy together with an enhancement of regional induction of MTs in the brain, which could be a helpful therapeutical approach.

6. Amyotrophic Lateral Sclerosis

Amyotrophic lateral sclerosis (ALS) is a neurodegenerative disease characterized by selective loss of motor neurons in the cerebral cortex, brain stem, and spinal cord, resulting in muscular atrophy (including respiratory muscles), complete paralysis, and death (Rowland and Shneider, 2001).

The exact etiology of the disease is unknown, but approximately 10% of ALS cases are caused by inherited mutations in the gene encoding Cu/Zn-superoxide dismutase 1 (SOD1), a cytosolic enzyme with a zinc and copper binding site, which catalyzes the conversion of superoxide radicals to hydrogen peroxide (Rosen *et al.*, 1993). The toxicity of SOD1 mutants seems not to be due to the loss of enzymatic activity, but rather to a toxic gain of function (Borchelt *et al.*, 1994; Liu *et al.*, 1998; Bruijn *et al.*, 2004). Several studies demonstrated that mutations in SOD1 altered its metal affinity or coordination (Goto *et al.*, 2000), and in particular decreased the affinity of SOD1 for zinc up to 50-fold compared to the wild-type form (Crow *et al.*, 1997) and increased the affinity for copper (Crow *et al.*, 1997; Lyons *et al.*, 1996). SOD1-decreased affinity for zinc led to an increment in nitrotyrosine formation and induced apoptosis in cultured motor neurons (Estevez *et al.*, 1999); while copper-increased affinity enhanced copper-mediated oxidative stress, which may lead to neuronal death (Said Ahmed *et al.*, 2000; Wiedau-Pazos *et al.*, 1996). Some evidence supports this hypothesis, since copper-chelating agents such as D-penicillamine and trientine extend survival in SOD1 transgenic mice (Andreassen *et al.*, 2001; Nagano *et al.*, 2003). In this context, the ability of MTs to maintain transition metal homeostasis (Park *et al.*, 2001; You *et al.*, 2002) has been strongly linked with the pathology. The involvement of MTs in ALS pathogenesis was also suggested by data showing MT-I/MT-II upregulation in astrocytes of the spinal cord from both ALS patients and mutant SOD1 transgenic mice (Blaauwgeers *et al.*, 1996; Gong and Elliott, 2000; Sillevis Smitt *et al.*, 1992). This elevated expression of MT-I/MT-II may be the consequence of perturbation of astrocytic function due to the expression of mutant SOD1 in astrocytes.

The information is scarce regarding the MT-III isoform. It has been reported that MT-III mRNA is increased in spinal motor neurons and eventually in astrocytes of mutant SOD1 mice (Gong and Elliott, 2000). Studies of crossing mutations between SOD1 transgenic mice and MT knockout mice led to a reduction in lifespan in both MT-I/MT-II knockout and MT-III knockout mice (Puttaparthi *et al.*, 2002), suggesting a protective role for all MT isoforms in the brain. Moreover, Gong and Elliot (2000) reported an increase in motor neuronal MT-III expression in transgenic SOD1 mice, even if this was not accompanied by an increase in this protein expression investigated with an immunocytochemical approach. Only recently, Tokuda *et al.* (2007) showed a significant increase of MT-III protein levels in the SOD mutant mice spinal cord; but unlike MT-I/MT-II, which became altered in the early stage of ASL, the MT-III protein increased at the end stage of the disease. According to Ono *et al.* (2006), the MT-III isoform seemed not to be associated with motor neuron death in ALS, and the authors suggested that the disease might be a systemic disorder to which the spinal cord is particularly susceptible. Surprisingly, some scientists reported that MT-III synthesis in transgenic SOD1 mice was also observed in glial cells in both gray and white matter, and was not restricted only to neurons (Gong and Elliott, 2000). This is consistent with previous findings which showed how, under certain experimental conditions, astrocytic MT-III expression could be induced (Masters *et al.*, 1994), and confirms that the inflammatory reaction occurring in ALS (Kushner *et al.*, 1991) seems to differ from that occurring in other neuropathological conditions (e.g. AD, PD).

Compelling evidence indicates that ALS may also be linked with free radical toxicity (Olanow, 1993; Przedborski *et al.*, 1996). Increased oxidative stress associated with metal cation imbalance can aid in the molecular and pathological events leading to ASL (reviewed in Sillevis Smitt *et al.*, 1994). Expression of mutant SOD1 in astrocytes might perturb astrocytic function in some way that eventually results in subsequent expression of MTs.

These experiments overall proposed a protective role for MTs against oxidative stress and metal toxicity. Epidemiological studies have indicated that sporadic ALS patients are more likely to have been exposed to heavy metals (Chancellor *et al.*, 1993), although not all studies agree (Morahan and Pamphlett, 2006) and no direct link has been found. Since one of the most important functions of MTs is to reduce cellular uptake of metals (Park *et al.*, 2001), some authors investigated whether deficiencies in MT isoforms may lead to a decrease in the cellular defense mechanisms against heavy metals. Morahan *et al.* (2007) considered the possibility of inappropriate silencing of MT genes occurring in ALS, but neither the MT-III isoform nor MT-II displayed genetic changes which can be correlated with the pathology (Hayashi *et al.*, 2006).

To confirm the key role played by MTs, recent findings suggested that they may serve an important role in peripheral nerve function and regeneration (Ceballos *et al.*, 2003). Furthermore, *in vitro* studies have demonstrated that MT-IIA can assist in neurite elongation and postinjury reactive neurite growth (Chung *et al.*, 2003). Moreover, Stankovic and Li (2006) showed that MT-I/MT-II deficiency can cause a significant reduction in neurofilament (NF) density in large myelinated axons of the murine phrenic nerve. NF dysfunction is a pathological hallmark of the disease (Piao *et al.*, 2003), and it seems reasonable that the oxidative stress mechanism, through the endogenous generation of ROS, could partly explain this phenomenon (Stankovic, 2005). Further investigations need to be done in order to establish the contribution of these factors to the development of ALS.

7. Prion Protein Disease

Prion disease (PrD) belongs to a group of fatal neurodegenerative disorders called transmissible spongiform encephalopathies (TSEs). Human forms include Gerstmann–Straussler–Scheinker (GSS) disease, kuru, fatal familial insomnia, and sporadic and variant forms of Creutzfeldt–Jakob disease (CJD) (Hanlon *et al.*, 2002). They are characterized by behavioral impairment, ataxia, and vacuolation of neurons and the neuropil (spongiosis). The common feature shared by these disorders is the accumulation in the CNS of an abnormally folded proteinase K-resistant isoform of the cellular prion protein (PrP^{Sc}), which is the posttranslationally modified form of the normal cellular protein (PrP^{C}). The two isoforms have different physical properties, since PrP^{C} forms a soluble monomer, while PrP^{Sc} forms insoluble aggregates which are amyloidogenic in character (Prusiner, 1998). The mechanism through which PrP^{C} is involved in the development of PrD is not clear, but a recent investigation proposed that PrP^{C} might provide neuroprotection (Brown, 2004) which is suppressed with PrP^{Sc} (Harris and True, 2006).

Thus far, no clear functions have been found for PrP, and there is little information on the connection between MTs and the pathology. It is generally accepted that their expression depends upon the disease duration. An increased level of MT mRNA has been reported in scrapie-infected hamster brain (Dandoy-Dron *et al.*, 1998), but studies regarding the human variants are scarce. Kawashima *et al.* (2000) showed, by immunohistochemistry and Western blot analysis, that MT-I/MT-II increased in the brain, especially in the astrocyte population, of CJD cases with a disease course shorter than 15 months compared to the controls. Activation of microglia has been recognized as a feature related to PrD (Rezaie and Lantos, 2001; Rezaie and Al-Sarraj, 2007). It is reasonable that alteration within the microglia through cellular accumulation of PrP^{Sc} could produce cellular damage because of the insoluble nature of the mutant isoform (Rezaie and Lantos, 2001). On the contrary, MT-III seemed to be decreased in the CJD brain with a long disease duration compared to the normal brain (Kawashima *et al.*, 2000). However, it is not established whether the different expressions of MTs could be involved in the abnormal deposition of PrP, such as kuru plaques, or vice versa whether the mutant PrP might influence the expression of MTs in astrocytes and neurons.

A link between prion protein disease and MTs could lie in the capacity of PrP^C to selectively bind Cu^{2+} (Brown *et al.*, 1997). Moreover, Cu^{2+} binding to a specific region of PrP^C is necessary for the normal folding (Miura *et al.*, 1996). This led to the hypothesis of a potential involvement of PrP^C in Cu^{2+} metabolism (Brown *et al.*, 1999; Whittal *et al.*, 2000; Millhauser, 2004). This link was first proposed by Pattison and Jebbett (1971) with the observation that the histopathology of mouse scrapie resembled that induced by cuprizone. Oxalic acid bis(cyclohexylidene hydrazide) is a copper chelator which induced spongiform encephalopathy and gliosis as a consequence of long-term treatment (Matsushima and Morell, 2001; Zatta *et al.*, 2005a). When chronically administered to mice, cuprizone was able to induce myelin loss; demyelination took place in deeper layers of the cortex and showed a pattern of rounded spots with variable diameter (Zatta *et al.*, 2005a).

It has been shown that PrP^C expression can alter Cu^{2+} uptake into cells and enhance Cu^{2+} incorporation into superoxide dismutase (SOD), and that PrP itself can act as an SOD (Brown *et al.*, 1999). Wells *et al.* (2006) proposed that the physiological binding of Cu^{2+} to normal PrP^C in the brain occurs only when the concentration of Cu^{2+} increased locally, for instance, during depolarization. It has been suggested that PrP^C might play an important role in reducing oxidative stress and that its enzymatic function depends on Cu^{2+}

incorporation (Brown *et al.*, 1999). Furthermore, copper itself has recently been shown to induce expression of cellular prion protein in primary cell cultures, upregulating the expression of this protein both at the cell surface and within intracellular compartments (Varela-Nallar *et al.*, 2006). A decreased capacity to chelate copper or the copper deficiency could lead to an increase in oxidative stress and consequently to cell death (Brown *et al.*, 1998).

PrP^C also has the capacity to bind divalent metal ions other than Cu^{2+}, for example, Mn^{2+} (Brown *et al.*, 2000). These findings stimulated the investigations on metal ion alteration in PrD human brains. Changes in the levels of Cu^{2+} and Mn^{2+} bound to PrP^{Sc} were found in sporadic CJD compared to PrP^C in normal subjects (Wong *et al.*, 2001a). Moreover, it was also observed that there was a decrease of up to 50% of Cu and a 10-fold increase in Mn in brain tissues of CJD compared to controls (Wong *et al.*, 2001b). In addition, antioxidant activity of purified PrP^{Sc} was widely reduced up to 85% in the sporadic CJD; this was in parallel with an increase in oxidative stress markers. These results suggest that metal ions might be potentially involved in the pathology (Thackray *et al.*, 2002), and therefore it is possible that metal imbalance could cause conversion of PrP^C into PrP^{Sc}. Thus, MTs could be proposed as a potential therapeutical approach for regulating divalent metal homeostasis, and may also represent a critical element in the diagnostic assessment of the disease course.

However, it is interesting to note that in bovine spongiform encephalopathy (BSE), the brain regions involved in the degeneration did not show any analytical alteration of Cu^{2+} level compared with healthy animals (unpublished data from our laboratory). Moreover, extraction of purified PrP from the frontal cortex of sporadic CJD variants revealed a significant reduction of copper bound to PrP compared to controls (Wong *et al.*, 2001b). This evidence creates uncertainties on the putative role of Cu in PrD (Zatta and Frank, 2007), and suggests further study to clarify this relevant issue.

8. Conclusion

The involvement of MTs in neurodegenerative disorders is far from being completely demonstrated, but is currently the subject of many investigations. Whether MTs are in part responsible for the etiologic processes leading to neurodegeneration or whether their alteration is an aspecific consequence of dysfunction of the CNS has to be proven. The role played by MTs in the detoxification of heavy metals and the distribution of essential metals is clearly established, but in recent years there has grown a wider appreciation of MT functions in responding to cellular stress, which is a consequence of normal brain function and can be enhanced in the presence of pathological events. Recent investigations in the mammalian brain have proposed a contribution of MTs in zinc metabolism, free radical scavenging, and regeneration following neurological injury. The wide presence of MTs within the CNS and the identification of a specific brain isoform, MT-III, are reasonable clues which support the hypothesis of this family of proteins having important neurological functions.

Despite the similarity among MT isomers, the roles of MT-I/MT-II are partly established, while information regarding the functions of MT-III is scarce and sometimes contrasting. Whereas research is still in its early stages, future experiments should provide a much greater understanding of the roles of MTs and may also suggest their involvement in the development of novel therapeutic strategies.

References

Adlard PA, West AK, Vickers JC. Increased density of metallothionein I/II-immunopositive cortical glial cells in the early stages of Alzheimer's disease. *Neurobiol Dis* 1998; **5**:349–355.

Amoureux MC, Van Gool D, Herrero MT, *et al.* Regulation of metallothionein-III (GIF) mRNA in the brain of patients with Alzheimer's disease is not impaired. *Mol Chem Neuropathol* 1997; **32**:101–121.

Andreassen OA, Dedeoglu A, Friedlich A, *et al.* Effects of an inhibitor of poly(ADP-ribose) polymerase, desmethylselegiline, trientine, and lipoic acid in transgenic ALS mice. *Exp Neurol* 2001; **168**:419–421.

Aschner M. Metallothionein (MT) isoforms in the central nervous system (CNS): Regional and cell-specific distribution and potential functions as an antioxidant. *Neurotoxicology* 1998; **19**:653–660.

Aschner M, Cherian MG, Klaassen CD, *et al.* Metallothioneins in brain — The role in physiology and pathology. *Toxical Appl Pharmacol* 1997; **142**:229–242.

Banati RB, Daniel SE, Blunt SB. Glial pathology but absence of apoptotic nigral neurons in long-standing Parkinson's disease. *Mov Disord* 1998; **13**:221–227.

Barnham KJ, Masters CL, Bush AI. Neurodegenerative diseases and oxidative stress. *Nat Rev Drug Discov* 2004; **3**:205–214.

Beal MF. Excitotoxicity and nitric oxide in Parkinson's disease pathogenesis. *Ann Neurol* 1998; **44**:110–114.

Belloso E, Henandez J, Giralt M, *et al.* Effect of stress on mouse and rat brain metallothionein I and II mRNA levels. *Neuroendocrinology* 1996; **64**:430–439.

Blaauwgeers HG, Vianney de Jong JM, Verspaget HW, *et al.* Enhanced superoxide dismutase-2 immunoreactivity of astrocytes and occasional neurons in amyotrophic lateral sclerosis. *J Neurol Sci* 1996; **140**:21–29.

Borchelt DR, Lee MK, Slunt HS, *et al.* Superoxide dismutase 1 with mutations linked to familial amyotrophic lateral sclerosis possesses significant activity. *Proc Natl Acad Sci USA* 1994; **91**:8292–8296.

Bray TM, Bettger WJ. The physiological role of zinc as an antioxidant. *Free Radic Biol Med* 1990; **8**:281–291.

Bringold U, Ghafourifar P, Richter R. Peroxynitrite formed by mitochondrial NO synthase promotes mitochondrial Ca^{2+} release. *Free Radic Biol Med* 2000; **29**:343–348.

Brown DR. Role of the prion protein in copper turnover in astrocytes. *Neurobiol Dis* 2004; **15**:534–543.

Brown DR, Hafiz F, Glassmith LL, *et al.* Consequences of manganese replacement of copper for prion protein function and proteinase resistance. *EMBO J* 2000; **19**:1180–1186.

Brown DR, Qin K, Herms JW, *et al.* The cellular prion protein binds copper *in vivo*. *Nature* 1997; **390**:684–687.

Brown DR, Schmidt B, Kretzschmar HA. Effects of the copper on survival of prion protein knockout neurons and glia. *J Neurochem* 1998; **70**:1686–1693.

Brown DR, Wong BS, Hafiz F, *et al.* Normal prion protein has an activity like that of superoxide dismutase. *Biochem J* 1999; **344**:1–5.

Bruijn LI, Miller TM, Cleveland DW. Unravelling the mechanisms involved in motor neuron degeneration in ALS. *Annu Rev Neurosci* 2004; **27**:723–749.

Cai L, Klein JB, Kang YJ. MT inhibits peroxynitrite-induced DNA and lipoprotein damage. *J Biol Chem* 2001; **275**:38957–38960.

Carrasco J, Giralt M, Molinero A, *et al.* Metallothionein (MT)-III: Generation of polyclonal antibodies, comparison with MT-I+II in the freeze lesioned rat brain and in a bioassay with astrocytes, and analysis of Alzheimer's disease brains. *J Neurotrauma* 1999; **16**:1115–1129.

Ceballos D, Lago N, Verdu E, *et al.* Role of metallothionein in peripheral nerve function and regeneration. *Cell Mol Life Sci* 2003; **60**:1209–1216.

Chancellor AM, Slattery JM, Fraser H, Warlow CP. Risk factors for motor neuron disease: A case-control study based on patients from the Scottish Motor Neuron Disease Register. *J Neurol Neurosurg Psychiatry* 1993; **56**:1200–1206.

Chen CF, Guo SS, Jia HY, *et al.* The role of endothelin-1 in Binswanger disease. *Chung Hua I Hsueh Tsa Chih (Taipei)* 1996; **58**:151–155.

Chuah MI, Getchell ML. Metallothionein in olfactory mucosa of Alzheimer's disease patients and apoE-deficient mice. *Neuroreport* 1999; **10**:1919–1924.

Chung RS, Vickers C, Meng IC, *et al.* Metallothionein-IIA promotes initial neurite elongation and postinjury reactive neurite growth and facilitates healing after focal cortical brain injury. *J Neurosci* 2003; **23**:3336–3342.

Cooper PN, Jackson M, Lennox G, *et al.* τ, ubiquitin and αB-crystallin immunocytochemistry define the principle causes of degenerative frontotemporal dementias. *Arch Neurol* 1995; **52**:1011–1015.

Crow P, Sampson JB, Zhuang Y, *et al.* Decreased zinc affinity of amyotrophic lateral sclerosis-associated superoxide dismutase mutants leads to enhanced catalysis of tyrosine nitration by peroxynitrite. *J Neurochem* 1997; **69**:1936–1944.

Cuajungco MP, Lees GJ. Zinc metabolism in the brain: Relevance to human neurodegenerative disorders. *Neurobiol Dis* 1997; **4**:137–169.

Dandoy-Dron F, Guillo F, Benboudjema L, *et al.* Gene expression in scrapie. *J Biol Chem* 1998; **273**:7691–7697.

Dawson VL, Dawson TM. Nitric oxide in neurodegeneration. *Prog Brain Res* 1998; **118**:215–229.

Dexter DT, Carayon A, Javoy-Agid F, *et al.* Alterations in the levels of iron, ferritin and other trace metals in Parkinson's disease and other neurodegenerative diseases affecting the basal ganglia. *Brain* 1991; **114**:1953–1975.

Duguid JR, Bohmont CW, Liu NG, Tourtellotte WW. Changes in brain gene expression shared by scrapie and Alzheimer's disease. *Proc Natl Acad Sci USA* 1989; **86**:7260–7264.

Ebadi M, Brown-Borg H, El Rafaey H, *et al.* Metallothionein-mediated neuroprotection in genetically engineered mouse models of Parkinson's disease. *Mol Brain Res* 2005a; **134**:67–75.

Ebadi M, Ramana Kumari MV, Hiramatsu M, *et al.* Metallothionein, neurotrophins and selegiline in providing neuroprotection in Parkinson's disease. *Restor Neurol Neurosci* 1998; **12**:103–111.

Ebadi M, Sharma S. Metallothioneins 1 and 2 attenuate peroxynitrite-induced oxidative stress in Parkinson disease. *Exp Biol Med (Maywood)* 2006; **231**:1576–1583.

Ebadi M, Sharma SK, Ghafourifar P, *et al.* Peroxynitrite in the pathogenesis of Parkinson's disease and the neuroprotective role of metallothioneins. *Methods Enzymol* 2005b; **396**:276–298.

El Ghazi I, Martin BL, Armitage IM. Metallothionein-3 is a component of a multiprotein complex in the mouse brain. *Exp Biol Med (Maywood)* 2006; **231**:1500–1506.

Erickson JC, Sewell AK, Jensen LT, *et al.* Enhanced neurotrophic activity in Alzheimer's disease cortex is not associated with down regulation of metallothionein-III, GIF. *Brain Res* 1994; **649**:297–304.

Esiri MM, Wilcock GK, Morris JH. Neuropathological assessment of the lesions of significance in vascular dementia. *J Neurol Neurosurg Psychiatry* 1997; **63**:749–753.

Estevez AG, Crow JP, Sampson JB, *et al.* Induction of nitric oxide-dependent apoptosis in motor neurons by zinc-deficient superoxide dismutase. *Science* 1999; **286**: 2498–2500.

Forsleff L, Schauss AG, Bier ID, Stuart S. Evidence of functional zinc deficiency in Parkinson's disease. *J Altern Complement Med* 1999; **5**:54–64.

Foster NL, Wilhelmsen K, Sima AAF, *et al.* Frontotemporal dementia and parkinsonism linked to chromosome 17: A consensus conference. *Ann Neurol* 1997; **41**:706–715.

Furuta A, Ishii N, Nishihara Y, Horie A. Medullary arteries in aging and dementia. *Stroke* 1991; **22**:442–446.

Gong YH, Elliott JL. Metallothionein expression is altered in a transgenic murine model of familial amyotrophic lateral sclerosis. *Exp Neurol* 2000; **162**:27–36.

Goto JJ, Zhu H, Sanchez RJ, *et al.* Loss of *in vitro* metal ion binding specificity in mutant copper-zinc superoxide dismutases associated with familial amyotrophic lateral sclerosis. *J Biol Chem* 2000; **275**:1007–1014.

Hanlon J, Monks E, Hughes C, *et al.* Metallothionein in bovine spongiform encephalopathy. *J Comp Pathol* 2002; **127**:280–289.

Harris DA, True HL. New insights into prion structure and toxicity. *Neuron* 2006; **40**:547–586.

Hayashi Y, Hashizume T, Wakida K, *et al.* Association between metallothionein genes polymorphisms and sporadic amyotrophic lateral sclerosis in a Japanese population. *Amyotroph Lateral Scler* 2006; **7**:22–26.

Hidalgo J, Aschner M, Zatta P, Vašàk M. Roles of the metallothionein family of proteins in the central nervous system. *Brain Res Bull* 2001; **55**:133–145.

Hidalgo J, Belloso E, Henandez J, *et al.* Role of glucocorticoids on rat brain metallothionein-I and -III response to stress. *Stress* 1997; **1**:231–240.

Hidalgo J, Penkowa M, Espejo C, *et al.* Expression of metallothionein-I, -II, and -III in Alzheimer disease and animal models of neuroinflammation. *Exp Biol Med (Maywood)* 2006; **231**:1450–1458.

Hirsch EC, Hunot S. Nitric oxide, glial cells and neuronal degeneration in parkinsonism. *Trends Pharmacol Sci* 2002; **21**:163–165.

Hussain S, Slikker W Jr, Ali SF. Role of metallothionein and other antioxidants in scavenging superoxide radicals and their possible role in neuroprotection. *Neurochem Int* 1996; **29**:145–152.

Irie Y, Keung WM. Metallothionein-III antagonizes the neurotoxic and neurotrophic effects of amyloid beta peptides. *Biochem Biophys Res Commun* 2001; **282**:416–420.

Jackson M, Lowe J. The new neuropathology of neurodegenerative frontotemporal dementia. *Acta Neuropathol* 1996; **91**:127–134.

Jellinger K, Neumayer E. Progressive subcorticale vasculäre encephalopathie Binswanger. Eine klinisch-neuropathologische studie. *Arch Psychiatr Nervenkr* 1964; **205**:523–554.

Jenner P. Oxidative mechanisms in nigral cell death in Parkinson's disease. *Mov Disord* 1998; **13**(Suppl 1):24–34.

Jimenéz-Jimenéz FJ, Molina JA, Anguilar MV, *et al.* Cerebrospinal fluid levels of transition metals in patients with Parkinson's disease. *J Neural Transm* 1998; **105**: 497–505.

Jorgensen OS, Balazs R. Plastic neuronal changes in Alzheimer's disease associated with activation of astrocytes and enhanced neurotrophic activity. In: Corain B, Iqbal K, Nicolini M, *et al.* (eds.), *Alzheimer's Disease: Advances in Clinical and Basic Research*, Wiley, Chichester, UK, 1993, pp. 189–198.

Kaji T, Yamamoto C, Tsubaki S, *et al.* Metallothionein induction by cadmium, cytokines, thrombin and endothelin-1 in cultured vascular endothelial cells. *Life Sci* 1993; **53**:1185–1191.

Kawashima T, Doh-ura K, Torisu M, *et al.* Differential expression of metallothioneins in human prion disease. *Dement Geriatr Cogn Disord* 2000; **11**:251–262.

Kushner PD, Stephenson DT, Wright S. Reactive astrogliosis is widespread in the subcortical white matter of amyotrophic lateral sclerosis brain. *J Neuropathol Exp Neurol* 1991; **50**:263–277.

Lee JY, Kim JH, Palmiter RD, Koh JY. Zinc released from metallothionein-III may contribute to hippocampal CA1 and thalamic neuronal death following acute brain injury. *Exp Neurol* 2003; **184**:337–347.

Liu G, Huang W, Moir RD, *et al.* Metal exposure and Alzheimer's pathogenesis. *J Struct Biol* 2006; **155**:45–51.

Liu Y, Brooks BR, Taniguchi N, Hartmann HA. CuZnSOD and MnSOD immunoreactivity in brain stem motor neurons from amyotrophic lateral sclerosis patients. *Acta Neuropathol* 1998; **95**:63–70.

Lovell MA, Robertson JD, Teesdale WJ, *et al.* Copper, iron and zinc in Alzheimer's disease senile plaques. *J Neurol Sci* 1998; **158**:47–52.

Lowe J. Degenerative non-Alzheimer dementias. *Brain Pathol* 1997; **7**:1047–1051.

Lowe J, Spillantini MG. Non-Alzheimer degenerative dementias. *Brain Pathol* 1998; **8**:295–297.

Lund and Manchester Groups. Clinical and neuropathological criteria for frontotemporal dementia. *J Neurol Neurosurg Psychiatry* 1994; **57**:416–418.

Lyons TJ, Liu H, Goto JJ, *et al.* Mutations in copper-zinc superoxide dismutase that cause amyotrophic lateral sclerosis alter the zinc binding site and the redox behavior of the protein. *Proc Natl Acad Sci USA* 1996; **93**:12240–12244.

Martin BL, Tokheim AM, McCarthy PT, *et al.* Metallothionein-3 and neuronal nitric oxide synthase levels in brains from the Tg2576 mouse model of Alzheimer's disease. *Mol Cell Biochem* 2006; **283**:129–137.

Masters BA, Quaife CJ, Erickson JC, *et al.* Metallothionein III is expressed in neurons that sequester zinc in synaptic vesicles. *J Neurosci* 1994; **14**:5844–5847.

Matsushima GK, Morell P. The neurotoxicant cuprizone as a model to study demyelination and remyelination in the central nervous system. *Brain Pathol* 2001; **11**:107–116.

Mattson MP. Pathways towards and away from Alzheimer's disease. *Nature* 2004; **430**:631–639.

McGeer PL, Itagaki S, Boyes BE, McGeer EG. Reactive microglia are positive for HLA-DR in the substantia nigra of Parkinson's and Alzheimer's disease brains. *Neurology* 1988; **38**:1285–1291.

Millhauser GL. Copper binding in the prion protein. *Acc Chem Res* 2004; **37**:79–85.

Mirza B, Hadberg P, Thomsen P, Moos T. The absence of reactive astrocytosis is indicative of a unique inflammatory process in Parkinson's disease. *Neuroscience* 2000; **95**:425–432.

Miu AC, Benga O. Aluminum and Alzheimer's disease: A new look. *J Alzheimers Dis* 2006; **10**:179–201.

Miura T, Hori-i A, Takeuchi H. Metal-dependent α-helix formation promoted by the glycine-rich octapeptide region of prion protein. *FEBS Lett* 1996; **396**:248–252.

Miyazaki I, Asanuma M, Higashi Y, *et al.* Age-related changes in expression of metallothionein-III in rat brain. *Neurosci Res* 2002; **43**:323–333.

Mocchegiani E, Giacconi R, Cipriano C, *et al.* Are zinc-bound metallothionein isoforms (I+II and III) involved in impaired thymulin production and thymic involution during ageing? *Immun Ageing* 2004; **12**:1–5.

Mogi M, Harada M, Narabayashi H, *et al.* Interleukin (IL)-1 beta, IL-2, IL-4, IL-6 and transforming growth factor-alpha levels are elevated in ventricular cerebrospinal fluid in juvenile parkinsonism and Parkinson's disease. *Neurosci Lett* 1996; **14**:13–16.

Morahan JM, Pamphlett R. Amyotrophic lateral sclerosis and exposure to environmental toxins: An Australian case-control study. *Neuroepidemiology* 2006; **27**:130–135.

Morahan JM, Yu B, Trent RJ, Pamphlett R. Are metallothionein genes silenced in ALS? *Toxicol Lett* 2007; **168**:83–87.

Nagano S, Fujii Y, Yamamoto T, *et al.* The efficacy of trientine or ascorbate alone compared to that of the combined treatment with these two agents in familial amyotrophic lateral sclerosis model mice. *Exp Neurol* 2003; **179**:176–180.

Nakajima K, Suzuki K. Immunochemical detection of metallothionein in brain. *Neurochem Int* 1995; **27**:73–87.

Naruse S, Igarashi S, Furuya T, *et al.* Structure of the human and mouse growth inhibitory factor-encoding genes. *Gene* 1994; **144**:283–287.

Olanow CW. A radical hypothesis for neurodegeneration. *Trends Neurosci* 1993; **16**:439–444.

Ono S, Endo Y, Tokuda E, *et al.* Upregulation of metallothionein-I mRNA expression in a rodent model for amyotrophic lateral sclerosis. *Biol Trace Elem Res* 2006; **113**:93–104.

Palmiter RD. Constitutive expression of metallothionein-III (MT-III), but not MT-I, inhibits growth when cells become zinc deficient. *Toxicol Appl Pharmacol* 1995; **135**:139–146.

Park JD, Liu Y, Klaassen CD. Protective effect of metallothionein against the toxicity of cadmium and other metals. *Toxicology* 2001; **163**:93–100.

Pattison IH, Jebbett JN. Histopathological similarities between scrapie and cuprizone toxicity in mice. *Nature* 1971; **230**:115–117.

Persechini A, McMillan K, Masters BS. Inhibition of nitric oxide synthase activity by Zn^{2+} ion. *Biochemistry* 1995; **34**:15091–15095.

Piao YS, Wakabayashi K, Kakita A, *et al.* Neuropathology with clinical correlations of sporadic amyotrophic lateral sclerosis: 102 autopsy cases examined between 1962 and 2000. *Brain Pathol* 2003; **13**:10–22.

Prusiner SB. Prions. *Proc Natl Acad Sci USA* 1998; **95**:13363–13383.

Przedborski S, Donaldson DM, Murphy PL, *et al.* Blood superoxide dismutase, catalase and glutathione peroxidase activities in familial and sporadic amyotrophic lateral sclerosis. *Neurodegeneration* 1996; **5**:57–64.

Puttaparthi K, Gitomer WL, Krishnan U, *et al.* Disease progression in a transgenic model of familial amyotrophic lateral sclerosis is dependent on both neuronal and non-neuronal zinc binding proteins. *J Neurosci* 2002; **22**:8790–8796.

Rasia RM, Bertoncini CW, Marsh D, *et al.* Structural characterization of copper(II) binding to alpha-synuclein: Insights into the bioinorganic chemistry of Parkinson's disease. *Proc Natl Acad Sci USA* 2005; **102**:4294–4299.

Rezaie P, Al-Sarraj S. Vacuolar degeneration affecting brain macrophages/microglia in variant CJD: A report on two cases. *Acta Neuropathol* 2007; **114**:651–658.

Rezaie P, Lantos PL. Microglia and the pathogenesis of spongiform encephalopathies. *Brain Res Rev* 2001; **35**:55–72.

Rojas P, Rojas-Castaneda J, Vigueras RM, *et al.* MPTP decreases MT-I mRNA in mouse striatum. *Neurochem Res* 2001; **25**:503–509.

Rosen DR, Siddique T, Patterson D, *et al.* Mutations in Cu/Zn superoxide dismutase gene are associated with familial amyotrophic lateral sclerosis. *Nature* 1993; **362**:59–62.

Rowland LP, Shneider NA. Amyotrophic lateral sclerosis. *N Engl J Med* 2001; **344**: 1688–1700.

Said Ahmed M, Hung WY, Zu JS, *et al.* Increased reactive oxygen species in familial amyotrophic lateral sclerosis with mutations in SOD1. *J Neurol Sci* 2000; **176**:88–94.

Sato M, Bremner I. Oxygen free radicals and metallothionein. *Free Radic Biol Med* 1993; **14**:325–337.

Sharma SK, Ebadi M. Metallothionein attenuates 3-morpholinosydnonimine (SIN-1)-induced oxidative stress in dopaminergic neurons. *Antioxid Redox Signal* 2003; **5**:251–264.

Shiraga H, Pfeiffer RF, Ebadi M. The effect of 6-hydroxydopamine and oxidative stress on the level of brain metallothionein. *Neurochem Int* 1993; **23**:561–566.

Sillevis Smitt PA, Blaauwgeers HG, Troost D, de Jong JM. Metallothionein immunoreactivity is increased in the spinal cord of patients with amyotrophic lateral sclerosis. *Neurosci Lett* 1992; **144**:107–110.

Sillevis Smitt PA, Mulder TP, Verspaget HW, *et al.* Metallothionein in amyotrophic lateral sclerosis. *Biol Signals* 1994; **3**:193–197.

Sogawa CA, Asanuma M, Sogawa N, *et al.* Localization, regulation, and function of metallothionein-III/growth inhibitory factor in the brain. *Acta Med Okayama* 2001; **55**:1–9.

Stankovic RK. Atrophy of large myelinated fibres in metallothionein-I, II knockout mice. *Cell Mol Neurobiol* 2005; **25**:943–953.

Stankovic RK, Li Z. Decreased neurofilament density in large myelinated axons of metallothionein-I, II knockout mice. *Neurosci Lett* 2006; **102**:1–6.

Streit WJ. Microglia and Alzheimer's disease pathogenesis. *J Neurosci Res* 2004; **77**:1–8.

Takahashi S, Abe T, Tohgi H. Problems in vascular dementia. *Nippon Ronen Igakkai Zasshi* 1996; **33**:158–163.

Tanoi Y, Okeda R, Budka H. Binswanger's encephalopathy: Serial sections and morphometry of the cerebral arteries. *Acta Neuropathol* 2000; **100**:347–355.

Thackray AM, Knight R, Haswell SJ, *et al.* Metal imbalance and compromised antioxidant function are early changes in prion disease. *Biochem J* 2002; **362**:253–258.

Tissot R, Costantinidis J, Richard J. *La Maladie de Pick.* Masson, Paris, 1975.

Tohgi H, Sasaki K, Chiba K, Nozaki Y. Effect of vinpocetine on oxygen release of haemoglobin and erythrocyte organic polyphosphate concentrations in patients with vascular dementia of the Binswanger type. *Arzneimittelforschung* 1990; **40**:640–643.

Tokuda E, Ono S, Ishige K, *et al.* Metallothionein proteins expression, copper and zinc concentrations, and lipid peroxidation level in a rodent model for amyotrophic lateral sclerosis. *Toxicology* 2007; **229**:33–41.

Tsuji S, Kobayashi H, Uchida Y, *et al.* Molecular cloning of human inhibitory factor cDNA and its down-regulation in Alzheimer's disease. *EMBO J* 1992; **11**:4843–4850.

Uchida Y, Takio K, Titani K, *et al.* The growth inhibitory factor that is deficient in the Alzheimer's disease brain is a 68 amino acid metallothionein-like protein. *Neuron* 1991; **7**:337–347.

Varela-Nallar L, Toledo EM, Larrondo FL, *et al.* Induction of cellular prion protein gene expression by copper in neurons. *Am J Physiol Cell Physiol* 2006; **290**:271–281.

Vickers J. A cellular mechanism for the neuronal changes underlying Alzheimer's disease. *Neuroscie*nce 1997; **78**:629–639.

Wells MA, Jackson GS, Jones S, *et al.* A reassessment of copper(II) binding in the full length prion protein. *Biochem J* 2006; **399**:435–444.

Whittal RM, Ball HL, Cohen FE, *et al.* Copper binding to octarepeat peptides of the prion protein monitored by mass spectrometry. *Protein Sci* 2000; **9**:332–343.

Wiedau-Pazos M, Goto JJ, Rabizadeh S, *et al.* Altered reactivity of superoxide dismutase in familial amyotrophic lateral sclerosis. *Science* 1996; **271**:515–518.

Wong BS, Brown DR, Pan T, *et al.* Oxidative impairment in scrapie-infected mice is associated with brain metals perturbations and altered antioxidant activities. *J Neurochem* 2001a; **79**:689–698.

Wong BS, Chen SG, Colucci M, *et al.* Aberrant metal binding by prion protein in human prion disease. *J Neurochem* 2001b; **78**:1400–1408.

You HJ, Oh DH, Choi CY, *et al.* Protective effect of metallothionein-III on DNA damage in response to reactive oxygen species. *Biochim Biophys Acta* 2002; **1573**:33–38.

Yu WH, Lukiw WJ, Bergeron C, *et al.* Metallothionein III is reduced in Alzheimer's disease. *Brain Res* 2001; **894**:37–45.

Yuguchi T, Kohmura E, Yamada K, *et al.* Expression of growth inhibitory factor mRNA following cortical injury. *J Neurotrauma* 1995; **12**:299–306.

Zambenedetti P, Giordano R, Zatta P. Metallothioneins are highly expressed in astrocytes and microcapillaries in Alzheimer's disease. *J Chem Neuroanat* 1998; **15**:21–26.

Zambenedetti P, Schmitt HP, Zatta P. Metallothionein I-II immunocytochemical reactivity in Binswanger's encephalopathy. *J Alzheimers Dis* 2002; **4**:459–466.

Zatta P, Frank A. Copper deficiency and neurological disorders in man and animals. *Brain Res Rev* 2007; **54**:19–33.

Zatta P, Raso M, Zambenedetti P, *et al.* Copper and zinc dismetabolism in the mouse brain upon chronic cuprizone treatment. *Cell Mol Life Sci* 2005a; **62**:1502–1513.

Zatta P, Raso M, Zambenedetti P, *et al.* Metallothionein-I-II expression in young and adult bovine pineal gland. *J Chem Neuroanat* 2006; **31**:124–129.

Zatta P, Zambenedetti P, Musicco M, Adorni F. Metallothionein-I-II and GFAP positivity in the brains from frontotemporal dementia patients. *J Alzheimers Dis* 2005b; **8**:109–116.

Chapter 4

METALLOTHIONEIN AND BRAIN INFLAMMATION

Juan Hidalgo

Most neurodegenerative diseases are characterized by a prominent inflammatory response (often called neuroinflammation), which consists mainly of elements of the innate immune response. Whether neuroinflammation is a cause or simply a consequence of the neuropathological events seen during neurodegenerative diseases is unclear, but it is undisputed that chronic inflammation normally leads to permanent scarring and tissue damage. In order to understand this complex biological response, it is essential to identify all factors involved in it. There is a clear consensus that a number of cytokines orchestrate the brain inflammatory response, in terms of both inducing and limiting it, but also control the expression of proteins important for coping with the potential tissue damage. One such protein is metallothionein (MT). Results obtained in *MT-1/MT-2*–null mice and in *MT-1*–overexpressing mice strongly suggest that these MT isoforms are important antioxidant, anti-inflammatory, and antiapoptotic proteins in the brain. Results in *MT-3*–null mice show a very different pattern, with no support for MT-1/MT-2–like functions; rather, MT-3 could be involved in neuronal sprouting and survival.

Keywords: Metallothionein-1/metallothionein-2; metallothionein-3; neuroinflammation; cytokines; IL-6; oxidative stress.

1. Introduction

Fifty years ago, an unusual cadmium-binding protein was isolated from horse kidney (Margoshes and Vallee, 1957). Due to its high content of metals and cysteine residues, this protein was named "metallothionein" (MT) (Kägi and Vallee, 1960; Kägi and Vallee, 1961). During these 50 years, an overwhelming flow of information on these proteins has been published, ranging from biochemical to physiological aspects of these proteins. It is clear that they constitute a superfamily of proteins, widely distributed in the animal kingdom but also in eukaryotic and prokaryotic microorganisms and plants (Hamer, 1986; Bremner, 1987a; Andrews, 2000; Vašák and Hasler, 2000; Ghoshal and Jacob, 2001; Coyle *et al.*, 2002).

Mammalian MTs belong to class I proteins, and display extensive genetic polymorphism, especially in ungulates and primates. In humans, the MT genes are tightly clustered in the q13 region of chromosome 16 (Karin *et al.*, 1984a; West *et al.*, 1990; Palmiter *et al.*, 1992; Quaife *et al.*, 1994), consisting of seven functional *MT-1* genes (*MT-1A, MT-B, MT-E, MT-F, MT-G, MT-H*, and *MT-X*) and a single gene encoding each of the other MT isoforms (namely, MT-2, MT-3, and MT-4). Mice have only one functional gene for each isoform, MT-1 to MT-4, all located on chromosome 8 (Cox and Palmiter, 1983; Searle *et al.*, 1984; Palmiter *et al.*, 1992; Quaife *et al.*, 1994). MT-1/MT-2 are expressed coordinately in most tissues including the central nervous system (CNS) (Searle *et al.*, 1984; Yagle and Palmiter, 1985; van Lookeren Campagne *et al.*, 2000), while MT-3 and MT-4 show a much more restricted tissue expression (basically, CNS and stratified squamous epithelia, respectively). It is generally agreed that metallothioneins play major roles in the body, despite the fact that their physiological functions remain elusive to some extent. In the last 10–15 years, MT research in the brain has contributed significantly to understand the putative physiological functions of these proteins, in part because transgenic mice have been extensively used.

2. Metallothionein-1/Metallothionein-2 Are Differentially Expressed in the Brain

Studies of *in situ* hybridization and immunohistochemistry have demonstrated that MT-1/MT-2 occur throughout the brain and spinal cord, and that the main cell expressing these MT isoforms is the astrocyte, especially the reactive astrocyte. Nevertheless, MT-1/MT-2 expression is also found in ependymal cells, epithelial cells of the choroid plexus, meningeal cells of the pia mater, and endothelial cells of blood vessels. Neurons appear to express MT-1/MT-2 to a much lower extent than astrocytes (Hidalgo *et al.*, 1994; Masters *et al.*, 1994b; Kramer *et al.*, 1996a; Kramer *et al.*, 1996b); while in the normal brain, oligodendrocytes and microglia are essentially devoid of MT-1/MT-2, but the latter cells do upregulate MT-1/MT-2 expression in response to injury (Vanguri, 1995; Vela *et al.*, 1997; Agulló *et al.*, 1998; Acarin *et al.*, 1999b; Penkowa *et al.*, 1999c). MT-3 was discovered as a putative factor important in Alzheimer's disease (Uchida *et al.*, 1991). In contrast to MT-1/MT-2, there is still significant uncertainty regarding the cellular source of MT-3 in the CNS; astrocytes and neurons have been suggested to be the main cellular sources (Kobayashi *et al.*, 1993; Uchida, 1993; Masters *et al.*, 1994b; Hidalgo *et al.*, 2001).

It is generally accepted that MT-1/MT-2 expression is highly inducible by a range of stimuli including metals, hormones, cytokines, oxidative agents, inflammation, and stress (Bremner, 1987b; Sato and Bremner, 1993). MT-1 and MT-2 expression is coordinately regulated in mice by metals and glucocorticoids (Searle *et al.*, 1984; Yagle and Palmiter, 1985). Metal-induced synthesis is mediated through the action of short *cis*-acting DNA sequences known as metal-responsive elements (MREs) — which are present in the promoter region of all mammalian MT genes (Karin *et al.*, 1984b; Radtke *et al.*, 1993) — and is mediated mainly by MTF-1, a zinc-sensitive *trans*-acting factor (Westin and Schaffner, 1988; Radtke *et al.*, 1993; Andrews, 2000). Similarly, glucocorticoid-responsive elements (GREs) are responsible for MT expression in response to glucocorticoids (Karin *et al.*, 1984b; Kelly *et al.*, 1997). MT expression is also regulated by antioxidant response elements (AREs), although some MREs also respond to oxidants, again MTF-1 being involved (Samson and Gedamu, 1998; Zhang *et al.*, 2003). MT induction by inflammatory factors is very likely to be influenced by cytokines such as IL-6 through STAT factors (Lee *et al.*, 1999).

Nevertheless, it is feasible that multiple signals participate in stress, inflammation, and oxidative stress induction of the MT genes. The regulation of MT-3 and MT-4 is poorly known.

3. Metallothionein-1/Metallothionein-2 Are Essential for Coping with Brain Damage

Brain MT-1/MT-2 levels have been consistently reported to be increased in Alzheimer's disease (Duguid *et al.*, 1989; Uchida, 1993; Adlard *et al.*, 1998; Zambenedetti *et al.*, 1998; Chuah and Getchell, 1999), Pick's disease (Duguid *et al.*, 1989), frontotemporal dementias (Zatta *et al.*, 2005), short-course Creutzfeldt–Jakob disease (Kawashima *et al.*, 2000), amyotrophic lateral sclerosis (Sillevis Smitt *et al.*, 1992; Sillevis Smitt *et al.*, 1994; Blaauwgeers *et al.*, 1996), multiple sclerosis (Lock *et al.*, 2002; Penkowa *et al.*, 2003c), and aging (Suzuki *et al.*, 1992). Presumably, the fact that MT-1/MT-2 expression is increased in these situations reflects a physiological response attempting to cope with the tissue damage known to occur in all of them. The therapeutic use of MT is still years ahead, but results obtained in animal models are very promising.

It is a well-known fact that MT-1/MT-2 expression is increased by inflammatory factors, since the early report 30 years ago that bacterial infection induces liver MT (Sobocinski *et al.*, 1978). Soon thereafter, it was shown that lipopolysaccharide (LPS) components of Gram-negative bacteria, the so-called endotoxins, could induce liver MT-1/MT-2 synthesis (Suzuki and Yamamura, 1980) by a mechanism which does not involve either heavy metals or glucocorticoids (Durnam *et al.*, 1984). Since agents such as turpentine oil, which also initiates inflammatory responses, were similarly found to induce MT-1/MT-2 synthesis (Sobocinski and Canterbury, 1982), it was soon suspected that cytokines could be mediating the effect of LPS or turpentine on MT-1/MT-2 levels. The exogenous administration of a number of cytokines, including IL-1α/β, IL-6, TNF-α, and IFN-γ, was found to increase liver MT-1/MT-2 levels (DiSilvestro and Cousins, 1984; Cousins and Leinart, 1988;

De *et al.*, 1990), strongly suggesting that these cytokines could mediate the induction of MT-1/MT-2 by LPS. Further support for this notion was obtained with the C3H/HeJ mice, a strain which has a mutation in the *Lps* gene that causes a very low cytokine response to endotoxin (Vogel, 1992), since these mice produced less hepatic MT-1/MT-2 than controls when injected with endotoxin, but responded equally well when administered cytokines (Maitani *et al.*, 1986). Brain MT-1/MT-2 are also upregulated by LPS, and presumably the same mechanisms operate (Searle *et al.*, 1984; Itano *et al.*, 1991; Palmiter *et al.*, 1992; Carrasco *et al.*, 1998a).

In more recent studies, it has been shown that MT-1/MT-2 expression is significantly affected whenever there is tissue damage, and even following subtle insults. For instance, brain MT-1/MT-2 are well known to be upregulated by a mainly psychogenic stress such as restraint (Hidalgo *et al.*, 1990), or by the administration of glutamate analogs (Dalton *et al.*, 1995; Hidalgo *et al.*, 1997; Montpied *et al.*, 1998; Acarin *et al.*, 1999b; Carrasco *et al.*, 2000b; Tang *et al.*, 2002), traumatic injury (Chung *et al.*, 2003), cryogenic injury (Penkowa *et al.*, 1999a; Penkowa *et al.*, 1999c), or stroke/ischemia (Neal *et al.*, 1996; van Lookeren Campagne *et al.*, 1999; van Lookeren Campagne *et al.*, 2000; Tang *et al.*, 2002; Trendelenburg *et al.*, 2002). Moreover, MT-1/MT-2 are also upregulated in animal models of Alzheimer's disease (Carrasco *et al.*, 2006), perinatal viral infections (Williams *et al.*, 2006), multiple sclerosis (Penkowa and Hidalgo, 2000; Espejo *et al.*, 2001; Penkowa *et al.*, 2001), and familial amyotrophic lateral sclerosis (Gong and Elliott, 2000; Fukada *et al.*, 2001; Nagano *et al.*, 2001; Puttaparthi *et al.*, 2002).

It is thus clear that MT-1/MT-2 may play an important role in the overall response of the brain to damage. Tissue injury elicits an inflammatory response that is thought to be orchestrated by a number of cytokines in a complex manner, as well as by oxidative stress (Halliwell, 1992; Coyle and Puttfarcken, 1993; Merrill and Benveniste, 1996; Smith *et al.*, 1996; Noseworthy, 1999; Barnham *et al.*, 2004; Andersen, 2004; Block *et al.*, 2007). Moreover, there is increasing experimental evidence that oxidative stress contributes significantly to the neuropathology of several adult neurodegenerative disorders as well as stroke, trauma, seizures, and neuronal degeneration, and that cytokines play a prominent role (Frei *et al.*, 1988; Hopkins and Rothwell, 1995; Rothwell and Hopkins, 1995; Gruol and Nelson, 1997; Muñoz-Fernández and Fresno, 1998; Noseworthy, 1999; Horner and Gage, 2000; Allan and Rothwell, 2001; Lucas *et al.*, 2006). That cytokines may cause significant brain damage has been clearly demonstrated by results obtained in transgenic mice which express in astrocytes cytokines such as interleukin-6 (IL-6), tumor necrosis factor-α (TNF-α), interleukin-3 (IL-3), and interferon-α (IFN-α) under the control of the glial fibrillary acidic protein (GFAP) gene promoter. Each of these transgenic mice showed a specific phenotype of cytokine-induced damage that resembled some of the most important neurological diseases in humans (Campbell *et al.*, 1993; Stalder *et al.*, 1998). Results obtained with microarrays and transgenic mice with null mutations have demonstrated that IL-6 (Poulsen *et al.*, 2005) and TNF-α (Quintana *et al.*, 2007) do play major roles in the response of the cortex to injury. These studies also demonstrated that both cytokines regulate MT-1/MT-2 expression, which is in line with results obtained from transgenic mice overexpressing either IL-6 (Hernández *et al.*, 1997; Carrasco *et al.*, 1998b) or TNF-α (Carrasco *et al.*, 2000a) and with the presence of STAT response elements in the MT promoter (Lee *et al.*, 1999). MT-1/MT-2 levels were also upregulated in GFAP-IL3 and GFAP-IFNα mice (Giralt *et al.*, 2001).

Taken together, the above studies strongly suggest a significant role of MT-1/MT-2 during neurodegenerative diseases and in response to brain injury and the ensuing neuroinflammation. In order to establish their putative role(s), the generation of genetically modified mice (Michalska and Choo, 1993; Palmiter *et al.*, 1993; Masters *et al.*, 1994a) has greatly potentiated this research. Mice overexpressing MT-1 were partially protected against mild focal cerebral ischemia and reperfusion, since the volume of affected tissue was smaller and the motor performance (3 weeks after the lesion) better (van Lookeren Campagne *et al.*, 1999). Conversely, MT-1/MT-2–null mice developed approximately threefold larger infarcts than wild-type mice and a significantly worse neurological outcome (Trendelenburg *et al.*, 2002), pointing to an essential role of MT-1/MT-2 to cope with ischemic damage of the brain. Other studies have clearly demonstrated a similar essential role to cope with damage elicited by kainic acid-induced seizures (Carrasco *et al.*, 2000b), the gliotoxin 6-aminonicotinamide (Penkowa *et al.*, 1999b; Penkowa *et al.*, 2002), dopaminergic neurotoxicity (Asanuma *et al.*, 2002; Xie *et al.*, 2004; Miyazaki *et al.*, 2007), mutated Cu,Zn-superoxide dismutase (Nagano *et al.*, 2001; Puttaparthi *et al.*, 2002), multiple sclerosis models (Penkowa *et al.*, 2001; Penkowa *et al.*, 2003b), traumatic brain injury (Penkowa *et al.*, 1999a; Penkowa *et al.*, 2000; Giralt *et al.*, 2002b; Natale *et al.*, 2004; Potter *et al.*, 2007), and transgenic IL-6–induced neuropathology (Giralt *et al.*, 2002a; Molinero *et al.*, 2003; Penkowa *et al.*, 2003a). Overall, the results obtained in these studies are compatible with a role of MT-1/MT-2 decreasing oxidative stress, inflammation, and apoptosis in the CNS, which is in accordance with results in other tissues (Lazo *et al.*, 1995; Kang *et al.*, 1997; Kondo *et al.*, 1997; Youn *et al.*, 2002).

The exact mechanisms through which MT-1/MT-2 elicit these neuroprotective effects are ill defined (Hidalgo *et al.*, 2001), and may include not only the putative roles of the intracellular protein but also those of the extracellular one: exogenously applied MT-2 protein mimics MT-1 transgenic overexpression, causing a significant clinical improvement in the animal model of multiple sclerosis experimental autoimmune encephalomyelitis (EAE) (Penkowa and Hidalgo, 2000; Pankowa and Hidalgo, 2003), and protecting against traumatic brain injury (Giralt *et al.*, 2002b; Chung *et al.*, 2003) and dopaminergic neurotoxicity (Xie *et al.*, 2004), which opens up exciting perspectives about the use of the MT family as therapeutical agents (Hidalgo *et al.*, 2001; Chung and West, 2004; Chung *et al.*, 2008). It is important to note that exogenous MT-1/MT-2 also show benefits in other diseases such as collagen-induced arthritis

(Youn *et al.*, 2002) and stress-induced stomach ulcers (Jiang *et al.*, 2005). It is very likely that all of these MT effects could be somehow linked to effects on the immune system (Lynes *et al.*, 2006), but more specific effects cannot be ruled out, at least in the brain. MT-1/MT-2 have potent effects on neurons *in vitro* that appear to be mediated at least in part through binding to receptors of the low-density lipoprotein receptor family (Ambjørn *et al.*, 2008; Fitzgerald *et al.*, 2007); one of these receptors, megalin, was previously suggested to be involved in MT-1/MT-2 uptake in kidney cells (Klassen *et al.*, 2004). Interestingly, a lipid raft-dependent endocytosis of MT has also been demonstrated in HepG2 cells (Hao *et al.*, 2007). It must be emphasized that a distant MT, *Drosophila* MTN, functions similarly to mammalian MT-1/MT-2 in the cryolesion model (Penkowa *et al.*, 2006b). Thus, both intracellular and extracellular MT-1/MT-2 must be taken into account regarding putative physiological functions. Finally, a microarray study comparing *MT-1/MT-2* knockout (KO) mice with wild-type (WT) mice clearly shows the complexity of challenge of understanding fully the roles of these proteins (Penkowa *et al.*, 2006a).

4. Metallothionein-3: A Challenging Protein

MT-3, originally named "growth inhibitory factor" (GIF), was discovered unexpectedly while pursuing putative mechanisms underlying the neuropathology of Alzheimer's disease (Uchida *et al.*, 1991). It was originally suggested that MT-3 is downregulated during AD, but unfortunately this is not a consistent finding (Uchida *et al.*, 1991; Tsuji *et al.*, 1992; Erickson *et al.*, 1994; Uchida, 1994; Amoureux *et al.*, 1997; Carrasco *et al.*, 1999; Yu *et al.*, 2001). Also, MT-3 expression increases or decreases in other human diseases such as Down syndrome (Arai *et al.*, 1997), Creutzfeldt–Jakob disease (Kawashima *et al.*, 2000), pontosubicular necrosis (Isumi *et al.*, 2000), Parkinson's disease, meningitis, and amyotrophic lateral sclerosis (Uchida, 1994). In line with the unclear type of response in these neurodegenerative diseases, in animal models of brain injury MT-3 expression also shows upregulation or downregulation depending on the model, time, etc., in sharp contrast to MT-1/MT-2. For instance, MT-3 expression has been shown to be increased by stab wounds (Anezaki *et al.*, 1995; Hozumi *et al.*, 1995; Hozumi *et al.*, 1996), kainic acid administration (Anezaki *et al.*, 1995), and mild hyperoxia (Gomi *et al.*, 2005); but decreased by cortical ablation of the somatosensory cortex (Yuguchi *et al.*, 1995a), facial nerve transection (Yuguchi *et al.*, 1995b), and middle cerebral artery occlusion (Inuzuka *et al.*, 1996). A biphasic response of MT-3 to CNS injury, with initial downregulation followed by upregulation, was observed in response to *N*-methyl-D-aspartate (NMDA) (Acarin *et al.*, 1999a) or to a cryolesion (Penkowa *et al.*, 1999c; Penkowa *et al.*, 2000). There are also reports of increased levels of MT-3 in astrocytes but a downregulation in neurons, such as in a model of perinatal viral infection (Williams *et al.*, 2006), which further complicates the understanding of the control of this protein.

As for MT-1/MT-2, the generation of genetically modified mice (Erickson *et al.*, 1995; Erickson *et al.*, 1997) could help significantly to understand the potential biological roles of MT-3. In normal conditions, MT-3–null mice do not show an appreciable phenotype (Erickson *et al.*, 1997; Carrasco *et al.*, 2003), as is the case for MT-1/MT-2. If challenged with the seizures elicited by the glutamate analog kainic acid, the MT-3 KO mice showed enhanced sensitivity, convulsing longer and having greater mortality than littermate controls, and showing increased neuronal death in the CA3 pyramidal cell layer of hippocampus; while transgenic mice overexpressing MT-3 showed the opposite trends, strongly suggesting a neuroprotective role of MT-3 (Erickson *et al.*, 1997). Consistent with these results, MT-3 was found to prevent glutamate neurotoxicity in primary cultures of cerebellar neurons (Montoliu *et al.*, 2000). Further studies demonstrated in fact that MT-3 has opposing effects in the CA3 and CA1 areas (Lee *et al.*, 2003). Other studies have suggested a neuroprotective role of MT-3 for motoneurons *in vivo*

(Puttaparthi *et al.*, 2002; Sakamoto *et al.*, 2003). In the cryolesion model, however, MT-3 deficiency did not affect the inflammatory response, oxidative stress, or apoptotic death (Carrasco *et al.*, 2003), in sharp contrast to what is observed in MT-1/MT-2–deficient mice (see above). However, MT-3 deficiency did increase the expression of some neurotrophins and other factors that may significantly affect neuronal survival and/or growth, such as GAP43 (Benowitz and Routtenberg, 1997; Bibel and Barde, 2000), which would support an inhibitory role of MT-3 in line with some *in vitro* bioassays (Uchida *et al.*, 1991; Erickson *et al.*, 1994; Sewell *et al.*, 1995; Chung *et al.*, 2002; Chung *et al.*, 2003; Chung and West, 2004). A report analyzing the response of MT-3–deficient mice in a peripheral nerve model supports the role of MT-3 as an inhibitory factor of neuronal sprouting *in vivo* (Ceballos *et al.*, 2003), since axonal regeneration was faster as substantiated electrophysiologically and histologically.

As for MT-1/MT-2, the use of recombinant proteins would facilitate further *in vivo* studies, including the putative therapeutic value and eventual use of MT-3. Indeed, it has already been demonstrated that the exogenous administration of MT-3 does have prominent effects on the injured cortex (Hozumi *et al.*, 2006; Penkowa *et al.*, 2006b). The first study showed, again remarkably, opposing effects of MT-3 depending on the amount applied directly onto the lesioned cortex in the stab wound model. In the second study, using the cryolesion model, the protein was administered intraperitoneally in the same amount as MT-1/MT-2, and the results were consistent with those observed in the corresponding null mice: no effect of MT-3 on the inflammatory response and oxidative stress, and a decreasing effect on several neurotrophins and other growth factors, in contrast to MT-1/MT-2 (Penkowa *et al.*, 2006b).

Thus, it might be envisaged that MT-3 could serve different functions in the CNS, promoting neuronal survival (perhaps only in specific neuronal populations such as motoneurons or hippocampus CA3 neurons) or death (hippocampus CA1 neurons) while inhibiting neuronal sprouting. The mechanisms underlying these effects remain to be established, despite much effort with *in vitro* studies, since significant discrepancies are observed in the literature regarding the role of MT-3 in oxidative stress defense [i.e. Uchida *et al.* (2002) vs. Chen *et al.* (2002) and Shi *et al.* (2003)], its putative biological partners [Kang *et al.* (2001) and Knipp *et al.* (2005) vs. Lahti *et al.* (2005)], and the essential sequence motifs that explain its unique features [Sewell *et al.* (1995), Romero-Isart *et al.* (2002), and Cai *et al.* (2006) vs. Ding *et al.* (2006)].

5. Conclusion

The results produced in the last 15 years demonstrate that the MT family of proteins is potentially very important in the neurobiology field, with a wide range of effects on the essential brain cells, astrocytes, and neurons, but also in microglia/macrophages. MT-1/MT-2 seem to differ from MT-3 physiologically as well as regarding their control, i.e. in the brain. A very interesting avenue is set by the fact that the exogenously applied proteins show comparable effects to the endogenous proteins, inviting therapeutic approaches in the near future.

Acknowledgments

The help of Ministerio de Ciencia y Tecnología and Feder SAF2005-00671 and European Comission FP6 Integrated Project Exgenesis (Ref. LSHM-CT-2004-005272) is acknowledged. Thanks are also given to the colleagues who have helped so much over the years.

References

Acarin L, Carrasco J, González B, *et al.* Expression of growth inhibitory factor (metallothionein-III) mRNA and protein following excitotoxic immature brain injury. *J Neuropathol Exp Neurol* 1999a; **58**:389–397.

Acarin L, González B, Hidalgo J, *et al.* Primary cortical glial reaction versus secondary thalamic glial response in the excitotoxically injured young brain: Astroglial response and metallothionein expression. *Neuroscience* 1999b; **92**:827–839.

Adlard PA, West AK, Vickers JC. Increased density of metallothionein I/II-immunopositive cortical glial cells in the early stages of Alzheimer's disease. *Neurobiol Dis* 1998; **5**:349–356.

Agulló L, García A, Hidalgo J. Metallothionein-I+II induction by zinc and copper in primary cultures of rat microglia. *Neurochem Int* 1998; **33**:237–242.

Allan SM, Rothwell NJ. Cytokines and acute neurodegeneration. *Nat Rev Neurosci* 2001; **2**:734–744.

Ambjørn M, Asmussen JW, Lindstam M, *et al.* Metallothionein and a peptide modeled after metallothionein, EmtinB, induce neuronal differentiation and survival through binding to receptors of the low density lipoprotein receptor family. *J Neurochem* 2008; **104**:21–37.

Amoureux MC, Van Gool D, Herrero MT, *et al.* Regulation of metallothionein-III (GIF) mRNA in the brain of patients with Alzheimer disease is not impaired. *Mol Chem Neuropathol* 1997; **32**:101–121.

Andersen JK. Oxidative stress in neurodegeneration: Cause or consequence? *Nat Med* 2004; **10**:S18–S25.

Andrews GK. Regulation of metallothionein gene expression by oxidative stress and metal ions. *Biochem Pharmacol* 2000; **59**:95–104.

Anezaki T, Ishiguro H, Hozumi I, *et al.* Expression of growth inhibitory factor (GIF) in normal and injured rat brains. *Neurochem Int* 1995; **27**:89–94.

Arai Y, Uchida Y, Takashima S. Developmental immunohistochemistry of growth inhibitory factor in normal brains and brains of patients with Down syndrome. *Pediatr Neurol* 1997; **17**:134–138.

Asanuma M, Miyazaki I, Higashi Y, *et al.* Aggravation of 6-hydroxydopamine-induced dopaminergic lesions in metallothionein-I and -II knock-out mouse brain. *Neurosci Lett* 2002; **327**:61–65.

Barnham KJ, Masters CL, Bush AI. Neurodegenerative diseases and oxidative stress. *Nat Rev Drug Discov* 2004; **3**:205–214.

Benowitz LI, Routtenberg A. GAP-43: An intrinsic determinant of neuronal development and plasticity. *Trends Neurosci* 1997; **20**:84–91.

Bibel M, Barde Y-A. Neurotrophins: Key regulators of cell fate and cell shape in the vertebrate nervous system. *Genes Dev* 2000; **14**:2919–2937.

Blaauwgeers HG, Anwar Chand M, van den Berg FM, *et al.* Expression of different metallothionein messenger ribonucleic acids in motor cortex, spinal cord and liver from patients with amyotrophic lateral sclerosis. *J Neurol Sci* 1996; **142**:39–44.

Block ML, Zecca L, Hong JS. Microglia-mediated neurotoxicity: Uncovering the molecular mechanisms. *Nat Rev Neurosci* 2007; **8**:57–69.

Bremner I. Interactions between metallothionein and trace elements. *Prog Food Nutr Sci* 1987a; **11**:1–37.

Bremner I. Nutritional and physiological significance of metallothionein. *Experientia Suppl* 1987b; **52**:81–107.

Cai B, Zheng Q, Teng XC, *et al.* The role of Thr5 in human neuron growth inhibitory factor. *J Biol Inorg Chem* 2006; **11**:476–482.

Campbell IL, Abraham CR, Masliah E, *et al.* Neurologic disease in transgenic mice by cerebral overexpression of interleukin 6. *Proc Natl Acad Sci USA* 1993; **90**:10061–10065.

Carrasco J, Adlard P, Cotman C, *et al.* Metallothionein-I and -III expression in animal models of Alzheimer disease. *Neuroscience* 2006; **143**:911–922.

Carrasco J, Giralt M, Molinero A, *et al.* Metallothionein (MT)-III: Generation of polyclonal antibodies, comparison with MT-I+II in the freeze lesioned rat brain and in a bioassay with astrocytes, and analysis of Alzheimer's disease brains. *J Neurotrauma* 1999; **16**:1115–1129.

Carrasco J, Giralt M, Penkowa M, *et al.* Metallothioneins are upregulated in symptomatic mice with astrocyte-targeted expression of tumor necrosis factor-α. *Exp Neurol* 2000a; **163**:46–54.

Carrasco J, Hernández J, Bluethmann H, Hidalgo J. Interleukin-6 and tumor necrosis factor-alpha type 1 receptor deficient mice reveal a role of IL-6 and TNF-alpha on brain metallothionein-I and -III regulation. *Mol Brain Res* 1998a; **57**:221–234.

Carrasco J, Hernández J, González B, *et al.* Localization of metallothionein-I and -III expression in the CNS of transgenic mice with astrocyte-targeted expression of interleukin 6. *Exp Neurol* 1998b; **153**:184–194.

Carrasco J, Penkowa M, Giralt M, *et al.* Role of metallothionein-III following central nervous system damage. *Neurobiol Dis* 2003; **13**:22–36.

Carrasco J, Penkowa M, Hadberg H, *et al.* Enhanced seizures and hippocampal neurodegeneration following kainic acid induced seizures in metallothionein-I+II deficient mice. *Eur J Neurosci* 2000b; **12**:2311–2322.

Ceballos D, Lago N, Verdú E, *et al.* Role of metallothioneins in peripheral nerve function and regeneration. *Cell Mol Life Sci* 2003; **60**:1209–1216.

Chen Y, Irie Y, Keung WM, Maret W. *S*-Nitrosothiols react preferentially with zinc thiolate clusters of metallothionein III through transnitrosation. *Biochemistry* 2002; **41**: 8360–8367.

Chuah MI, Getchell ML. Metallothionein in olfactory mucosa of Alzheimer's disease patients and apoE-deficient mice. *Neuroreport* 1999; **10**:1919–1924.

Chung RS, Hidalgo J, West A. New insight into the molecular pathways of metallothionein-mediated neuroprotection and regeneration. *J Neurochem* 2008; **104**:14–20.

Chung RS, Vickers JC, Chuah MI, *et al.* Metallothionein-III inhibits initial neurite formation in developing neurons as well as postinjury, regenerative neurite sprouting. *Exp Neurol* 2002; **178**:1–12.

Chung RS, Vickers JC, Chuah MI, West AK. Metallothionein-IIA promotes initial neurite elongation and postinjury reactive neurite growth and facilitates healing after focal cortical brain injury. *J Neurosci* 2003; **23**:3336–3342.

Chung RS, West AK. A role for extracellular metallothioneins in CNS injury and repair. *Neuroscience* 2004; **123**:595–599.

Cousins RJ, Leinart AS. Tissue-specific regulation of zinc metabolism and metallothionein genes by interleukin 1. *FASEB J* 1988; **2**:2884–2890.

Cox DR, Palmiter RD. The metallothionein-I gene maps to mouse chromosome 8: Implications for human Menkes' disease. *Hum Genet* 1983; **64**:61–64.

Coyle J, Puttfarcken P. Oxidative stress, glutamate, and neurodegenerative disorders. *Science* 1993; **262**:689–695.

Coyle P, Philcox JC, Carey LC, Rofe AM. Metallothionein: The multipurpose protein. *Cell Mol Life Sci* 2002; **59**:627–647.

Dalton T, Pazdernik TL, Wagner J, *et al.* Temporalspatial patterns of expression of metallothionein-I and -III and other stress related genes in rat brain after kainic acid-induced seizures. *Neurochem Int* 1995; **27**:59–71.

De SK, McMaster MT, Andrews GK. Endotoxin induction of murine metallothionein gene expression. *J Biol Chem* 1990; **265**:15267–15274.

Ding ZC, Teng XC, Cai B, *et al.* Mutation at Glu23 eliminates the neuron growth inhibitory activity of human metallothionein-3. *Biochem Biophys Res Commun* 2006; **349**:674–682.

DiSilvestro RA, Cousins RJ. Glucocorticoid independent mediation of interleukin-1 induced changes in serum zinc and liver metallothionein levels. *Life Sci* 1984; **35**:2113–2118.

Duguid JR, Bohmont CW, Liu NG, Tourtellotte WW. Changes in brain gene expression shared by scrapie and Alzheimer disease. *Proc Natl Acad Sci USA* 1989; **86**:7260–7264.

Durnam DM, Hoffman JS, Quaife CJ, *et al.* Induction of mouse metallothionein-I mRNA by bacterial endotoxin is independent of metals and glucocorticoid hormones. *Proc Natl Acad Sci USA* 1984; **81**:1053–1056.

Erickson JC, Hollopeter G, Thomas SA, *et al.* Disruption of the metallothionein-III gene in mice: Analysis of brain zinc, behavior, and neuron vulnerability to metals, aging, and seizures. *J Neurosci* 1997; **17**:1271–1281.

Erickson JC, Masters BA, Kelly EJ, *et al.* Expression of human metallothionein-III in transgenic mice. *Neurochem Int* 1995; **27**:35–41.

Erickson JC, Sewell AK, Jensen LT, *et al.* Enhanced neurotrophic activity in Alzheimer's disease cortex is not associated with down-regulation of metallothionein-III (GIF). *Brain Res* 1994; **649**:297–304.

Espejo C, Carrasco J, Hidalgo J, *et al.* Differential expression of metallothioneins in the CNS of mice with experimental autoimmune encephalomyelitis. *Neuroscience* 2001; **105**:1055–1065.

Fitzgerald M, Nairn P, Bartlett CA, *et al.* Metallothionein-IIA promotes neurite growth via the megalin receptor. *Exp Brain Res* 2007; **183**:171–180.

Frei K, Leist T, Meager A, *et al.* Production of B cell stimulatory factor-2 and interferon gamma in the central nervous system during viral meningitis and encephalitis. *J Exp Med* 1988; **168**:449–453.

Fukada K, Nagano S, Satoh M, *et al.* Stabilization of mutant Cu/Zn superoxide dismutase (SOD1) protein by coexpressed wild SOD1 protein accelerates the disease progression in familial amyotrophic lateral sclerosis mice. *Eur J Neurosci* 2001; **14**:2032–2036.

Ghoshal K, Jacob ST. Regulation of metallothionein gene expression. *Prog Nucleic Acid Res Mol Biol* 2001; **66**:357–384.

Giralt M, Carrasco J, Penkowa M, *et al.* Astrocyte-targeted expression of interleukin-3 and interferon-α causes specific changes in metallothionein expression in the brain. *Exp Neurol* 2001; **168**:334–346.

Giralt M, Penkowa M, Hernández J, *et al.* Metallothionein-1+2 deficiency increases brain pathology in transgenic mice with astrocyte-targeted expression of interleukin 6. *Neurobiol Dis* 2002a; **9**:319–338.

Giralt M, Penkowa M, Lago N, *et al.* Metallothionein-1+2 protect the CNS after a focal brain injury. *Exp Neurol* 2002b; **173**:114–128.

Gomi F, Nakano S, Uchida Y. Tolerance of aged rat brains to mild hyperoxia: Possible involvement of higher GIF content. *Brain Res* 2005; **1033**:113–116.

Gong YH, Elliott JL. Metallothionein expression is altered in a transgenic murine model of familial amyotrophic lateral sclerosis. *Exp Neurol* 2000; **162**:27–36.

Gruol DL, Nelson TE. Physiological and pathological roles of interleukin-6 in the central nervous system. *Mol Neurobiol* 1997; **15**:307–339.

Halliwell B. Reactive oxygen species and the central nervous system. *J Neurochem* 1992; **59**:1609–1623.

Hamer DH. Metallothionein. *Annu Rev Biochem* 1986; **55**:913–951.

Hao Q, Hong SH, Maret W. Lipid raft-dependent endocytosis of metallothionein in HepG2 cells. *J Cell Physiol* 2007; **210**:428–435.

Hernández J, Molinero A, Campbell IL, Hidalgo J. Transgenic expression of interleukin 6 in the central nervous system regulates brain metallothionein-I and -III expression in mice. *Brain Res Mol Brain Res* 1997; **48**:125–131.

Hidalgo J, Aschner M, Zatta P, Vašák M. Roles of the metallothionein family of proteins in the central nervous system. *Brain Res Bull* 2001; **55**:133–145.

Hidalgo J, Borras M, Garvey JS, Armario A. Liver, brain, and heart metallothionein induction by stress. *J Neurochem* 1990; **55**:651–654.

Hidalgo J, Castellano B, Campbell IL. Regulation of brain metallothioneins. *Curr Top Neurochem* 1997; **1**:1–26.

Hidalgo J, García A, Oliva AM, *et al.* Effect of zinc, copper and glucocorticoids on metallothionein levels of cultured neurons and astrocytes from rat brain. *Chem Biol Interact* 1994; **93**:197–219.

Hopkins S, Rothwell N. Cytokines and the nervous system. I: Expression and recognition. *Trends Neurosci* 1995; **18**:83–88.

Horner PJ, Gage FH. Regenerating the damaged central nervous system. *Nature* 2000; **407**:963–970.

Hozumi I, Inuzuka T, Hiraiwa M, *et al.* Changes of growth inhibitory factor after stab wounds in rat brain. *Brain Res* 1995; **688**:143–148.

Hozumi I, Inuzuka T, Ishiguro H, *et al.* Immunoreactivity of growth inhibitory factor in normal rat brain and after stab wounds — An immunocytochemical study using confocal laser scan microscope. *Brain Res* 1996; **741**:197–204.

Hozumi I, Uchida Y, Watabe K, *et al.* Growth inhibitory factor (GIF) can protect from brain damage due to stab wounds in rat brain. *Neurosci Lett* 2006; **395**:220–223.

Inuzuka T, Hozumi I, Tamura A, *et al.* Patterns of growth inhibitory factor (GIF) and glial fibrillary acidic protein relative level changes differ following left middle cerebral artery occlusion in rats. *Brain Res* 1996; **709**:151–153.

Isumi H, Uchida Y, Hayashi T, *et al.* Neuron death and glial response in pontosubicular necrosis. The role of the growth inhibition factor. *Clin Neuropathol* 2000; **19**:77–84.

Itano Y, Noji S, Koyama E, *et al.* Bacterial endotoxin-induced expression of metallothionein genes in rat brain, as revealed by *in situ* hybridization. *Neurosci Lett* 1991; **124**:13–16.

Jiang P, Chang L, Pan CS, *et al.* Protective role of metallothionein in stress-induced gastric ulcer in rats. *World J Gastroenterol* 2005; **11**:2739–2743.

Kägi JHR, Vallee BL. Metallothionein: A cadmium- and zinc-containing protein from equine renal cortex. *J Biol Chem* 1960; **235**:3460–3465.

Kägi JHR, Vallee BL. Metallothionein: A cadmium- and zinc-containing protein from equine renal cortex. II. Physicochemical properties. *J Biol Chem* 1961; **236**:2435–2442.

Kang QH, Chen QL, Ren HW, Ru BG. Growth inhibitory factor (GIF) directly interacts with G protein-Rab3A. *Prog Biochem Biophys* 2001; **28**:880–884.

Kang YJ, Chen Y, Yu A, *et al.* Overexpression of metallothionein in the heart of transgenic mice suppresses doxorubicin cardiotoxicity. *J Clin Invest* 1997; **100**:1501–1506.

Karin M, Eddy RL, Henry WM, *et al.* Human metallothionein genes are clustered on chromosome 16. *Proc Natl Acad Sci USA* 1984a; **81**:5494–5498.

Karin M, Haslinger A, Holtgreve H, *et al.* Characterization of DNA sequences through which cadmium and glucocorticoid hormones induce human metallothionein-IIA gene. *Nature* 1984b; **308**:513–519.

Kawashima T, Doh-ura K, Torisu M, *et al.* Differential expression of metallothioneins in human prion diseases. *Dement Geriatr Cogn Disord* 2000; **11**:251–262.

Kelly EJ, Sandgren EP, Brinster RL, Palmiter RD. A pair of adjacent glucocorticoid response elements regulate expression of two mouse metallothionein genes. *Proc Natl Acad Sci USA* 1997; **94**:10045–10050.

Klassen RB, Crenshaw K, Kozyraki R, *et al.* Megalin mediates renal uptake of heavy metal metallothionein complexes. *Am J Physiol Renal Physiol* 2004; **287**:F393–F403.

Knipp M, Meloni G, Roschitzki B, Vašák M. Zn_7 metallothionein-3 and the synaptic vesicle cycle: Interaction of metallothionein-3 with the small GTPase Rab3A. *Biochemistry* 2005; **44**:3159–3165.

Kobayashi H, Uchida Y, Ihara Y, *et al.* Molecular cloning of rat growth inhibitory factor cDNA and the expression in the central nervous system. *Mol Brain Res* 1993; **19**:188–194.

Kondo Y, Rusnak J, Hoyt D, *et al.* Enhanced apoptosis in metallothionein null cells. *Mol Pharmacol* 1997; **52**:195–201.

Kramer KK, Liu J, Choudhuri S, Klaassen CD. Induction of metallothionein mRNA and protein in murine astrocyte cultures. *Toxicol Appl Pharmacol* 1996a; **136**:94–100.

Kramer KK, Zoelle JT, Klaassen CD. Induction of metallothionein mRNA and protein in primary murine neuron cultures. *Toxicol Appl Pharmacol* 1996b; **141**:1–7.

Lahti DW, Hoekman JD, Tokheim AM, *et al.* Identification of mouse brain proteins associated with isoform 3 of metallothionein. *Protein Sci* 2005; **14**:1151–1157.

Lazo JS, Kondo Y, Dellapiazza D, *et al.* Enhanced sensitivity to oxidative stress in cultured embryonic cells from transgenic mice deficient in metallothionein I and II genes. *J Biol Chem* 1995; **270**:5506–5510.

Lee DK, Carrasco J, Hidalgo J, Andrews GK. Identification of a signal transducer and activator of transcription (STAT) binding site in the mouse metallothionein-I promoter involved in interleukin-6–induced gene expression. *Biochem J* 1999; **337**:59–65.

Lee JY, Kim JH, Palmiter RD, Koh JY. Zinc released from metallothionein-iii may contribute to hippocampal CA1 and thalamic neuronal death following acute brain injury. *Exp Neurol* 2003; **184**:337–347.

Lock C, Hermans G, Pedotti R, *et al.* Gene-microarray analysis of multiple sclerosis lesions yields new targets validated in autoimmune encephalomyelitis. *Nat Med* 2002; **8**:500–508.

Lucas SM, Rothwell NJ, Gibson RM. The role of inflammation in CNS injury and disease. *Br J Pharmacol* 2006; **147**(Suppl 1):S232–S240.

Lynes MA, Zaffuto K, Unfricht DW, *et al.* The physiological roles of extracellular metallothionein. *Exp Biol Med* 2006; **231**:1548–1554.

Maitani T, Saito Y, Fujimaki H, Suzuki KT. Comparative induction of hepatic zinc-thionein and increase in tissue calcium by bacterial endotoxin in endotoxin-sensitive (C3H/HeN) and endotoxin-resistant (C3H/HeJ) mice. *Toxicol Lett* 1986; **30**:181–187.

Margoshes M, Vallee BL. A cadmium protein from equine kidney cortex. *J Am Chem Soc* 1957; **79**:4813–4814.

Masters BA, Kelly EJ, Quaife CJ, *et al.* Targeted disruption of metallothionein I and II genes increases sensitivity to cadmium. *Proc Natl Acad Sci USA* 1994a; **91**:584–588.

Masters BA, Quaife CJ, Erickson JC, *et al.* Metallothionein III is expressed in neurons that sequester zinc in synaptic vesicles. *J Neurosci* 1994b; **14**:5844–5857.

Merrill JE, Benveniste EN. Cytokines in inflammatory brain lesions: Helpful and harmful. *Trends Neurosci* 1996; **19**:331–338.

Michalska AE, Choo KH. Targeting and germ-line transmission of a null mutation at the metallothionein I and II loci in mouse. *Proc Natl Acad Sci USA* 1993; **90**:8088–8092.

Miyazaki I, Asanuma M, Hozumi H, *et al.* Protective effects of metallothionein against dopamine quinone-induced dopaminergic neurotoxicity. *FEBS Lett* 2007; **581**:5003–5008.

Molinero A, Penkowa M, Hernández J, *et al.* Metallothionein-I overexpression decreases brain pathology in transgenic mice with astrocyte-targeted expression of interleukin 6. *J Neuropathol Exp Neurol* 2003; **62**:315–328.

Montoliu C, Monfort P, Carrasco J, *et al.* Metallothionein-III prevents glutamate and nitric oxide neurotoxicity in primary cultures of cerebellar neurons. *J Neurochem* 2000; **75**:266–273.

Montpied P, de Bock F, Baldy Moulinier M, Rondouin G. Alterations of metallothionein II and apolipoprotein J mRNA levels in kainate-treated rats. *Neuroreport* 1998; **9**:79–83.

Muñoz-Fernández MA, Fresno M. The role of tumour necrosis factor, interleukin 6, interferon-gamma and inducible nitric oxide synthase in the development and pathology of the nervous system. *Prog Neurobiol* 1998; **56**:307–340.

Nagano S, Satoh M, Sumi H, *et al.* Reduction of metallothioneins promotes the disease expression of familial amyotrophic lateral sclerosis mice in a dose-dependent manner. *Eur J Neurosci* 2001; **13**:1363–1370.

Natale JE, Knight JB, Cheng Y, *et al.* Metallothionein I and II mitigate age-dependent secondary brain injury. *J Neurosci Res* 2004; **78**:303–314.

Neal JW, Singhrao SK, Jasani B, Newman GR. Immunocytochemically detectable metallothionein is expressed by astrocytes in the ischaemic human brain. *Neuropathol Appl Neurobiol* 1996; **22**:243–247.

Noseworthy JH. Progress in determining the causes and treatment of multiple sclerosis. *Nature* 1999; **399**(6738 Suppl):A40–A47.

Palmiter RD, Findley SD, Whitmore TE, Durnam DM. MT-III, a brain-specific member of the metallothionein gene family. *Proc Natl Acad Sci USA* 1992; **89**:6333–6337.

Palmiter RD, Sandgren EP, Koeller DM, Brinster RL. Distal regulatory elements from the mouse metallothionein locus stimulate gene expression in transgenic mice. *Mol Cell Biol* 1993; **13**:5266–5275.

Penkowa M, Cáceres M, Borup R, *et al.* Novel roles for metallothionein-I+II (MT-I+II) in defense responses, neurogenesis and tissue restoration after traumatic brain injury: Insights from global gene expression profiling in wildtype and MT-I+II knockout mice. *J Neurosci Res* 2006a; **84**:1452–1474.

Penkowa M, Camats J, Giralt M, *et al.* Metallothionein-I overexpression alters brain inflammation and stimulates brain repair in transgenic mice with astrocyte-targeted interleukin-6 expression. *Glia* 2003a; **42**:287–306.

Penkowa M, Carrasco J, Giralt M, *et al.* CNS wound healing is severely depressed in metallothionein I- and II-deficient mice. *J Neurosci* 1999a; **19**:2535–2545.

Penkowa M, Carrasco J, Giralt M, *et al.* Altered central nervous system cytokine-growth factor expression profiles and angiogenesis in metallothionein-I+II deficient mice. *J Cerebral Blood Flow Metab* 2000; **20**:1174–1189.

Penkowa M, Espejo C, Martínez-Cáceres EM, *et al.* Altered inflammatory response and increased neurodegeneration in metallothionein I+II deficient mice during experimental autoimmune encephalomyelitis. *J Neuroimmunol* 2001; **119**:248–260.

Penkowa M, Espejo C, Martínez-Cáceres EM, *et al.* Increased demyelination and axonal damage in metallothionein I+II-deficient mice during experimental autoimmune encephalomyelitis. *Cell Mol Life Sci* 2003b; **60**:185–197.

Penkowa M, Espejo C, Ortega-Aznar A, *et al.* Metallothionein expression in the central nervous system of multiple sclerosis patients. *Cell Mol Life Sci* 2003c; **60**:1258–1266.

Penkowa M, Giralt M, Camats J, Hidalgo J. Metallothionein 1+2 protect the CNS during neuroglial degeneration induced by 6-aminonicotinamide. *J Comp Neurol* 2002; **444**:174–189.

Penkowa M, Giralt M, Moos T, *et al.* Impaired inflammatory response to glial cell death in genetically metallothionein-I- and -II-deficient mice. *Exp Neurol* 1999b; **156**: 149–164.

Penkowa M, Hidalgo J. Metallothionein I+II expression and their role in experimental autoimmune encephalomyelitis. *Glia* 2000; **32**:247–263.

Penkowa M, Hidalgo J. Treatment with metallothionein prevents demyelination and axonal damage and increases oligodendrocyte precursors and tissue repair during experimental autoimmune encephalomyelitis (EAE). *J Neurosci Res* 2003; **72**:574–586.

Penkowa M, Moos T, Carrasco J, *et al.* Strongly compromised inflammatory response to brain injury in interleukin-6–deficient mice. *Glia* 1999c; **25**:343–357.

Penkowa M, Tió L, Giralt M, *et al.* Specificity and divergence in the neurobiological effects of different metallothioneins after brain injury. *J Neurosci Res* 2006b; **83**:974–984.

Potter EG, Cheng Y, Knight JB, *et al.* Metallothionein I and II attenuate the thalamic microglial response following traumatic axotomy in the immature brain. *J Neurotrauma* 2007; **24**:28–42.

Poulsen CB, Penkowa M, Borup R, *et al.* Brain response to traumatic brain injury in wild-type and interleukin-6 knockout mice: A microarray analysis. *J Neurochem* 2005; **92**:417–432.

Puttaparthi K, Gitomer WL, Krishnan U, *et al.* Disease progression in a transgenic model of familial amyotrophic lateral sclerosis is dependent on both neuronal and non-neuronal zinc binding proteins. *J Neurosci* 2002; **22**:8790–8796.

Quaife CJ, Findley SD, Erickson JC, *et al.* Induction of a new metallothionein isoform (MT-IV) occurs during differentiation of stratified squamous epithelia. *Biochemistry* 1994; **33**:7250–7259.

Quintana A, Molinero A, Florit S, *et al.* Diverging mechanisms for TNF-alpha receptors in normal mouse brains and in functional recovery after injury: From gene to behavior. *J Neurosci Res* 2007; **85**:2668–2685.

Radtke F, Heuchel R, Georgiev O, *et al.* Cloned transcription factor MTF-1 activates the mouse metallothionein I promoter. *EMBO J* 1993; **12**:1355–1362.

Romero-Isart N, Jensen LT, Zerbe O, *et al.* Engineering of metallothionein-3 neuroinhibitory activity into the inactive isoform metallothionein-1. *J Biol Chem* 2002; **277**:37023–37028.

Rothwell NJ, Hopkins SJ. Cytokines and the nervous system II: Actions and mechanisms of action. *Trends Neurol Sci* 1995; **18**:130–136.

Sakamoto T, Kawazoe Y, Uchida Y, *et al.* Growth inhibitory factor prevents degeneration of injured adult rat motoneurons. *Neuroreport* 2003; **14**:2147–2151.

Samson SL, Gedamu L. Molecular analyses of metallothionein gene regulation. *Prog Nucleic Acid Res Mol Biol* 1998; **59**:257–288.

Sato M, Bremner I. Oxygen free radicals and metallothionein. *Free Radic Biol Med* 1993; **14**:325–337.

Searle PF, Davison BL, Stuart GW, *et al.* Regulation, linkage, and sequence of mouse metallothionein I and II genes. *Mol Cell Biol* 1984; **4**:1221–1230.

Sewell AK, Jensen LT, Erickson JC, *et al.* Bioactivity of metallothionein-3 correlates with its novel beta domain sequence rather than metal binding properties. *Biochemistry* 1995; **34**:4740–4747.

Shi Y, Wang W, Mo J, *et al.* Interactions of growth inhibitory factor with hydroxyl and superoxide radicals. *Biometals* 2003; **16**:383–389.

Sillevis Smitt PA, Blaauwgeers HG, Troost D, de Jong JM. Metallothionein immunoreactivity is increased in the spinal cord of patients with amyotrophic lateral sclerosis. *Neurosci Lett* 1992; **144**:107–110.

Sillevis Smitt PA, Mulder TP, Verspaget HW, *et al.* Metallothionein in amyotrophic lateral sclerosis. *Biol Signals* 1994; **3**:193–197.

Smith MA, Perry G, Richey PL, *et al.* Oxidative damage in Alzheimer's. *Nature* 1996; **382**:120–121.

Sobocinski PZ, Canterbury WJ Jr. Hepatic metallothionein induction in inflammation. *Ann NY Acad Sci* 1982; **389**:354–367.

Sobocinski PZ, Canterbury WJ Jr, Mapes CA, Dinterman RE. Involvement of hepatic metallothioneins in hypozincemia associated with bacterial infection. *Am J Physiol* 1978; **234**:E399–E406.

Stalder AK, Carson MJ, Pagenstecher A, *et al.* Late-onset chronic inflammatory encephalopathy in immune-competent and severe combined immune-deficient (SCID) mice with astrocyte-targeted expression of tumor necrosis factor. *Am J Pathol* 1998; **153**:767–783.

Suzuki K, Nakajima K, Kawaharada U, *et al.* Metallothionein in the human brain. *Acta Histochem Cytochem* 1992; **25**:617–622.

Suzuki KT, Yamamura M. Induction of hepatic zinc-thionein in rat by endotoxin. *Biochem Pharmacol* 1980; **29**:2260.

Tang Y, Lu A, Aronow BJ, *et al.* Genomic responses of the brain to ischemic stroke, intracerebral haemorrhage, kainate seizures, hypoglycemia, and hypoxia. *Eur J Neurosci* 2002; **15**:1937–1952.

Trendelenburg G, Prass K, Priller J, *et al.* Serial analysis of gene expression identifies metallothionein-II as major neuroprotective gene in mouse focal cerebral ischemia. *J Neurosci* 2002; **22**:5879–5888.

Tsuji S, Kobayashi H, Uchida Y, *et al.* Molecular cloning of human growth inhibitory factor cDNA and its down-regulation in Alzheimer's disease. *EMBO J* 1992; **11**:4843–4850.

Uchida Y. Growth inhibitory factor in brain. In: Suzuki KT, Imura N, Kimura M (eds.), *Metallothionein III*, Birkhäuser Verlag, Basel, 1993, pp. 315–328.

Uchida Y. Growth-inhibitory factor, metallothionein-like protein, and neurodegenerative diseases. *Biol Signals* 1994; **3**:211–215.

Uchida Y, Gomi F, Masumizu T, Miura Y. Growth inhibitory factor prevents neurite extension and death of cortical neurons caused by high oxygen exposure through hydroxyl radical scavenging. *J Biol Chem* 2002; **277**:32353–32359.

Uchida Y, Takio K, Titani K, *et al.* The growth inhibitory factor that is deficient in the Alzheimer's disease brain is a 68 amino acid metallothionein-like protein. *Neuron* 1991; **7**:337–347.

van Lookeren Campagne M, Thibodeaux H, van Bruggen N, *et al.* Evidence for a protective role of metallothionein-1 in focal cerebral ischemia. *Proc Natl Acad Sci USA* 1999; **96**:12870–12875.

van Lookeren Campagne M, Thiobodeaux H, van Bruggen N, *et al.* Increased binding activity at an antioxidant-responsive element in the metallothionein-1 promoter and rapid induction of metallothionein-1 and -2 in response to cerebral ischemia and reperfusion. *J Neurosci* 2000; **20**:5200–5207.

Vanguri P. Interferon-gamma-inducible genes in primary glial cells of the central nervous system: Comparisons of astrocytes with microglia and Lewis with brown Norway rats. *J Neuroimmunol* 1995; **56**:35–43.

Vašák M, Hasler DW. Metallothioneins: New functional and structural insights. *Curr Opin Chem Biol* 2000; **4**:177–183.

Vela JM, Hidalgo J, González B, Castellano B. Induction of metallothionein in astrocytes and microglia in the spinal cord from the myelin-deficient jimpy mouse. *Brain Res* 1997; **767**:345–355.

Vogel SN. The Lps gene: Insights into the genetic and molecular basis of LPS responsiveness and macrophage differentiation. In: Beutler B (ed.), *Tumor Necrosis Factors: The Molecules and Their Emerging Role in Medicine*, Raven Press, New York, 1992, pp. 485–513.

West AK, Stallings R, Hildebrand CE, *et al.* Human metallothionein genes: Structure of the functional locus at 16q13. *Genomics* 1990; **8**:513–518.

Westin G, Schaffner W. A zinc-responsive factor interacts with a metal-regulated enhancer element (MRE) of the mouse metallothionein-I gene. *EMBO J* 1988; **7**:3763–3770.

Williams BL, Yaddanapudi K, Kirk CM, *et al.* Metallothioneins and zinc dysregulation contribute to neurodevelopmental damage in a model of perinatal viral infection. *Brain Pathol* 2006; **16**:1–14.

Xie T, Tong L, McCann UD, *et al.* Identification and characterization of metallothionein-1 and -2 gene expression in the context of (+/-)3,4-methylenedioxymethamphetamine-induced toxicity to brain dopaminergic neurons. *J Neurosci* 2004; **24**:7043–7050.

Yagle MK, Palmiter RD. Coordinate regulation of mouse metallothionein I and II genes by heavy metals and glucocorticoids. *Mol Cell Biol* 1985; **5**:291–294.

Youn J, Hwang SH, Ryoo ZY, *et al.* Metallothionein suppresses collagen-induced arthritis via induction of TGF-beta and down-regulation of proinflammatory mediators. *Clin Exp Immunol* 2002; **129**:232–239.

Yu WH, Lukiw WJ, Bergeron C, *et al.* Metallothionein III is reduced in Alzheimer's disease. *Brain Res* 2001; **894**:37–45.

Yuguchi T, Kohmura E, Yamada K, *et al.* Expression of growth inhibitory factor mRNA following cortical injury. *J Neurotrauma* 1995a; **12**:299–306.

Yuguchi T, Kohmura E, Yamada K, *et al.* Changes in growth inhibitory factor mRNA expression compared with those in c-jun mRNA expression following facial nerve transection. *Mol Brain Res* 1995b; **28**:181–185.

Zambenedetti P, Giordano R, Zatta P. Metallothioneins are highly expressed in astrocytes and microcapillaries in Alzheimer's disease. *J Chem Neuroanat* 1998; **15**:21–26.

Zatta P, Zambenedetti P, Musicco M, Adorni F. Metallothionein-I-II and GFAP positivity in the brains from frontotemporal dementia patients. *J Alzheimers Dis* 2005; **8**:109–116.

Zhang B, Georgiev O, Hagmann M, *et al.* Activity of metal-responsive transcription factor 1 by toxic heavy metals and H_2O_2 *in vitro* is modulated by metallothionein. *Mol Cell Biol* 2003; **23**:8471–8485.

Chapter 5

METALLOTHIONEIN AND AUTISM

Sarah E. Owens, Marshall L. Summar and Michael Aschner

Autism is a complex disorder with both genetic and environmental factors. Metal disturbances have been reported in autistic individuals. In particular, exposure to ethylmercury (EtHg) in the vaccine preservative thimerosal and exposure to methylmercury (MeHg) through fish consumption have been implicated as environmental contributors to the autism spectrum disorder (ASD) phenotype. Metallothioneins (MTs) are small sulfhydryl (−SH)-rich metal-binding proteins that have important functions in metal homeostasis and protection against metal-induced toxicity. MT1 and MT2 have a high affinity for toxic heavy metals and are induced following mercury (Hg) exposure. It has been suggested that altered or dysfunctional MTs could enhance susceptibility to Hg-induced toxicity, and that alterations in Hg metabolism may contribute to the neurodevelopmental phenotypes present in ASD. One proposed treatment of autism is to attempt to restore MT function; however, there is no evidence in the peer-reviewed literature that MT restoration is an effective treatment for autistic symptoms. To date, there is no evidence for the efficacy of MT in ameliorating MeHg- or EtHg-induced neurotoxicity. Currently, there is nothing in the literature that suggests altered MT homeostasis is a contributing factor to the development of autism.

Keywords: Metallothionein; autism; metals; mercury; ethylmercury; methylmercury.

1. Introduction

This chapter serves to review the current literature examining the relationship between autism and metallothionein (MT). We will briefly define the neurodevelopmental features associated with the autism spectrum disorder (ASD). This will be followed by general features of mercury neurotoxicity, a critical overview of studies suggesting metal disturbances in autism, general structural and functional properties of MTs, and MTs and neurodegenerative disorders.

2. Autism

Autism is a common neurodevelopmental disorder affecting more than 1.5 million children and adults in the United States (Yeargin-Allsopp *et al.*, 2003). Characteristic symptoms include repetitive patterns of behavior, impairment in social interaction, and disrupted verbal or nonverbal communication (Table 1). These clinical manifestations are usually evident by the age of 3 years, and represent both prenatal and postnatal developmental aberrations.

The *Diagnostic and Statistical Manual of Mental Disorders, Fourth Edition* (DSM-IV) lists autistic disorder (autism) as one of several pervasive developmental disorders. The diagnostic criteria for

Table 1 Phenotypic manifestations of the three core symptoms present in autism.

Three Core Symptoms of Autism	Phenotypic Manifestations
Atypical social behavior	Retreat into isolation Playing alone "in their own world"
Disrupted verbal or nonverbal communication	Loss or lack of speech around 18 months of age Little or no eye contact, little interest in faces Loss or lack of gestures such as pointing or waving
Unusual patterns of highly restricted interests and repetitive behaviors	Fixation on a word, an object, or an activity (i.e. stacking objects, lining things up, and putting them in a certain order) Repetitive speech or actions (i.e. hand flapping, spinning in circles) Sensory integration difficulty (i.e. unusual reactions to the way things look, feel, smell, taste, or sound) Development of obsessions or routines around food (i.e. colors, textures, types of food)

autism require six characteristic symptoms to be present from three different categories of the DSM-IV (American Psychiatric Association, 2000). The term "autism spectrum disorder" (ASD) refers collectively to autism, Asperger's syndrome, and pervasive developmental disorder, not otherwise specified (PDD-NOS) when these related conditions are difficult to differentiate (Lord and Volkmar, 2002).

Prevalence estimates of autism in the general population have increased gradually over the past 20 years (Chakrabarti and Fombonne, 2001; Fombonne, 2002; Fombonne, 2003; Yeargin-Allsopp *et al.*, 2003), but the causative factors of autism are still poorly understood. Recent studies suggest a marked increase in the incidence of this disorder over the last few decades, approximating that 1 in 150 children in the United States have ASD (CDC, 2007).

Autism is a complex disorder with both genetic and environmental components (Rapin, 1997). Multiple genetic loci have been proposed to impact susceptibility and the overall autistic phenotype, but few surpass the suggestive and significant linkage threshold values (Vorstman *et al.*, 2006). The evidence for a strong genetic influence on autism is overwhelming (Folstein and Rosen-Sheidley, 2001), but the inability of researchers to identify the underlying genetic variations suggests that multiple genes are acting independently or interacting with each other and/or with environmental factors to raise the risk of autism (Folstein and Rosen-Sheidley, 2001).

Autistic brains exhibit gross morphological changes such as smaller neuronal size but increased cell density in the limbic system, decreased numbers of cerebellar Purkinje cells, enlarged brain size, and age-dependent changes in neuronal size in the inferior olive and cerebellar nuclei (Palmen *et al.*, 2004). Interestingly, Purkinje cells are particularly vulnerable to heavy metal-induced toxicity, such as MeHg, as well as other types of insults including ischemia, hypoxia, viral infections, and toxins (Welsh *et al.*, 2002). Similar symptoms have been observed in both inorganic mercury (Hg) poisoning and autism, including social impairments, movement disorders, speech difficulties, repetitive behaviors, and sensory abnormalities (Bernard *et al.*, 2001), with a higher prevalence of affected males than females (Bernard *et al.*, 2002).

Although a prenatal origin is suspected for neuropathological defects in autism, one report (Kern, 2003) suggested that some children develop autism after postnatal neuronal cell death or brain damage resulting from injuries associated with heavy metal exposure. This could explain why autistic symptoms do not appear in some children until 2 years after birth (Institute of Medicine, 2001). Notably, morphologic variability is observed within autism groups (Pickett and London, 2005), and polymorphisms might account for this. Although speculative, it has been suggested that individual genetic susceptibility to heavy metals, combined with prenatal exposure to Hg, could result in the early appearance of autism symptoms; whereas postnatal exposure to Hg could explain a delay.

3. Mercury

In the US, approximately 8% of the women of childbearing age have Hg levels exceeding Environmental Protection Agency (EPA) limits (Schober *et al.*, 2003). Hg is highly toxic, especially during central nervous system (CNS) development. Inhaled Hg vapor deposits in the CNS, and organic Hg (e.g. MeHg, EtHg) is readily transported across the blood-brain barrier (BBB). Prenatal exposure to organic Hg is especially damaging, as it rapidly crosses both the placenta and the BBB. MeHg is efficiently transported across the BBB and appears in brain tissue 5 minutes after intravenous injection or within 4 hours of ingestion (Sager *et al.*, 1982; Thomas and Smith, 1982; Hirayama, 1985). Although a mature adult may be resistant to the toxic insults of what is considered "acceptable levels" of MeHg, the fetus is highly susceptible (Counter and Buchanan, 2004). This increased level of MeHg exposure exacerbates the toxicity hazard to the developing fetus, and introduces a health risk to pregnant mothers.

Environmental as well as occupational and domestic Hg exposures represent growing health concerns (Counter and Buchanan,

2004). Hg toxicity can result from exposure to a variety of sources. The most prevalent source of Hg is from organic MeHg, ingested through fish consumption. In this scenario, Hg vapor is released into the atmosphere by the natural degassing of the earth's crust. It then returns to earth contained in rainwater. Methanogenic bacteria present in sediments of fresh and ocean water then produce MeHg, which humans ingest through consumption of fish. In contrast, EtHg is found at high concentrations ($\sim$50%) in the preservative thimerosal, which was administered in some childhood vaccinations. It has been removed from childhood vaccines in the United States, but is still widely used in developing countries. Finally, inhalation of elemental Hg vapor occurs in some occupational settings such as industry, dentistry (amalgam tooth fillings), and extraction of gold. Metallic Hg in dental amalgams can release both Hg vapor and divalent mercury (Hg^{2+}) into surrounding tissues. Topically applied skin creams, infant teething powders, and contaminated food are also routes of inorganic Hg exposure (Casarett, 2001; Counter and Buchanan, 2004).

While much is known about MeHg pharmacokinetics, data on EtHg are scarce. Due to a larger alkyl chain compared to MeHg, EtHg passes through the BBB more slowly, but it decomposes into inorganic Hg faster (Magos, 2003). Many of the theories concerning organic Hg toxicity suggest that it is the inorganic form that causes the damage. EtHg-treated rats have a greater concentration of Hg (organic and inorganic combined) in blood and less in the brain compared to rats treated with identical doses of MeHg; however, the proportion of inorganic Hg (to organic Hg) after EtHg treatments is higher in both blood and brain (Magos, 2003). These studies were confirmed in infant monkeys (Burbacher *et al.*, 2005).

EtHg appears to behave like MeHg, with fecal excretion accounting for most of the elimination from the body. The absorption rate and initial distribution volume of total mercury are also reported to be generally similar after EtHg injections and oral MeHg exposure (Burbacher *et al.*, 2005). In other words, peak total blood mercury levels after a single exposure to either EtHg or MeHg are very similar, implying that the organic mercury compounds behave similarly in the early hours after exposure. However, the blood half-time of EtHg is shorter than that of MeHg (~7 days vs. 19 days, respectively) (Pichichero *et al.*, 2002; Magos, 2003; Burbacher *et al.*, 2005), and thus EtHg should clear faster from the body than MeHg. Because EtHg decomposes much faster than MeHg, the total amount of Hg in the brain upon exposure to MeHg is greater when compared to exposure to EtHg (Burbacher *et al.*, 2005).

4. Critical Overview of Studies Suggesting Metal Disturbances in Austism

There are multiple reports of metal disturbances associated with autism (Shearer *et al.*, 1982; Chauhan *et al.*, 2004; Fido and Al-Saad, 2005). It has been suggested that autistic children have inherent differences in metabolism compared to normal children, possibly a consequence of genetic differences, which result in increased body burdens of Hg observed in children with autistic disorders in comparison to normal children (Bradstreet *et al.*, 2003; James *et al.*, 2004). It has also been reported that autistic children have significantly higher concentrations of toxic trace elements such as lead, mercury, and uranium in their hair than age-matched controls (Fido and Al-Saad, 2005), while other studies report lower concentrations of these metals.

A study by Holmes *et al.* (2003) found statistically significant lower concentrations of mercury in first baby haircut samples of autistic children compared to normal controls. The authors hypothesized that this indicates a decreased ability to excrete mercury into the hair, and thus mercury is being sequestered in the brain and other tissues. However, a national study by the U.S. Centers for Disease Control and Prevention (CDC) reported that the mean hair mercury concentration in normal U.S. children who consume high quantities of fish in their diet was 0.40 ppm (with a 95% confidence interval of 0.24–0.55) (McDowell *et al.*, 2004). When comparing this to data reported in the Holmes study, it appears that the mean hair mercury concentration in autistic children (0.47 ppm) was actually within the normal range, while the mean mercury concentration reported in the hair of nonautistic controls (3.63 ppm) was 10 times greater than the CDC mean. Two other studies which investigated the hypothesis that mercury deposition in the hair differs between autistic and nonautistic individuals (Ip *et al.*, 2004; Kern *et al.*, 2007) both found that levels of mercury in the hair were not statistically different between autistic children and nonautistic controls exposed to similar amounts of mercury.

A study by Bradstreet *et al.* (2003) examined mercury excretion in urine in 221 children with ASD after chelation with DMSA, and reported that postchelation urine mercury levels were three times greater than those of controls. The authors claimed that this suggests that ASD children retain a greater amount of mercury in their tissues due to insufficient excretion mechanisms. Although the difference observed was statistically significant, the authors also acknowledged that it could simply indicate a higher current exposure to mercury instead of an increased body burden of mercury. In addition, no prechelation mercury levels were given for comparison, making it exceedingly difficult to ascertain whether autism is associated with increased body burdens of Hg.

It has also been reported that autistic children have increased levels of zinc and copper (Shearer *et al.*, 1982). A study of 503 ASD patients (comprised of 318 with autism, 23 with Asperger's syndrome, and 162 with PDD) found that their copper:zinc ratio was significantly higher compared to healthy controls ($p < 0.0001$) (Walsh, 2002). The author claimed that an altered copper:zinc ratio contributes to a variety of symptoms including emotional instability, neurotransmitter imbalance, and impairment of hippocampus and amygdala function. However, a recent report which examined the level of zinc and mercury in baby teeth of children with ASD determined that they had 2.1 times higher levels of mercury, but similar levels of zinc, compared to normal controls (Adams *et al.*, 2007).

Furthermore, in a study examining 19 children (15 with autism, 3 PDD, 1 with disintegrative disorder), it was reported that reduced levels of the metal-binding proteins transferrin (iron-binding) and ceruloplasmin (copper-binding) were correlated with the loss of previously acquired language skills in autistic kids, providing evidence for abnormal iron and copper metabolism in autism (Chauhan *et al.*, 2004).

There are numerous examples of diseases other than autism in which metal levels are altered. Increased aluminum, copper, and zinc are present in Alzheimer's disease (Liu *et al.*, 2005). Menkes and Wilson's diseases result in copper insufficiency and copper accumulation, respectively (Daniel *et al.*, 2004). Copper, manganese, and iron levels are altered in prion disease (Wong *et al.*, 2001). Iron overload has been observed in the brains of patients with Alzheimer's disease (increased iron in Lewy bodies), Parkinson's disease, and Huntington's disease (Moos and Morgan, 2004). Since MTs function in regulating levels of the metals mentioned above, it is possible that polymorphisms in MT genes may also explain the altered metal content in these other disorders.

5. Ethylmercury and Autism

Ethylmercury thiosalicylate (chemical structure, $C_9H_9HgNaO_2S$), also known under the trade name thimerosal (American Academy of Pediatrics, 1999), was introduced as a preservative in vaccines in the 1930s, after a series of studies in several animal species and humans provided evidence for its safety and effectiveness (Powell and Jamieson, 1931). Thimerosal in concentrations of 0.001% (1 part in 100 000) to 0.01% (1 part in 10 000) has been shown to be effective in clearing a broad spectrum of pathogens. A vaccine containing 0.01% thimerosal as a preservative contains 50 μg of thimerosal per 0.5-mL dose or approximately 25 μg of mercury per 0.5-mL dose (U.S. FDA, 2007). After approximately 70 years of safe practice and a long record of effectiveness in preventing bacterial and fungal contamination of vaccines with only minor local reactions at the site of injection, in 2001 the use of thimerosal was questioned as a potential toxic hazard to infants (Ball *et al.*, 2001). Though it is still used in developing countries, where the advantages of multiple-use vials outweigh thimerosal's putative toxicity (World Health Organization, 2002), it was removed from the US market in 2001.

The Word Health Organization (WHO, 1990), the US Environmental Protection Agency (EPA; online, 2007), the US Agency for Toxic Substances and Disease Registry (ATSDR, 1999), and the US Food and Drug Administration (US FDA; online, 2007) have assessed the risk associated with MeHg in diet and have published recommendations for safe exposure to this metal. These recommendations encompass a safety margin, and range from 0.7 μg MeHg/kg of body weight per week (EPA) to 3.3 μg MeHg/kg of body weight per week (WHO). The range of recommendations reflects varying safety margins, differing emphases placed on various sources of data, the different missions of the agencies, and the population that the guideline is intended to protect. All guidelines, however, fall within the same order of magnitude. If applied to a female infant in the lowest fifth percentile of weight between birth and 14 weeks, the period during which most infant vaccines are administered, these guidelines translate to limits of safe total MeHg exposure of 34 μg and 159 μg, as per the EPA and WHO safe exposure limits, respectively. An infant generally receives three doses of DTP vaccine or a total of 75 μg of EtHg during the first 14 weeks of life (US FDA; online, 2007). If the hepatitis B vaccine is added to the immunization schedule during the first 14 weeks of life, the maximum exposure to EtHg is 112.5 μg. If *Haemophilus influenzae* type b conjugate (Hib) vaccine is added during the same time, the total EtHg dose reaches 187.5 μg. Thus, some infants receiving vaccines according to the recommended schedule will receive doses of mercury exceeding the cutoff levels established by regulatory agencies. However, as mentioned above, and as will be illustrated in detail below, the application of

MeHg risk assessment guidelines to thimerosal (or EtHg) exposure assumes that EtHg and MeHg are identically distributed in the body, specifically in the central nervous system (CNS), reaching equal concentrations at sensitive sites and exerting similar toxic sequelae. This assumption is invalid as it is refuted by existing scientific evidence.

Most human exposures to EtHg are in the form of thimerosal, and tissue disposition patterns of mercury in experimental animals after equivalent doses of either EtHg chloride or thimerosal are the same (Suzuki *et al.*, 1973). Accordingly, it appears that the thiosalicylic acid anion attached to EtHg in the thimerosal plays no role in influencing the fate of EtHg in the body. Thus, thimerosal rapidly dissociates to release EtHg (Reader and Lines, 1983; Tan and Parkin, 2000), which is the active species of concern.

Conclusions on the toxicity of EtHg (or thimerosal) are predominantly drawn from analogies to the organomercurial, MeHg. While the scientific literature supports the concept that MeHg is a potent and well-known developmental neurotoxin, the assertion that thimerosal in vaccines leads to developmental abnormalities is hypothetical and unsubstantiated, resting on indirect and incomplete information, primarily from analogies with MeHg. This approach is not surprising, as until recently there was sparse information on the disposition of EtHg as compared to MeHg. However, results from the few studies that have provided a direct comparison between these compounds [reviewed by Clarkson *et al.* (2003), Magos (2003), Clarkson and Magos (2006), and Magos and Clarkson (2006)] have established that extrapolation of EtHg's disposition and toxic potential from the MeHg literature is flawed, as distinct differences exist with respect to the pharmacokinetic behavior of the two organomercurials. Simply stated, MeHg (and even more so, metallic or Hg vapor) is not a suitable reference for evaluating EtHg toxicity.

A recent review critically examined 12 original studies on the link between autism and vaccines containing EtHg and concluded that, because of methodological flaws, none could establish this association (Parker *et al.*, 2005). As this review is not meant to be exhaustive, we refer the reader to Parker *et al.* (2005) for more detailed information on the studies. Based on current evidence, the Institute of Medicine (IOM) released a final report from the Immunization Safety Review Committee in 2004 that "favors the rejection of a causal relationship between thimerosal-containing vaccines and autism" (IOM, 2004). However, genetic susceptibility to abnormal Hg metabolism is offered as a plausible explanation for autism in some children, with the recommendation that this possibility be vigorously pursued. A genetically susceptible population of children may exhibit increased accumulation of Hg and/or decreased brain Hg excretion, which alters several key biochemical pathways — for example, apoptosis and DNA metabolism — leading to autism (Bradstreet, 2004). The report further stated that abnormal Hg metabolism may be a causal factor of autism or a comorbid condition, and further research on the topic was encouraged (Institute of Medicine, 2004).

Recently, the possible association between early vaccines containing thimerosal and neuropsychological performance was assessed in 1047 children between 7 and 10 years old (Thompson *et al.*, 2007). Mercury exposure during the prenatal period, the neonatal period (birth to 28 days), and the first 7 months of life was evaluated for association to 42 different neuropsychological outcomes. Out of all the tests performed, only a few were significantly associated with mercury exposure from thimerosal, and these associations exhibited both positive and negative effects equally. For example, an increased performance for one measure of language but a decreased measure of attention and executive functioning was associated with prenatal exposure to mercury. Mercury exposure during the neonatal period was associated with worse performance on one measure of speech articulation, yet improved performance on one measure of fine motor coordination. Finally, better performance on one measure of fine motor coordination and on one measure of attention was associated with mercury exposure during the first 7 months of life (Thompson *et al.*, 2007). With this mix of results, the authors concluded that their study does not support an association between early exposure to mercury in the vaccine preservative thimerosal and deficits in neuropsychological functioning in children who are 7–10 years old.

6. Metallothioneins

MT1 and MT2 have a high affinity for toxic heavy metals and are induced following Hg exposure (Aschner, 1996), presumably to protect essential cellular functions and enhance survival. MT1 and MT2 can serve as potent antioxidants, and both are upregulated in response to other metals including Cd^{2+}, Cu^{2+}, Ag^{2+}, Hg^{2+}, MeHg, and Zn^{2+} (Casarett, 2001), thus attenuating metal-induced reactive oxygen species (ROS) damage (Maret, 2000).

Transgenic mice with deletions of the MT-1 and MT-2 genes (MT1/MT2-null mice) do not exhibit any obvious atypical phenotypes at normal metal concentrations. However, they are more susceptible to metal-induced toxicity and environmental stress.

Prolonged exposure to low-level mercury vapor, at concentrations relevant to human exposure levels, resulted in altered neurobehavioral functions in mice in two behavioral tests evaluated. Significantly enhanced locomotion in the open-field test (which examines the response to a novel environmental stimulus) and decreased performance in the passive avoidance test (which examines the retention of a learned response) were not only observed in control mice, but were slightly more severe in MT-null mice. MT1- and MT2-null mice appeared to be more susceptible to Hg toxicity, suggesting that MT1 and MT2 can protect against neurobehavioral deficits which result from metal-induced toxicity (Yoshida *et al.*, 2004).

Recently, the role of MTs and cognitive function were examined in MT1/MT2-null mice. No metal challenge was presented during these studies, so it was assumed that an altered MT system would perturb normal trace metal metabolism and consequently affect normal brain function and development. Spatial learning and memory were tested using a win-shift task in an eight-arm radial maze. In this test, food cups were positioned at the end of each of the arms and were filled only once; the mouse had to remember where it had already explored in prior sessions in order to know which remaining arms still contained a reward. The MT1/MT2-null mice exhibited a poorer rate of learning when trained, as well as severe choice accuracy impairment which did not lessen with additional training. Interestingly, nicotine treatment has been shown to improve working memory, and when applied at a sufficient dose it significantly improved memory performance in MT1/MT2-null mice (Levin *et al.*, 2006).

Polymorphisms have been identified in some of the MT genes. Although association studies have examined the relationship between MT single nucleotide polymorphisms (SNPs) and a variety of factors including aging (Cipriano *et al.*, 2006), carotid artery stenosis (Giacconi *et al.*, 2007), cellular response to cadmium (Kita *et al.*, 2006), amyotrophic lateral sclerosis (ALS) (Hayashi *et al.*, 2006), and type 2 diabetes (Giacconi *et al.*, 2005), no studies to date have examined the relationship (if any) between genetic variants in MT genes and individual susceptibility to mercury exposure.

Polymorphisms in MT genes may prevent cells from mounting an optimal response to oxidative stress and may also decrease protection mechanisms initiated by MTs, rendering individuals more susceptible to Hg exposure. Whether decreased expression or "quality" of MTs could lead to an increasing body burden of metal that might explain cases of regressive autism, where parents report that their children regress into autistic phenotypes shortly after receiving childhood vaccines containing thimerosal, has yet to be determined.

7. MT and Autism

If autistic individuals have an increased vulnerability to mercury exposure, it is presumed that they harbor an increased body burden of the metal. As discussed earlier, efforts to examine the mercury content in the hair and urine of autistic children have produced conflicting results. However, findings of the same levels of Hg in the body (and even the brain) in autistic and normal children do not refute a potential role for Hg as a contributing factor in the development and progression of autism. Even if brain Hg concentrations are the same in both groups of children, it is still possible that the distribution of Hg could still be altered, especially if autistic children cannot mount an optimal response to Hg via MT induction (Aschner and Walker, 2002).

It has been suggested that some autistic individuals are more susceptible to heavy metal exposure due to variations in genes which attenuate metal toxicity, such as metallothioneins. MTs are critical in protecting the body against heavy metal toxicity, and are upregulated upon exposure to toxic metals such as Hg, Pb, and Cd. Elevated toxic metals are present in 92% of autism cases (Walsh, 2002). In addition, MTs function in zinc and copper homeostasis. As mentioned earlier, a study of 503 ASD patients found that their copper:zinc ratio was significantly higher compared to healthy controls ($p < 0.0001$) (Walsh, 2002). Collectively, this implies impairment in proper MT functioning in autistic children. Altered MTs could result in decreased ability to guard against heavy metals such as mercury, as well as misregulation of essential metals in the body. However, it is not known whether MTs actually cause the difference in metal levels, or if metal imbalance is merely a consequence of having autism.

Heavy metals such as mercury are able to initiate an autoimmune reaction (El-Fawall, 1999; Hultman and Hansson-Georgiadis, 1999; Bigazzi, 2000). Therefore, it has been postulated that autistic children might upregulate MT and anti-MT as an indication of metal-induced immune response. A study was conducted using MT and anti-MT as biomarkers of detoxification response to mercury derived from vaccines in 62 children with autism and 60 normal children (Singh and Hanson, 2006). All children had received their full immunization schedule, and the authors performed the study based on the assumption that participating children had been exposed to mercury only through the thimerosal preservative in the vaccines, but not any other source. Serum profiles of MT protein and anti-MT were generated by ELISA assays. MT protein was found in all 122 children, but there was not a significant difference in its concentration between autistic and normal groups. There was also no effect of serum dilution, as the levels of antibodies to MT were present at the same amounts in autistic and normal children. Finally, the distribution of antibodies to MT1, MT2, and MT3 proteins in autistic children was not significantly different from normal controls. The upregulation of MT protein and antibodies to MT could be considered biomarkers if mercury exposure was correlated with an autoimmune response in autism. However, as no significant difference was observed in serum profiles of MT protein and anti-MT between autistic and normal children, this study does not support a link between mercury in the preservative thimerosal and autoimmunity in autism.

Caveats of this study were later brought to attention (Guzzi, 2007). The timing between the last vaccination and the evaluation of serum MT was not indicated for individuals tested ($\sim$7 days is the half-life of EtHg in blood). In addition, since details of fish consumption (a common source of mercury) were not given, it is unknown if children were truly only exposed to mercury from vaccines. It is also unclear whether EtHg induces MTs as MeHg does. However, it is likely that EtHg is more potent than MeHg, because exposure to EtHg is associated with an increased generation of inorganic mercury compared to MeHg exposure, and the former is known to be a stronger inducer of MTs. Finally, it would have been extremely informative if urine MTs in both groups had been measured, but these data are lacking.

In the authors' response to these caveats (Singh, 2007), they stated that since the study was retrospective, they did not have information regarding the time lapsed between the last inoculation and blood draw. Attempts to gather information retrospectively on fish consumption revealed that the children did not consume fish, according to the parents. The authors admitted that they do not know if EtHg induces changes in MT at the molecular level, and stated that this needs to be examined.

A recent paper explored the hypothesis that MTs are not adequately upregulated in autistic individuals upon exposure to heavy metals, such as mercury found in the preservative thimerosal (49.6% EtHg). If true, this would render autistic individuals more susceptible to heavy metal exposure. It is known that white blood cells, when treated with mercury, express metallothionein (Yamada and Koizumi, 1991). The authors examined the effect of thimerosal challenge in lymphocyte cultures derived from 11 children (5 autistic individuals, 5 nonautistic siblings, 1 sibling with PDD) representing three multiplex families (Walker *et al.*, 2006). Total RNA was isolated and gene expression was analyzed by high-density microarray after exposure to either zinc or thimerosal. At least nine different MT transcripts were significantly upregulated in cultured lymphocytes from both autistic and nonautistic individuals exposed to 150 μM zinc. In contrast, cultured lymphocytes from both autistic and nonautistic individuals exposed to 10 μM thimerosal did not upregulate MT transcripts, but rather heat shock transcripts. Genes involved in neurodegeneration and apoptosis pathways were upregulated, while several metabolic and cell-signaling pathways were downregulated, in response to thimerosal. Although the genomic responses to the two metals were quite distinct, no difference in responses to either challenge by zinc or challenge by thimerosal was observed between the autistic and nonautistic siblings.

8. MT as Treatment

One suggested treatment of autism is to attempt to restore MT function. There have been anecdotal reports on this type of treatment on the Web (Pfeiffer Treatment Center, 2007), but there is no evidence in the peer-reviewed literature that MT restoration is an effective treatment for autistic symptoms nor is it clear that any of the suggested treatments actually meet the stated goals of increasing MT expression.

MTs have been suggested as treatment for other neurodegenerative disorders such as Alzheimer's disease, multiple sclerosis (MS), and amyotrophic lateral sclerosis (ALS) (Hozumi *et al.*, 2004; Penkowa, 2006; Stankovic *et al.*, 2007). MTs could be a possible pharmacological agent for the successful treatment of MS. It has been shown that treatment with exogenous MT can inhibit experimental autoimmune encephalomyelitis (EAE) and delay its progression. MT1 and MT2 are commonly found to reduce cell death, and to promote brain repair and angiogenesis. Application of exogenous MT1 and MT2 to neurons produced a neurotrophic effect. In addition, MT1 and MT2 promote neuron survival and axon outgrowth (Chung *et al.*, 2003). Based on these findings, MT1 and MT2 are suggested to be new neurotherapeutic drug targets for treating neurodegenerative diseases (Stankovic *et al.*, 2007).

To date, there is no evidence for the efficacy of MT in ameliorating MeHg- or EtHg-induced neurotoxicity. There is evidence that MT-null and MT-overexpressing mice are more and less susceptible, respectively, to Hg, but further research is needed.

9. Significance

The factors influencing autism are still poorly understood. It is generally assumed that Hg toxicity occurs when a "safe" level of exposure has been surpassed. In the context of genetic susceptibility, however, even normal Hg levels could be implicated in the etiology of autism (Hattis *et al.*, 2002). Thus, Hg deposition in the CNS could be altered as a result of unbalanced transport (increased accumulation versus release), or by the inability of MT genes that normally combat the metal's toxic effects to function optimally. Whether there is a genetically susceptible subset of children who develop autism, for which environmental factors are important contributors to the expression of the disease phenotype, is undocumented. Identifying the genetic polymorphisms of MTs and other metal metabolism genes could help elucidate the etiology of autism, particularly by determining the extent of individual susceptibility to neurotoxicity caused by various Hg species.

MTs are synthesized not only in response to divalent metal cations, but also in response to glucocorticoids (stress signal), cytokines, ROS, and endotoxins. Furthermore, MTs are associated with a number of diseases (Simpkins, 2000) and are implicated in aging and neurodegenerative brain disorders. Interestingly, MT1 and MT2 have been shown to be neuroprotective in animal models of familial ALS (Nagano *et al.*, 2001) and multiple sclerosis (Espejo *et al.*, 2001). In addition, MT levels are also elevated in cancer (Schmid *et al.*, 1993; Zelger *et al.*, 1993). Recently, it has been proposed that MT is the "danger signal" that indicates cellular damage has occurred in order to mount an active immune response (Yin *et al.*, 2005). MT in the extracellular environment facilitates the movement of white blood cells to the site of inflammation (Yin *et al.*, 2005). The discovery that MTs mediate leukocyte chemotaxis implies that these proteins could be associated with autoimmune disease and toxicant exposure.

As there is compelling evidence of metal imbalances in other neurodegenerative diseases such as Alzheimer's disease, Parkinson's disease, prion disease, Menkes disease, and Wilson's disease, identification of the polymorphisms in metal transporters responsible for regulating these metals would provide invaluable information regarding the role of individual genetically determined susceptibility to metal toxicity and how it relates to these disorders.

10. Conclusions

Currently, there is nothing in the literature that suggests that altered MT homeostasis is a contributing factor to the development of autism. Further research needs to bridge the current gap between the areas of neurogenetics and environmental toxicology. MT-null mice have been challenged with mercury vapor and MeHg, but experiments need to be performed where MT-null mice are challenged with EtHg and compared to normal mice. It would also be interesting to see if MT gene variants explain the altered metal content observed in neurodegenerative diseases. Characterizing individual genetic susceptibility to heavy metal toxicity may have broad implications for numerous diseases, while further advancing our understanding of the complexities of the etiology of autism.

Acknowledgments

This work was supported by grants T32 ES007028, NIEHS R01 07331, and T32 MH065782.

References

Adams JB, Romdalvik J, Ramanujam VM, Legator MS. Mercury, lead, and zinc in baby teeth of children with autism versus controls. *J Toxicol Environ Health A* 2007; **70**:1046–1051.

Agency for Toxic Substances and Disease Registry (ATSDR). Toxicological profile for mercury CAS#: 7439-97-6. Atlanta, GA, 1999. Available at http://www.atsdr.cdc.gov/toxprofiles/tp46.html/.

American Academy of Pediatrics (AAP) and U.S. PHS. Joint statement of the American Academy of Pediatrics and the United States Public Health Service. *Pediatrics* 1999; **104**:568–569.

American Psychiatric Association. *Diagnostic and Statistical Manual of Mental Disorders*. American Psychiatric Association, Washington, D.C., 2000.

Aschner M. The functional significance of brain metallothioneins. *FASEB J* 1996; **10**:1129–1136.

Aschner M, Walker SJ. The neuropathogenesis of mercury toxicity. *Mol Psychiatry* 2002; **7**(Suppl 2):S40–S41.

Ball L, Ball R, Pratt R. An assessment of thimerosal use in childhood vaccines. *Pediatrics* 2001; **107**:1147–1154.

Bernard S, Enayati A, Redwood L, *et al.* Autism: A novel form of mercury poisoning. *Med Hypotheses* 2001; **56**:462–471.

Bernard S, Enayati A, Roger H, *et al.* The role of mercury in the pathogenesis of autism. *Mol Psychiatry* 2002; **7**(Suppl 2):S42–S43.

Bigazzi PE. *Autoimmunity Induced by Metals.* Lippincott Williams & Wilkins Press, Philadelphia, 2000.

Bradstreet J. Presentation to the Immunization Safety Review Committee. Rising Incidence of Autism: Association with Thimerosal. Cambridge, MA, 2004.

Bradstreet J, Geier DA, Kartzinel JJ, *et al.* A case-control study of mercury burden in children with autistic spectrum disorders. *J Am Physicians Surg* 2003; **8**:76–79.

Burbacher TM, Shen DD, Liberato N, *et al.* Comparison of blood and brain mercury levels in infant monkeys exposed to methylmercury or vaccines containing thimerosal. *Environ Health Perspect* 2005; **113**:1015–1021.

Casarett D. *Toxicology: The Basic Science of Poisons*. McGraw-Hill, New York, 2001.

CDC (Centers for Disease Control and Prevention). 2007.

Chakrabarti S, Fombonne E. Pervasive developmental disorders in preschool children. *JAMA* 2001; **285**:3093–3099.

Chauhan A, Chauhan V, Brown WT, Cohen I. Oxidative stress in autism: Increased lipid peroxidation and reduced serum levels of ceruloplasmin and transferrin — The antioxidant proteins. *Life Sci* 2004; **75**:2539–2549.

Chung RS, Vickers JC, Chuah MI, West AK. Metallothionein-IIA promotes initial neurite elongation and postinjury reactive neurite growth and facilitates healing after focal cortical brain injury. *J Neurosci* 2003; **23**:3336–3342.

Cipriano C, Malavolta M, Costarelli L, *et al.* Polymorphisms in MT1a gene coding region are associated with longevity in Italian Central female population. *Biogerontology* 2006; **7**:357–365.

Clarkson TW, Magos L. The toxicology of mercury and its chemical compounds. *Crit Rev Toxicol* 2006; **36**:609–662.

Clarkson TW, Magos L, Myers GJ. The toxicology of mercury — Current exposures and clinical manifestations. *N Engl J Med* 2003; **349**:1731–1737.

Counter SA, Buchanan LH. Mercury exposure in children: A review. *Toxicol Appl Pharmacol* 2004; **198**:209–230.

Daniel KG, Harbach RH, Guida WC, Dou QP. Copper storage diseases: Menkes, Wilsons, and cancer. *Front Biosci* 2004; **9**:2652–2662.

El-Fawall HA, Watreman SJ, De Feo A, Shamy MY. Neuroimmunotoxicology: Humoral assessment of neurotoxicity and autoimmune mechanisms. *Environ Health Perspect* 1999; **107**:767–775.

Espejo C, Carrasco J, Hidalgo J, *et al.* Differential expression of metallothioneins in the CNS of mice with experimental autoimmune encephalomyelitis. *Neuroscience* 2001; **105**:1055–1065.

Fido A, Al-Saad S. Toxic trace elements in the hair of children with autism. *Autism* 2005; **9**:290–298.

Folstein SE, Rosen-Sheidley B. Genetics of autism: Complex aetiology for a heterogeneous disorder. *Nat Rev Genet* 2001; **2**:943–955.

Fombonne E. Epidemiological trends in rates of autism. *Mol Psychiatry* 2002; **7**(Suppl 2):S4–S6.

Fombonne E. The prevalence of autism. *JAMA* 2003; **289**:87–89.

Giacconi R, Cipriano C, Muti E, *et al.* Novel −209A/G MT2A polymorphism in old patients with type 2 diabetes and atherosclerosis: Relationship with inflammation (IL-6) and zinc. *Biogerontology* 2005; **6**:407–413.

Giacconi R, Muti E, Malavolta M, *et al.* The +838 C/G MT2A polymorphism, metals, and the inflammatory/immune response in carotid artery stenosis in elderly people. *Mol Med* 2007; **13**:388–395.

Guzzi G. Of vaccine-autism, thimerosal, and metallothioneins. *Pediatr Allergy Immunol* 2007; **18**:88; author reply 89.

Hattis D, Baird S, Goble R. A straw man proposal for a quantitative definition of the RfD. *Drug Chem Toxicol* 2002; **25**:403–436.

Hayashi Y, Hashizume T, Wakida K, *et al.* Association between metallothionein genes polymorphisms and sporadic amyotrophic lateral sclerosis in a Japanese population. *Amyotroph Lateral Scler* 2006; **7**:22–26.

Hirayama K. Effects of combined administration of thiol compounds and methylmercury chloride on mercury distribution in rats. *Biochem Pharmacol* 1985; **34**:2030–2032.

Holmes AS, Blaxill MF, Haley BE. Reduced levels of mercury in first baby haircuts of autistic children. *Int J Toxicol* 2003; **22**:277–285.

Hozumi I, Asanuma M, Yamada M, Uchida Y. Metallothioneins and neurodegenerative diseases. *J Health Sci* 2004; **50**:323–331.

Hultman P, Hansson-Georgiadis H. Methyl mercury-induced autoimmunity in mice. *Toxicol Appl Pharmacol* 1999; **154**:203–211.

Institute of Medicine. *Immunization Safety Review: Measles-Mumps-Rubella Vaccine and Autism.* Stratton K, Gable A, Shetty P, McCormick M (eds.), National Academy Press, Washington, D.C., 2001.

Institute of Medicine. *Immunization Safety Review: Vaccines and Autism.* National Academy of Sciences, Washington, D.C., 2004.

Ip P, Wong V, Ho M, *et al.* Mercury exposure in children with autistic spectrum disorder: Case-control study. *J Child Neurol* 2004; **19**:431–434.

James SJ, Cutler P, Melnyk S, *et al.* Metabolic biomarkers of increased oxidative stress and impaired methylation capacity in children with autism. *Am J Clin Nutr* 2004; **80**:1611–1617.

Kern JK. Purkinje cell vulnerability and autism: A possible etiological connection. *Brain Dev* 2003; **25**:377–382.

Kern JK, Grannemann BD, Trivedi MH, Adams JB. Sulfhydryl-reactive metals in autism. *J Toxicol Environ Health A* 2007; **70**:715–721.

Kita K, Miura N, Yoshida M, *et al.* Potential effect on cellular response to cadmium of a single-nucleotide A $\rightarrow$ G polymorphism in the promoter of the human gene for metallothionein IIA. *Hum Genet* 2006; **120**:553–560.

Levin ED, Perraut C, Pollard N, Freedman JH. Metallothionein expression and neurocognitive function in mice. *Physiol Behav* 2006; **87**:513–518.

Liu G, Garrett MR, Men P, *et al.* Nanoparticle and other metal chelation therapeutics in Alzheimer disease. *Biochim Biophys Acta* 2005; **1741**:246–252.

Lord C, Volkmar F. Genetics of childhood disorders: XLII. Autism, part 1: Diagnosis and assessment in autistic spectrum disorders. *J Am Acad Child Adolesc Psychiatry* 2002; **41**:1134–1136.

Magos L. Neurotoxic character of thimerosal and the allometric extrapolation of adult clearance half-time to infants. *J Appl Toxicol* 2003; **23**:263–269.

Magos L, Clarkson TW. Overview of the clinical toxicity of mercury. *Ann Clin Biochem* 2006; **43**:257–268.

Maret W. The function of zinc metallothionein: A link between cellular zinc and redox state. *J Nutr* 2000; **130**:1455S–1458S.

McDowell MA, Dillon CF, Osterloh J, *et al.* Hair mercury levels in U.S. children and women of childbearing age: Reference range data from NHANES 1999–2000. *Environ Health Perspect* 2004; **112**:1165–1171.

Moos T, Morgan EH. The metabolism of neuronal iron and its pathogenic role in neurological disease: Review. *Ann NY Acad Sci* 2004; **1012**:14–26.

Nagano S, Satoh M, Sumi H, *et al.* Reduction of metallothioneins promotes the disease expression of familial amyotrophic lateral sclerosis mice in a dose-dependent manner. *Eur J Neurosci* 2001; **13**:1363–1370.

Palmen SJ, van Engeland H, Hof PR, Schmitz C. Neuropathological findings in autism. *Brain* 2004; **127**:2572–2583.

Parker S, Todd J, Schwartz B, Pickering L. Thimerosal-containing vaccines and autistic spectrum disorder: A critical review of published original data. *Pediatrics* 2005; **115**:200.

Penkowa M. Metallothioneins are multipurpose neuroprotectants during brain pathology. *FEBS J* 2006; **273**:1857–1870.

Pfeiffer Treatment Center. Available at http://www.hriptc.org/autism_protocol.html/. 2007.

Pichichero ME, Cernichiari E, Lopreiato J, Treanor J. Mercury concentrations and metabolism in infants receiving vaccines containing thiomersal: A descriptive study. *Lancet* 2002; **360**:1737–1741.

Pickett J, London E. The neuropathology of autism: A review. *J Neuropathol Exp Neurol* 2005; **64**:925–935.

Powell H, Jamieson WA. Merthiolate as a germicide. *Am J Hyg* 1931; **13**:296–310.

Rapin I. Autism. *N Engl J Med* 1997; **337**:97–104.

Reader M, Lines CB. Decomposition of thimerosal in aqueous solution and its determination by high-performance liquid chromatography. *J Pharm Sci* 1983; **72**:1406–1409.

Sager PR, Doherty RA, Rodier PM. Effects of methylmercury on developing mouse cerebellar cortex. *Exp Neurol* 1982; **77**:179–193.

Schmid KW, Ellis IO, Gee JM, *et al.* Presence and possible significance of immunocytochemically demonstrable metallothionein over-expression in primary invasive ductal carcinoma of the breast. *Virchows Arch A Pathol Anat Histopathol* 1993; **422**:153–159.

Schober SE, Sinks TH, Jones RL, *et al.* Blood mercury levels in US children and women of childbearing age, 1999–2000. *JAMA* 2003; **289**:1667–1674.

Shearer TR, Larson K, Neuschwander J, Gedney B. Minerals in the hair and nutrient intake of autistic children. *J Autism Dev Disord* 1982; **12**:25–34.

Simpkins CO. Metallothionein in human disease. *Cell Mol Biol (Noisy-le-grand)* 2000; **46**:465–488.

Singh VK. Thimerosal is unrelated to autoimmune autism. *Pediatr Allergy Immunol* 2007; **18**:89.

Singh VK, Hanson J. Assessment of metallothionein and antibodies to metallothionein in normal and autistic children having exposure to vaccine-derived thimerosal. *Pediatr Allergy Immunol* 2006; **17**:291–296.

Stankovic RK, Chung RS, Penkowa M. Metallothioneins I and II: Neuroprotective significance during CNS pathology. *Int J Biochem Cell Biol* 2007; **39**:484–489.

Suzuki T, Takemoto T, Kashiwazaki H, Miyama T. Metabolic fate of ethylmercury salts in man and animal. In: Miller M, Clarkson T (eds.), *Mercury, Mercurials and Mercaptans*, Charles C. Thomas Publishing, Springfield, IL, 1973, pp. 209–233.

Tan M, Parkin JE. Route of decomposition of thiomersal (thimerosal). *Int J Pharm* 2000; **208**:23–34.

Thomas DJ, Smith JC. Effects of coadministered low-molecular-weight thiol compounds on short-term distribution of methyl mercury in the rat. *Toxicol Appl Pharmacol* 1982; **62**:104–110.

Thompson WW, Price C, Goodson B, *et al.*, Early thimerosal exposure and neuropsychological outcomes at 7 to 10 years. *N Engl J Med* 2007; **357**:1281–1292.

U.S. FDA (Food and Drug Administration). http://www.fda.gov/CbER/vaccine/thimerosal.htm/. Accessed 2007.

Vorstman JA, Staal WG, van Daalen E, *et al.* Identification of novel autism candidate regions through analysis of reported cytogenetic abnormalities associated with autism. *Mol Psychiatry* 2006; **11**:1, 18–28.

Walker SJ, Segal J, Aschner M. Cultured lymphocytes from autistic children and non-autistic siblings up-regulate heat shock protein RNA in response to thimerosal challenge. *Neurotoxicology* 2006; **27**:685–692.

Walsh W. Metallothionein promotion therapy in autism spectrum disorders. In: Rimland B (ed.), *DAN! (Defeat Autism Now!) Spring 2002 Conference Practitioner Training*, Autism Research Institute, San Diego, CA, 2002.

Welsh JP, Yuen G, Placantonakis DG, *et al.* Why do Purkinje cells die so easily after global brain ischemia? Aldolase C, EAAT4, and the cerebellar contribution to posthypoxic myoclonus. *Adv Neurol* 2002; **89**:331–359.

Wong BS, Chen SG, Colucci M, *et al.* Aberrant metal binding by prion protein in human prion disease. *J Neurochem* 2001; **78**:1400–1408.

World Health Organization (WHO). Environmental Health Criteria 101: Methylmercury. International Program on Chemical Safety, World Health Organization, Geneva, 1990.

World Health Organization (WHO). Recommendations from the Strategic Advisory Group of Experts. *Wkly Epidemiol Rec* 2002; **77**:305–316.

Yamada H, Koizumi S. Metallothionein induction in human peripheral blood lymphocytes by heavy metals. *Chem Biol Interact* 1991; **78**:347–354.

Yeargin-Allsopp M, Rice C, Karapurkar T, *et al.* Prevalence of autism in a US metropolitan area. *JAMA* 2003; **289**:49–55.

Yin X, Knecht DA, Lynes MA. Metallothionein mediates leukocyte chemotaxis. *BMC Immunol* 2005; **6**:21.

Yoshida M, Watanabe C, Satoh M, *et al.* Susceptibility of metallothionein-null mice to the behavioral alterations caused by exposure to mercury vapor at human-relevant concentration. *Toxicol Sci* 2004; **80**:69–73.

Zelger B, Hittmair A, Schir M, *et al.* Immunohistochemically demonstrated metallothionein expression in malignant melanoma. *Histopathology* 1993; **23**:257–263.

Chapter 6

THE ROLE OF METALLOTHIONEIN AND ASTROCYTE–NEURON INTERACTIONS IN INJURY TO THE CNS

Samantha J. Fung, Roger S. Chung and Adrian West

The physiological role of metallothionein (MT) has been a topic of growing interest, particularly with regard to a potential therapeutic application in trauma of the central nervous system (CNS). An increasing number of studies describe the protective, regenerative, and anti-inflammatory properties of MT-I and MT-II isoforms (MT-I/MT-II) in the context of *in vitro* and animal models, using, for example, MT-I/MT-II null, overexpressing, or injected mice following induced CNS trauma or disease. MT-I/MT-II respond to trauma by upregulation, and may have roles in metal ion homeostasis and free radical scavenging. Notably, a direct action of MT-I/MT-II on neurons has been shown using *in vitro* models, whereby the application of exogenous MT-I/MT-II directly increases neurite outgrowth of young neurons and regeneration of injured, mature neurons. The expression and putative functions of MT within the injured CNS will be addressed within this chapter, with particular regard to the MT-I/MT-II isoforms that display neuroprotective and regenerative properties. Intriguingly, a further member of the MT family, MT-III, shows high homology to MT-I/MT-II, yet has a contrasting effect on neuron growth and survival in some models.

Keywords: Central nervous system (CNS); metallothionein; expression; astrocyte; neuron; injury.

1. Introduction

Mammalian metallothioneins (MTs) are small proteins with a molecular mass of 6–7 kDa. Four mammalian MT isoforms have been characterized from multigene families, MT-I to MT-IV (Palmiter, 1998), of which MT-I, MT-II, and MT-III are expressed within the mammalian central nervous system (CNS) (Hidalgo *et al.*,

2001). The final mammalian MT isoform, MT-IV, is expressed in squamous epithelial tissues (Quaife *et al.*, 1994) and remains relatively unstudied. The composition of all MT proteins is unique in its cysteine-rich nature and, in the case of mammalian MTs, cysteine residues constitute approximately one third of the 60–68 amino acids (see, for example, Kägi and Vallee, 1961; Huang *et al.*, 1977; Uchida *et al.*, 1991; Palmiter *et al.*, 1992). It is this cysteine-rich nature that allows metal–sulphur bonds to form between cysteine residues and metal ions to give the characteristic metal-binding property of MT (Hamer, 1986).

MT-I and MT-II are highly homologous in amino acid sequence; for example, human and mouse MT-I and MT-II both consist of 61 amino acids, of which 20 are cysteine residues in conserved motifs (e.g. human MT-IA and MT-IIA) (Huang *et al.*, 1977; Kissling and Kägi, 1977; Huang *et al.*, 1981; Kägi *et al.*, 1984; Miles *et al.*, 2000). Due to their striking homology in structure as well as similarities in expression and function (see, for example, Penkowa *et al.*, 2006b), MT-I and MT-II are often grouped together and are herein referred to as MT-I/MT-II. The MT-III isoform is homologous to MT-I/MT-II; for example, human MT-III displays 70% sequence homology to human MT-IIA and totals 68 amino acids, containing two insertions relative to the MT-IIA isoform — a threonine at residue 5 and a six-amino-acid acidic insertion in the C-terminus (Uchida *et al.*, 1991). Further to this, MT-III contains a Cys(6)-Pro-Cys-Pro motif adjacent to the threonine insertion when compared to the Cys(5)-Ser-Cys-Ala motif in MT-IIA (Uchida *et al.*, 1991; Palmiter *et al.*, 1992; Tsuji *et al.*, 1992; Kobayashi *et al.*, 1993; Chung *et al.*, 2002a).

2. MT Expression

Within the CNS, MT-I/MT-II are present primarily in astrocytes of gray matter and some astrocytes within white matter (Nishimura *et al.*, 1992; Blaauwgeers *et al.*, 1993; Holloway *et al.*, 1997), and in the nucleus dentatus and granular layer of the cerebellum

(Nakajima and Suzuki, 1995). Ependymal and choroid plexus cells as well as astrocytic end-feet surrounding blood vessels also express MT-I/MT-II (Nakajima and Suzuki, 1995). MT-I/MT-II are not generally thought to be expressed in neurons of the CNS, although some reports have identified MT-I/MT-II in neurons (for example, Skabo *et al.*, 1997; Campagne *et al.*, 1999), with expression restricted to minor and specific populations of neurons such as olfactory neurons (Skabo *et al.*, 1997) or cortical pyramidal neurons of mice when MT-I is overexpressed (Campagne *et al.*, 1999).

MT-III, also termed "growth inhibitory factor" due to its effect on neuron survival and neurite outgrowth in neuronal cultures (Uchida *et al.*, 1991; Tsuji *et al.*, 1992; Erickson *et al.*, 1994; Irie and Keung, 2001; Chung *et al.*, 2002b), is expressed mainly in the CNS (Uchida *et al.*, 1991; Palmiter *et al.*, 1992; Tsuji *et al.*, 1992; Sogawa *et al.*, 2001). Although first identified within astrocytes of the gray matter (Uchida *et al.*, 1991) and reported in glial cells in the cerebellum, neocortex, hippocampus, striatum, brain stem, and spinal cord (Uchida *et al.*, 1991; Uchida, 1994; Yamada *et al.*, 1996; Sogawa *et al.*, 2001), MT-III is nevertheless believed to be predominantly expressed in neurons. MT-III mRNA is found in neurons in the entire brain with the exception of fiber tracts (Kim *et al.*, 2003), particularly in the dentate gyrus (Choudhuri *et al.*, 1995) and CA1 and CA3 regions of the hippocampus, cerebral cortex, amygdala, and zinc-containing synaptic vesicles (Masters *et al.*, 1994). The MT-III protein has been detected in neurons of the dentate gyrus and CA1–3 regions of the hippocampus, cerebral cortex, thalamus, and olfactory bulb by immunohistochemistry (Yanagitani *et al.*, 1999; Lee *et al.*, 2003).

3. MT-I/MT-II Expression in Response to Injury

The genes encoding MT-I/MT-II are highly inducible and their expression is upregulated as a result of a variety of stimuli including heavy metals (e.g. Carter *et al.*, 1984; Yagle and Palmiter, 1985;

Culotta and Hamer, 1989), oxidative stress (e.g. Dalton *et al.*, 1994), glucocorticoids (Karin *et al.*, 1984; Yagle and Palmiter, 1985; Kelly *et al.*, 1997), and injury (e.g. Carrasco *et al.*, 1999; Penkowa *et al.*, 1999b; Campagne *et al.*, 1999; Trendelenburg *et al.*, 2002; Chung *et al.*, 2004). In particular, in the CNS following *in vivo* focal lesion in the neocortex of rats, MT-I/MT-II upregulation can be detected within days in astrocytes bordering the lesion tract (Chung *et al.*,

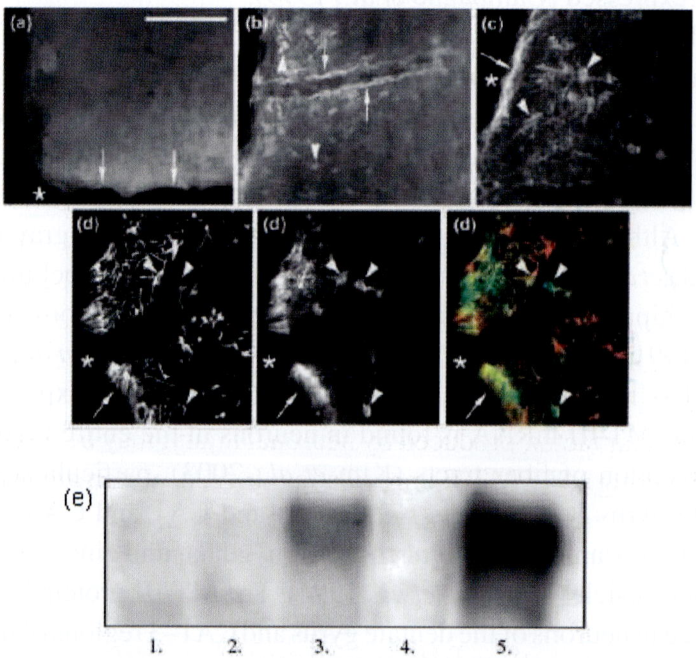

Fig. 1 Expression of MT-I/MT-II following focal lesion to the rat neocortex. (a) MT-I/MT-II labeling is absent 1 day postinjury. Arrows indicate the location of the lesion. At (b) 4 and (c) 7 days postinjury, MT-I/MT-II immunoreactivity is localized to glia-like cells (arrowheads), cells and processes surrounding blood vessels [arrows in (b)], and processes along the pial surface [arrow in (c)]. (d) Double-labeling immunohistochemistry for MT-I/MT-II (green; arrowheads) and for the astrocytic marker glial fibrillary acidic protein (GFAP, red) shows that cells immunopositive for MT-I/MT-II possess a cell body and/or processes immunopositive for GFAP. The pial surface is indicated (arrow). (e) Western blot analysis was performed for rat cortex proteins (1, uninjured; 2, 1 day postinjury; 3, 7 days postinjury; 4, 14 days postinjury; and 5, sheep liver MT-I/MT-II). Immunoreactivity for MT-I/MT-II was present in a single band for the sheep liver sample and at 7 days postinjury in the rat brain. Figure reprinted from Chung *et al.* (2004), with permission from Blackwell Publishing.

2004). Expression then spreads from the injury site to surrounding astrocytes and decreases to basal levels around 14 days postlesion (Chung *et al.*, 2004) (Fig. 1). MT-I/MT-II are also upregulated in astrocytes following cryolesion (Carrasco *et al.*, 1999; Penkowa *et al.*, 1999a), ischemia (Campagne *et al.*, 1999; Trendelenburg *et al.*, 2002), administration of the gliotoxin 6-aminonicotinamide (6-AN) (Penkowa *et al.*, 1999b; Penkowa *et al.*, 2002a; Penkowa *et al.*, 2002b), and kainic acid treatment (Carrasco *et al.*, 2000; Kim *et al.*, 2003); and have also been reported in microglia and macrophages at the lesion site (Vela *et al.*, 1997; Carrasco *et al.*, 1998; Penkowa *et al.*, 2002a).

MT-I/MT-II induction can be stimulated by several acute phase cytokines such as interleukin-1 (IL-1) and interleukin-6 (IL-6) (Karin *et al.*, 1985; Schroeder and Cousins, 1990; Kikuchi *et al.*, 1993) and tumor necrosis factor-α (TNF-α) (Sato *et al.*, 1992). In astrocytes, this appears to be promoted in part by neuronal injury, as demonstrated by an upregulation of MT-I/MT-II in neuron-astrocyte cocultures when medium from a neuron culture injured with a scalpel blade was applied (Chung *et al.*, 2004). Thus, it was hypothesized that an extracellular factor, produced by neuronal injury, may be responsible for triggering MT-I/MT-II upregulation. Indeed, glutamate was identified as a factor that may play a part in the upregulation of astrocytic MT-I/MT-II (Chung *et al.*, 2004).

4. Expression of MT-III in Response to Injury

Various reports exist concerning the expression of MT-III following injury. For example, Anezaki *et al.* (1995) reported the upregulation of MT-III in reactive astrocytes following stab wound injury and kainic acid injection. It is most likely that MT-III expression is dynamic in response to CNS injury. Several reports have suggested that MT-III is reduced 1 day postinjury, and then increases in the following 3–4 days; MT-III then returns to basal levels before increasing again 2–4 weeks postinjury, for example, in the case of a stab wound (Hozumi *et al.*, 1995; Yuguchi *et al.*, 1995a). Yuguchi *et al.* (1995b) reported downregulation of MT-III mRNA following transection of the facial nerve and a delayed increase in MT-III mRNA several weeks after injury in the perilesioned area where gliosis predominantly occurs (Yuguchi *et al.*, 1995a). This may indicate the upregulation of MT-III in reactive astrocytes (Carrasco *et al.*, 1999). MT-III upregulation has also been reported in astrocytes in hypoxic conditions (Tanji *et al.*, 2003).

5. MT-I/MT-II Promotes Wound Healing

CNS injury in MT-I/MT-II null mice demonstrates an important role for MT-I/MT-II in the regenerative and wound healing response (as reviewed by Hidalgo *et al.*, 2001; West *et al.*, 2004; Penkowa, 2006). Penkowa and colleagues (1999c) first established that MT-I/MT-II null mutant mice showed a deficiency in wound healing following cortical cryolesion. Since this pivotal report, models such as traumatic brain injury (Natale *et al.*, 2004; Potter *et al.*, 2007), kainic acid-induced seizure (Carrasco *et al.*, 2000), cryolesion (Penkowa *et al.*, 1999a; Penkowa *et al.*, 2000; Penkowa *et al.*, 2006b), ischemia (Trendelenberg *et al.*, 2002), induction of experimental autoimmune encephalomyelitis (EAE, the mouse model of human multiple sclerosis) (Penkowa *et al.*, 2001; Penkowa *et al.*, 2003b), the SOD1 G93A transgenic mouse model of amyotrophic lateral sclerosis (ALS) (Nagano *et al.*, 2001; Puttaparthi *et al.*, 2002), the α-synuclein knockout/SIN-1 model of Parkinson's disease (Ebadi and Sharma, 2006), and 6-AN treatment (Penkowa *et al.*, 1999b) have been used to cause physical or excitatory lesions in the CNS of MT-I/MT-II null mutant mice. In mice lacking endogenous MT-I/MT-II, such lesions universally result in increased degeneration of brain tissue and poor recovery compared to wild-type animals (for an excellent review, see Penkowa (2006)).

Following trauma, MT-I/MT-II null mutant mice exhibit increased apoptotic cell death and angiogenesis, and increased oxidative stress (Penkowa *et al.*, 1999a; Penkowa *et al.*, 2000; Penkowa *et al.*, 2003b;

Carrasco *et al.*, 2000; Nagano *et al.*, 2001; Giralt *et al.*, 2002b; Natale *et al.*, 2004). Processes associated with inflammation are also enhanced, including infiltration of macrophages, microglia, lymphocytes, and progenitors; secretion of proinflammatory cytokines such as IL-1β, IL-6, and TNF-α; and attenuated upregulation of growth factors such as basic fibroblast growth factor (bFGF), transforming growth factor-β1 (TGF-β1), vascular endothelial growth factor (VEGF), and neurotrophin-3 (NT-3) (Penkowa *et al.*, 1999a; Penkowa *et al.*, 1999b; Penkowa *et al.*, 2000; Penkowa *et al.*, 2001; Penkowa *et al.*, 2003b; Penkowa *et al.*, 2006b; Carrasco *et al.*, 2000; Giralt *et al.*, 2002a; Giralt *et al.*, 2002b; Potter *et al.*, 2007). The response of reactive astrocytes is also impaired and glial scar formation is delayed (Penkowa *et al.*, 1999a; Penkowa *et al.*, 1999b; Penkowa *et al.*, 2000; Penkowa *et al.*, 2001).

Conversely, transgenic MT-I/MT-II overexpressing mice (TgMT) are more resistant to trauma than wild-type mice, and experience attenuated pathology and inflammation as well as augmented astrogliosis and wound healing (Campagne *et al.*, 1999; Giralt *et al.*, 2002b; Penkowa *et al.*, 2002a; Penkowa *et al.*, 2003a; Penkowa *et al.*, 2004; Penkowa *et al.*, 2005; Molinero *et al.*, 2003; Ebadi and Sharma, 2006). Moreover, when exogenous MT-I/MT-II are administered to 6-AN treated, cryolesioned, EAE, or focal cortical needlestick models of injury, there is a similar improved outcome (Penkowa and Hidalgo, 2000; Penkowa and Hidalgo, 2001; Giralt *et al.*, 2002b; Penkowa *et al.*, 2002a; Penkowa *et al.*, 2006b; Chung *et al.*, 2003). In the case of EAE, demyelination is also reduced (Penkowa *et al.*, 2003b; Penkowa *et al.*, 2006a; Penkowa and Hidalgo, 2003), and recruitment of neuroglial precursors can be observed in response to cryolesion of TgMT mice and in MT-I/MT-II treatment of EAE (Penkowa *et al.*, 2003b; Penkowa *et al.*, 2006a; Penkowa and Hidalgo, 2003). This presents strong evidence for a protective and regenerative action of MT-I/MT-II in CNS injury through protective and trophic actions on neurons, anti-inflammatory properties, and a beneficial astrocytic response.

MT-III null mutant mice differ in their response to injury. Cortical cryolesion performed on MT-III null mutant mice appeared to increase GAP-43 and neurotrophin expression, indicative of an increase in sprouting in the absence of MT-III expression (Carrasco *et al.*, 2003). Moreover, in the peripheral nervous system, MT-III null mutant mice experienced rapid recovery compared to wild-type mice following a peripheral nerve crush lesion, supporting Uchida's original suggestion that MT-III is inhibitory to neuron growth (Ceballos *et al.*, 2003). Indeed, MT-III may contribute to neuronal death in the CA1 region of the hippocampus and the thalamus through the release of zinc following kainic acid-induced seizure (Lee *et al.*, 2003). However, exogenous MT-III reduces lesion size following stab wound to the rat brain in a dose-dependent manner, though excessive amounts of MT-III result in a larger lesion size (Hozumi *et al.*, 2005). Administration of an adenoviral vector encoding MT-III to the lesion site resulted in an increase in the amount of MAP-2 and GAP-43 (Hozumi *et al.*, 2005), and administration of the same vector promoted the survival of injured facial motoneurons (Sakamoto *et al.*, 2003). Thus, applied MT-III may also promote CNS wound healing in some paradigms. This effect is likely to be concentration-dependent and might be attributed to roles such as free radical scavenging (Hozumi *et al.*, 2005).

6. MT-I/MT-II Act Directly on Neurons

Thus, in the CNS, MT-I/MT-II have been implicated in increased regeneration and reactive sprouting of neurons due to deficiencies in these processes in MT-I/MT-II null mutant mouse injury models, while the contribution of MT-III is less clear. The application of exogenous MT-I/MT-II to a focal injury of the rat neocortex results in accelerated healing of the lesion and, importantly, leads to the presence of numerous SMI312-positive reactive sprouts in the lesion site compared to vehicle-treated controls (Chung *et al.*, 2003). Using cultured rat cortical neurons, we have demonstrated that application of MT-I/MT-II results in a direct regenerative effect on neurons.

MT-I/MT-II increased the rate of neurite growth in *in vitro* models, including models of injured cortical neurons (Chung *et al.*, 2003). Furthermore Kohler *et al.* (2003) reported MT-I/MT-II to have the same effect on hippocampal and dopaminergic neurons; and recently Fitzgerald *et al.* (2007) and Ambjørn *et al.* (2007) reported this phenomenon in retinal ganglion cells and cerebellar granule neurons, respectively, suggesting a generic effect of MT-I/MT-II on neurons.

7. Model of Extracellular MT-I/MT-II Action in CNS Injury

Taken together, MT-I/MT-II, expressed primarily in astrocytes, are protective to neurons and promote neurite regeneration in the injured CNS. This has led to our proposal of an extracellular action of MT-I/MT-II. We hypothesize that, following their astrocytic upregulation, MT-I/MT-II are released into the extracellular environment, where they then act on neurons to increase regenerative sprouting (Chung and West, 2004) (Fig. 2). Based on their amino acid sequence, MT-I/MT-II are not considered to contain any motifs consistent with traditional secretion pathways (Palmiter *et al.*, 1992), yet increasing evidence suggests that MT-I/MT-II have an extracellular presence and are secreted by various cell types in physiological and pathological situations and upon specific stimulation. This is also consistent with other stress proteins, being released through nonclassical secretory pathways that do not involve processing through the endoplasmic reticulum and Golgi apparatus (reviewed by Lynes *et al.*, 2006). Some proinflammatory cytokines such as IL-1α and IL-1β (Suttles *et al.*, 1990; Prudovsky *et al.*, 2003) — as well as stress response proteins such as heat shock protein 70 (Hsp70) (Gastpar *et al.*, 2005), FGF-1, and bFGF (Jackson *et al.*, 1992; Engling *et al.*, 2002; Prudovsky *et al.*, 2003; Taverna *et al.*, 2003) — are released by nonclassical pathways such as through lipid raft domains, membrane blebbing, and direct translocation from cytoplasm to extracellular space (Lynes *et al.*, 2006).

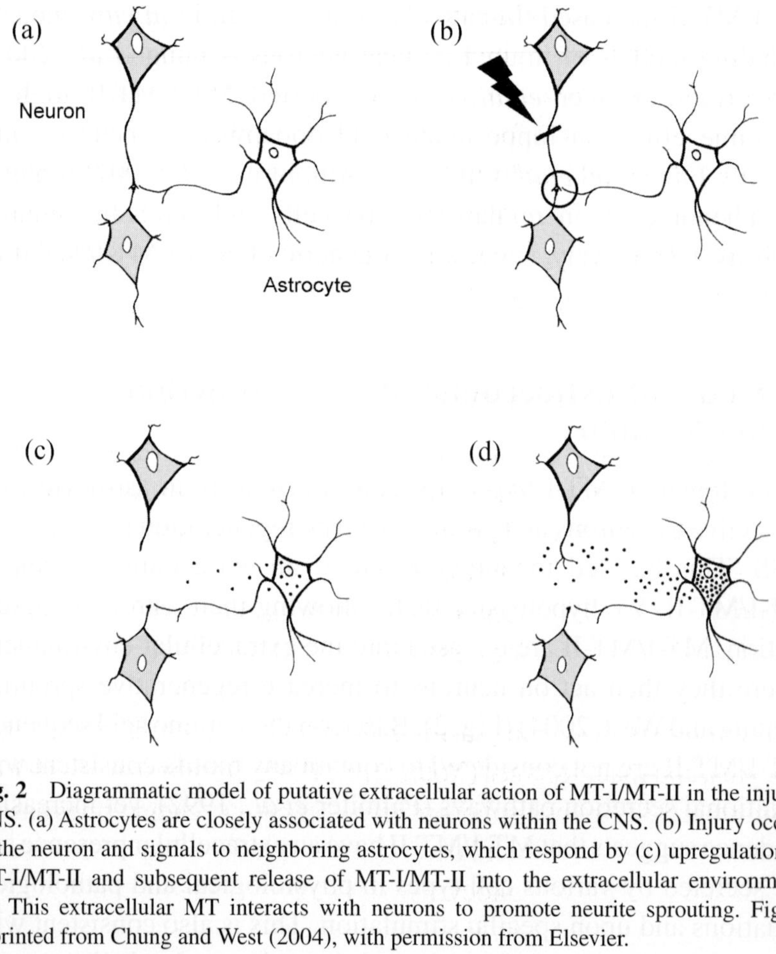

Fig. 2 Diagrammatic model of putative extracellular action of MT-I/MT-II in the injured CNS. (a) Astrocytes are closely associated with neurons within the CNS. (b) Injury occurs to the neuron and signals to neighboring astrocytes, which respond by (c) upregulation of MT-I/MT-II and subsequent release of MT-I/MT-II into the extracellular environment. (d) This extracellular MT interacts with neurons to promote neurite sprouting. Figure reprinted from Chung and West (2004), with permission from Elsevier.

Extracellular MT has been reported physiologically in various *in vivo* and *in vitro* models. We have noted extracellular MT-I/MT-II following injury to the rat CNS *in vivo* (unpublished data), and MT has also been reported surrounding cells in kainic acid-induced excitotoxicity of MT-I overexpressing mice (Penkowa *et al.*, 2005). Specific MT-I/MT-II release is also indicated by the presence of extracellular MT-I/MT-II in the medium of cultured preadipocytes during differentiation (Trayhurn *et al.*, 2000); and extracellular MT-I/MT-II were recently detected in the medium of cultured astrocytes following stimulation with a combination of zinc and IL-1α (Chung *et al.*, 2008), a treatment shown by Kikuchi *et al.* (1993) to induce MT accumulation on the surface of astrocytoma cells. There is also further physiological evidence for extracellular MT-I/MT-II *in vivo* — and thus MT-I/MT-II release from cells — in serum (Mehra and Bremner, 1983; Garvey, 1984; Hidalgo *et al.*, 1988), bile (Sato and Bremner, 1984), pancreatic fluid (De Lisle *et al.*, 1996), and other extracellular fluids (as reviewed by Lynes *et al.*, 2006).

8. Putative Interaction of MT and Neurons

Therefore, the proposition that MT-I/MT-II are found outside cells is reasonable, and evidence currently suggests that administration of exogenous MT-I/MT-II directly alters the growth of neurons in culture (Chung *et al.*, 2003; Kohler *et al.*, 2003; Fitzgerald *et al.*, 2007; Ambjørn *et al.*, 2008). One question that remains to be addressed in this model is, how does extracellular MT-I/MT-II interact with neurons?

MT interacts with the plasma membrane of immune cells, lymphocytes, and macrophages (Youn *et al.*, 1995; Borghesi *et al.*, 1996). Furthermore, MT may display properties similar to chemokines based on homology to the beta and delta cytokines CCL-17 and CX3CL-1, and on stimulation of leukocyte chemotaxis (Yin *et al.*, 2005). This chemotactic response was inhibited through the use of G-coupled-protein receptor antagonists, suggesting the intriguing possibility of a leukocytic G-protein-coupled MT receptor (Yin *et al.*, 2005). In kidney cell lines, MT has been shown to interact with the scavenger receptor megalin, which has been identified as contributing to MT-I/MT-II uptake in the kidney through an endocytic process (Erfurt *et al.*, 2003; Klassen *et al.*, 2004; Wolff *et al.*, 2006). Indeed, preliminary data from our laboratory suggest that MT-IIA is internalized by neurons (unpublished data) and that ectopic expression of human MT-IIA–enhanced GFP fusion protein by neurons increases neurite outgrowth, suggesting action from an intracellular location.

Megalin expression has been demonstrated in retinal ganglion cells and astrocytic processes (Fitzgerald *et al.*, 2007), in cerebellar granule neurons (Ambjørn *et al.*, 2008), and in cultured cortical neurons and regenerating neurites (unpublished observations). Direct binding of MT-I/MT-II to megalin (Klassen *et al.*, 2004; Ambjørn *et al.*, 2008) and the related low-density-lipoprotein (LDL) receptor-related protein (LRP1; reviewed by Qiu *et al.*, 2006; May *et al.*, 2007; Ambjørn *et al.*, 2008) has been demonstrated using surface plasmon resonance. Indeed, LRP1 is implicated in neurite outgrowth elicited by apolipoprotein E3; and this response was inhibited with a competitive ligand for such LDL receptors as megalin and LRP1, receptor-associated protein (RAP) or anti-LRP1 antibodies, to attenuate apolipoprotein E3-mediated neurite outgrowth (for example, Holtzman *et al.*, 1995; Narita *et al.*, 1997; Qiu *et al.*, 2004). Moreover, a recent report from Fitzgerald *et al.* (2007) suggested that neurite outgrowth from retinal ganglion cells is mediated, at least in part, by megalin following antibody pretreatment to the receptor prior to MT-IIA treatment. Ambjørn *et al.* (2008) also reported that inhibition of megalin and LRP1 with the competitive ligand RAP attenuates MT-I/MT-II–mediated neurite outgrowth and survival in cerebellar granule neurons, and data from our laboratory suggested that siRNA knockdown of megalin inhibits neurite outgrowth from cortical neurons (unpublished data).

9. Signaling Through LDL Receptors — A Putative Mechanism of MT-I/MT-II Activity?

Megalin (Zou *et al.*, 2004; Wicher *et al.*, 2006) and LRP1 (May *et al.*, 2002) undergo regulated intramembrane proteolysis (RIP), in which the extracellular domain is shed through matrix metalloproteases and cleavage by γ-secretase or γ-secretase-like activity (May *et al.*, 2002; Zou *et al.*, 2004). The cytoplasmic domain may then be translocated within the cell (May *et al.*, 2002; Zou *et al.*, 2004), including to the nucleus (Wicher *et al.*, 2006), which may result in a signaling event. The C-terminus of megalin contains motifs that imply interaction with intracellular adaptor and signaling molecules, such as Src homology 2 and 3 domains, and NPxY motifs (Saito *et al.*, 1994; Hjalm *et al.*, 1996). The NPxY motif is also common to LRP1 (Herz and Strickland, 2001); and activation of LRP1 promotes phosphorylation of the mitogen-activated protein kinase (MAPK), extracellular signal-regulated kinase 1/2 (ERK1/2) (Qiu *et al.*, 2004). Activation of the MAPK pathway and cAMP response-element binding (CREB) proteins as well as protein kinase B (PKB) by MT-I/MT-II has been demonstrated (Ambjørn *et al.*, 2008), indicating that MT-I/MT-II activates intracellular signaling pathways, possibly through interaction with LDL receptors (Fig. 3).

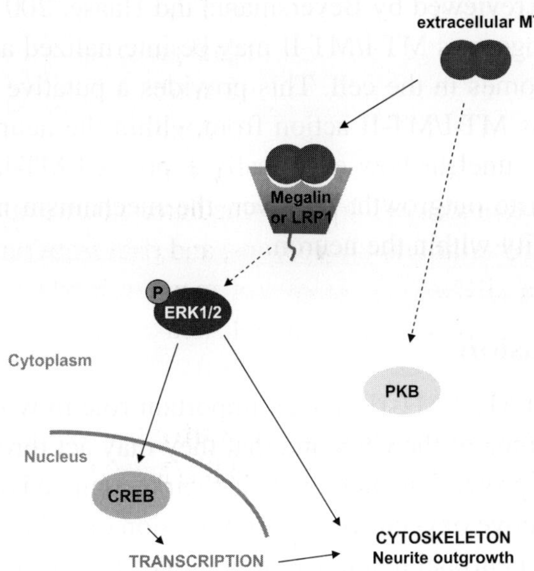

Fig. 3 Putative model of MT-I/MT-II interaction with megalin. Soluble, extracellular MT-I/MT-II interacts with the low-density-lipoprotein receptor megalin and/or LRP1 on the neuronal surface. Megalin/LRP1 may be activated by binding of extracellular MT-I/MT-II to initiate a signaling cascade through the MAPK/ERK pathway to activate transcription factors and impact cytoskeletal elements to influence neurite outgrowth. Activation of PKB has also been demonstrated in response to MT-I/MT-II.

In addition to signaling through LDL receptors, Hao *et al.* (2007) suggested that MT-II endocytosis may provide a route of metal ion delivery in HepG2 hepatocellular carcinoma cells; this paradigm may also be important in the regenerative capability of MT-I/MT-II. In this situation, exogenous MT-II loaded with metal ions is taken into endosomes and fuses with lysosomes, and the change in oxidative environment in this compartment results in release of metals (Austin *et al.*, 2005). Metals were detected within the cytoplasm of MT-II–treated cells (Hao *et al.*, 2007), and the proton-coupled divalent metal transporter (DMT1) was implicated in transport of metals from the endosome/lysosome (Abouhamed *et al.*, 2007). An analogous delivery of zinc to neuronal cytoplasm may have an effect on signaling pathways including MAPK/ERK (Huang *et al.*, 1999), protein kinase C (PKC) (Csermely *et al.*, 1988), and transcription factors (Smirnova *et al.*, 2000) (reviewed by Beyersmann and Haase, 2001); and suggests that exogenous MT-I/MT-II may be internalized and act from within endosomes in the cell. This provides a putative mechanism for exogenous MT-I/MT-II action from within the neuron (Fig. 3). It is currently unclear how ectopically expressed MT-I/MT-II may result in neurite outgrowth; however, the mechanism may involve zinc availability within the neuron.

10. Conclusion

It is clear that MT-I/MT-II play an important role in wound healing following trauma to the CNS and that they may act through several routes to give protection, including via their recognized roles in scavenging of reactive oxygen species, suppression of inflammation, and promotion of neuronal regeneration. In a physiological context, this primarily astrocytic protein appears to be able to be released from astrocytes, with the subsequent effect on neuronal regeneration potentially mediated through LDL receptors. In light of their neuroprotective and regenerative actions, it is conceivable that MT-I/MT-II may provide a promising therapeutic tool in CNS trauma.

Acknowledgments

The authors thank the Australian Research Council (DP0556630 and LP0774820), the National Health and Medical Research Council (Peter Doherty Fellowship to RSC), the Clive & Vera Rammaciotti Foundation, and the Alzheimer's Australia Research Institute. We also thank our colleagues in the Neurorepair Group lab; and our international collaborators Prof. Juan Hidalgo, Assoc. Prof. Milena Penkowa, Prof. Rannar Sillard, and Prof. Peep Palumaa.

References

Abouhamed M, Wolff N, Lee W-K, *et al.* Knockdown of endosomal/lysosomal divalent metal transporter 1 by RNA interference prevents cadmium-metallothionein-1 cytotoxicity in renal proximal tubule cells. *Am J Physiol Renal Physiol* 2007; **293**:F705–F712.

Ambjørn M, Asmussen J, Lindstam M, *et al.* Metallothionein and a peptide modelled after metallothionein, EmtinB, induce neuronal differentiation and survival through binding to receptors of the low density lipoprotein receptor family. *J Neurochem* 2008; **104**:21–37.

Anezaki T, Ishiguro H, Hozumi I, *et al.* Expression of growth inhibitory factor (GIF) in normal and injured rat brains. *Neurochem Int* 1995; **27**:89–94.

Austin C, Wen X, Gazzard L, *et al.* Oxidizing potential of endosomes and lysosomes limits intracellular cleavage of disulfide-based antibody–drug conjugates. *Proc Natl Acad Sci USA* 2005; **102**:17987–17992.

Beyersmann D, Haase H. Functions of zinc in signaling, proliferation and differentiation of mammalian cells. *Biometals* 2001; **14**:331–341.

Blaauwgeers H, Sillevis Smitt P, De Jong J, Troost D. Distribution of metallothionein in the human central nervous system. *Glia* 1993; **8**:62–70.

Borghesi L, Youn J, Olson E, Lynes M. Interactions of metallothionein with murine lymphocytes: Plasma membrane binding and proliferation. *Toxicology* 1996; **108**:129–140.

Campagne M, Thibodeaux H, van Bruggen N, *et al.* Evidence for a protective role of metallothionein-1 in focal cerebral ischemia. *Proc Natl Acad Sci USA* 1999; **96**: 12870–12875.

Carrasco J, Giralt M, Molinero A, *et al.* Metallothionein (MT)-III: Generation of polyclonal antibodies, comparison with MT-I+II in the freeze lesioned rat brain and in a bioassay with astrocytes, and analysis of Alzheimer's disease brains. *J Neurotrauma* 1999; **16**:1115–1129.

Carrasco J, Hernandez J, Gonzalez B, *et al.* Localization of metallothionein-I and -III expression in the CNS of transgenic mice with astrocyte-targeted expression of interleukin 6. *Exp Neurol* 1998; **153**:184–194.

Carrasco J, Penkowa M, Giralt M, *et al.* Role of metallothionein-III following central nervous system damage. *Neurobiol Dis* 2003; **13**:22–36.

Carrasco J, Penkowa M, Hadberg H, *et al.* Enhanced seizures and hippocampal neurodegeneration following kainic acid-induced seizures in metallothionein-I+II–deficient mice. *Eur J Neurosci* 2000; **12**:2311–2322.

Carter A, Felber B, Walling M, *et al.* Duplicated heavy metal control sequences of the mouse metallothionein-I gene. *Proc Natl Acad Sci USA* 1984; **81**:7392–7396.

Ceballos D, Lago N, Verdu E, *et al.* Role of metallothioneins in peripheral nerve function and regeneration. *Cell Mol Life Sci* 2003; **60**:1209–1216.

Choudhuri S, Kramer K, Berman N, *et al.* Constitutive expression of metallothionein genes in mouse brain. *Toxicol Appl Pharmacol* 1995; **131**:144–154.

Chung R, Adlard P, Dittmann J, *et al.* Neuron-glia communication: Metallothionein expression is specifically up-regulated by astrocytes in response to neuronal injury. *J Neurochem* 2004; **88**:454–461.

Chung R, Holloway A, Eckhardt B, *et al.* Sheep have an unusual variant of the brain-specific metallothionein, metallothionein-III. *Biochem J* 2002a; **365**:323–328.

Chung RS, Penkowa M, Dittmann J, *et al.* Redefining the role of metallothionein within the injured brain: Extracellular metallothioneins play an important role in the astrocyte-neuron response to injury. *J Biol Chem* 2008; **283**:15349–15358.

Chung R, Vickers J, Chuah M, West A. Metallothionein-IIA promotes initial neurite elongation and postinjury reactive neurite growth and facilitates healing after focal cortical brain injury. *J Neurosci* 2003; **23**:3336–3342.

Chung R, West A. A role for extracellular metallothioneins in CNS injury and repair. *Neuroscience* 2004; **123**:595–599.

Chung RS, Vickers JC, Chuah MI, *et al.* Metallothionein-III inhibits initial neurite formation in developing neurons as well as postinjury, regenerative neurite sprouting. *Exp Neurol* 2002b; **178**:1–12.

Csermely P, Szamel M, Rasch K, Somogyi J. Zinc can increase the activity of protein kinase C and contributes to its binding to plasma membranes in T lymphocytes. *J Biol Chem* 1988; **263**:6487–6490.

Culotta V, Hamer D. Fine mapping of a mouse metallothionein gene metal response element. *Mol Cell Biol* 1989; **9**:1376–1380.

Dalton T, Palmiter R, Andrews G. Transcriptional induction of the mouse metallothionein-I gene in hydrogen peroxide-treated Hepa cells involves a composite major late transcription factor/antioxidant response element and metal response promoter elements. *Nucleic Acids Res* 1994; **22**:5016–5023.

De Lisle R, Sarras MJ, Hidalgo J, Andrews G. Metallothionein is a component of exocrine pancreas secretion: Implications for zinc homeostasis. *Am J Physiol* 1996; **271**:C1103–C1110.

Ebadi M, Sharma S. Metallothioneins 1 and 2 attenuate peroxynitrite-induced oxidative stress in Parkinson disease. *Exp Biol Med* 2006; **231**:1576–1583.

Engling A, Backhaus R, Stegmayer C, *et al.* Biosynthetic FGF-2 is targeted to non-lipid raft microdomains following translocation to the extracellular surface of CHO cells. *J Cell Sci* 2002; **115**:3619–3631.

Erfurt C, Roussa E, Thevenod F. Apoptosis by Cd^{2+} or CdMT in proximal tubule cells: Different uptake routes and permissive role of endo/lysosomal CdMT uptake. *Am J Physiol Cell Physiol* 2003; **285**:1367–1376.

Erickson JC, Sewell AK, Jensen LT, *et al.* Enhanced neurotrophic activity in Alzheimer's disease cortex is not associated with down-regulation of metallothionein-III (GIF). *Brain Res* 1994; **649**:297–304.

Fitzgerald M, Nairn P, Bartlett C, *et al.* Metallothionein-IIA promotes neurite growth via the megalin receptor. *Exp Brain Res* 2007; **183**:171–180.

Garvey J. Metallothionein: Structure/antigenicity and detection/quantitation in normal physiological fluids. *Environ Health Perspect* 1984; **54**:117–127.

Gastpar R, Gehrmann M, Bausero M, *et al.* Heat shock protein 70 surface-positive tumor exosomes stimulate migratory and cytolytic activity of natural killer cells. *Cancer Res* 2005; **65**:5238–5247.

Giralt M, Penkowa M, Hernandez J, *et al.* Metallothionein-1+2 deficiency increases brain pathology in transgenic mice with astrocyte-targeted expression of interleukin 6. *Neurobiol Dis* 2002a; **9**:319–338.

Giralt M, Penkowa M, Lago N, *et al.* Metallothionein-1+2 protect the CNS after a focal brain injury. *Exp Neurol* 2002b; **173**:114–128.

Hamer D. Metallothionein. *Annu Rev Biochem* 1986; **55**:913–951.

Hao Q, Hong S-H, Maret W. Lipid raft-dependent endocytosis of metallothionein in HepG2 cells. *J Cell Physiol* 2007; **210**:428–435.

Herz J, Strickland D. LRP: A multifunctional scavenger and signaling receptor. *J Clin Invest* 2001; **108**:779–784.

Hidalgo J, Aschner M, Zatta P, Vasak M. Roles of the metallothionein family of proteins in the central nervous system. *Brain Res Bull* 2001; **55**:133–145.

Hidalgo J, Giralt M, Garvey J, Armario A. Physiological role of glucocorticoids on rat serum and liver metallothionein in basal and stress conditions. *Am J Physiol* 1988; **254**:E71–E78.

Hjalm G, Murray E, Crumley G, *et al.* Cloning and sequencing of human gp330, a Ca^{2+}-binding receptor with potential intracellular signalling properties. *Eur J Biochem* 1996; **239**:132–137.

Holloway A, Stennard F, Dziegielewska K, *et al.* Localisation and expression of metallothionein immunoreactivity in the developing sheep brain. *Int J Dev Neurosci* 1997; **15**:195–203.

Holtzman D, Pitas R, Kilbridge J, *et al.* Low density lipoprotein receptor-related protein mediates apolipoprotein E-dependent neurite outgrowth in a central nervous system-derived neuronal cell line. *Proc Natl Acad Sci USA* 1995; **92**:9480–9484.

Hozumi I, Inuzuka T, Hiraiwa M, *et al.* Changes of growth inhibitory factor after stab wounds in rat brain. *Brain Res* 1995; **688**:143–148.

Hozumi I, Uchida Y, Watabe K, *et al.* Growth inhibitory factor (GIF) can protect from brain damage due to stab wounds in rat brain. *Neurosci Lett* 2005; **395**:220–223.

Huang I-Y, Kimura M, Hata A, *et al.* Complete amino acid sequence of mouse liver metallothionein-II. *J Biochem* 1981; **89**:1839–1845.

Huang I-Y, Yoshida A, Tsunoo H, Nakajima H. Mouse liver metallothioneins. Complete amino acid sequence of metallothionein-I. *J Biol Chem* 1977; **252**:8217–8221.

Huang J-S, Mukherjee J, Chung T, *et al.* Extracellular calcium stimulates DNA synthesis in synergism with zinc, insulin and insulin-like growth factor I in fibroblasts. *Eur J Biochem* 1999; **266**:943–951.

Irie Y, Keung WM. Metallothionein-III antagonizes the neurotoxic and neurotrophic effects of amyloid beta peptides. *Biochem Biophys Res Commun* 2001; **282**:416–420.

Jackson A, Friedman S, Zhan X, *et al.* Heat shock induces the release of fibroblast growth factor 1 from NIH 3T3 cells. *Proc Natl Acad Sci USA* 1992; **89**:10691–10695.

Kägi J, Vallee B. Metallothionein: A cadmium and zinc-containing protein from equine renal cortex. *J Biol Chem* 1961; **236**:2435–2442.

Kägi J, Vašák M, Lerch K, *et al.* Structure of mammalian metallothionein. *Environ Health Perspect* 1984; **54**:93–103.

Karin M, Haslinger A, Holtgreve H, *et al.* Characterisation of DNA sequences through which cadmium and glucocorticoid hormones influence human metallothionein-IIA gene. *Nature* 1984; **308**:513–519.

Karin M, Imbra R, Heguy A, Wong G. Interleukin 1 regulates human metallothionein gene expression. *Mol Cell Biol* 1985; **5**:2866–2869.

Kelly EJ, Sandgren EP, Brinster RL, Palmiter RD. A pair of adjacent glucocorticoid response elements regulate expression of two mouse metallothionein genes. *Proc Natl Acad Sci USA* 1997; **95**:10045–10050.

Kikuchi Y, Irie M, Kasahara T, *et al.* Induction of metallothionein in a human astrocytoma cell line by interleukin-1 and heavy metals. *FEBS Lett* 1993; **317**:22–26.

Kim D, Kim E, Kim C, *et al.* Differential regulation of metallothionein-I, II and III mRNA expression in the rat brain following kainic acid treatment. *Neuroreport* 2003; **14**:679–682.

Kissling M, Kägi J. Primary structure of human hepatic metallothionein. *FEBS Lett* 1977; **82**:247–250.

Klassen R, Crenshaw K, Kozyraki R, *et al.* Megalin mediates renal uptake of heavy metal metallothionein complexes. *Am J Physiol Renal Physiol* 2004; **287**:F393–F403.

Kobayashi H, Uchida Y, Ihara Y, *et al.* Molecular cloning of rat growth inhibitory factor cDNA and the expression in the central nervous system. *Mol Brain Res* 1993; **19**:188–194.

Kohler L, Berezin V, Bock E, Penkowa M. The role of metallothionein II in neuronal differentiation and survival. *Brain Res* 2003; **992**:128–136.

Lee J-Y, Kim J-H, Palmiter R, Koh J-Y. Zinc released from metallothionein-III may contribute to hippocampal CA1 and thalamic neuronal death following acute brain injury. *Exp Neurol* 2003; **184**:337–347.

Lynes M, Zaffuto K, Unfricht D, *et al.* The physiological roles of extracellular metallothionein. *Exp Biol Med* 2006; **231**:1548–1554.

Masters BA, Quaife CJ, Erickson JC, *et al.* Metallothionein III is expressed in neurons that sequester zinc in synaptic vesicles. *J Neurosci* 1994; **14**:5844–5857.

May P, Reddy Y, Herz J. Proteolytic processing of low density lipoprotein receptor-related protein mediates regulated release of its intracellular domain. *J Biol Chem* 2002; **277**:18736–18743.

May P, Woldt E, Matz R, Boucher P. The LDL receptor-related protein (LRP) family: An old family of proteins with new physiological functions. *Ann Med* 2007; **39**:219–228.

Mehra R, Bremner I. Development of a radioimmunoassay for rat liver metallothionein-I and its application to the analysis of rat plasma and kidneys. *Biochem J* 1983; **213**:459–465.

Miles A, Hawksworth G, Beattie J, Rodilla V. Induction, regulation, degradation, and biological significance of mammalian metallothioneins. *Crit Rev Biochem Mol Biol* 2000; **35**:35–70.

Molinero A, Penkowa M, Hernandez J, *et al.* Metallothionein-I overexpression decreases brain pathology in transgenic mice with astrocyte-targeted expression of interleukin-6. *J Neuropathol Exp Neurol* 2003; **62**:315–328.

Nagano S, Satoh M, Sumi H, *et al.* Reduction of metallothioneins promotes the disease expression of familial amyotrophic lateral sclerosis mice in a dose-dependent manner. *Eur J Neurosci* 2001; **13**:1363–1370.

Nakajima K, Suzuki K. Immunochemical detection of metallothionein in brain. *Neurochem Int* 1995; **27**:73–87.

Narita M, Bu G, Holtzman D, Schwartz A. The low-density lipoprotein receptor-related protein, a multifunctional apolipoprotein E receptor, modulates hippocampal neurite development. *J Neurochem* 1997; **68**:587–595.

Natale J, Knight J, Cheng Y, *et al.* Metallothionein I and II mitigate age-dependent secondary brain injury. *J Neurosci Res* 2004; **78**:303–314.

Nishimura N, Nishimura H, Ghaffar A, Tohyama C. Localization of metallothionein in the brain of rat and mouse. *J Histochem Cytochem* 1992; **40**:309–315.

Palmiter R. The elusive function of metallothioneins. *Proc Natl Acad Sci USA* 1998; **95**:8428–8430.

Palmiter RD, Findley SD, Whitmore TE, Durnam DM. MT-III, a brain-specific member of the metallothionein gene family. *Proc Natl Acad Sci USA* 1992; **89**:6333–6337.

Penkowa M. Metallothioneins are multipurpose neuroprotectants during brain pathology. *FEBS J* 2006; **273**:1857–1870.

Penkowa M, Caceres M, Borup R, *et al.* Novel roles for metallothionein-I+II (MT-I+II) in defense responses, neurogenesis, and tissue restoration after traumatic brain injury: Insights from global gene expression profiling in wild-type and MT-I+II knockout mice. *J Neurosci Res* 2006a; **84**:1452–1474.

Penkowa M, Camats J, Giralt M, *et al.* Metallothionein-I overexpression alters brain inflammation and stimulates brain repair in transgenic mice with astrocyte-targeted interleukin-6 expression. *Glia* 2003a; **42**:287–306.

Penkowa M, Carrasco J, Giralt M, *et al.* CNS wound healing is severely depressed in metallothionein-I and II–deficient mice. *J Neurosci* 1999a; **19**:2535–2545.

Penkowa M, Carrasco J, Giralt M, *et al.* Altered central nervous system cytokine-growth factor expression profiles and angiogenesis in metallothionein-I+II–deficient mice. *J Cereb Blood Flow Metab* 2000; **20**:1174–1189.

Penkowa M, Espejo C, Martinez-Carceres E, *et al.* Altered inflammatory response and increased neurodegeneration in metallothionein I+II–deficient mice during experimental autoimmune encephalomyelitis. *J Neuroimmunol* 2001; **119**:248–260.

Penkowa M, Espejo C, Martinez-Carceres E, *et al.* Increased demyelination and axonal damage in metallothionein-I+II–deficient mice during experimental autoimmune encephalomyelitis. *Cell Mol Life Sci* 2003b; **60**:185–197.

Penkowa M, Florit S, Giralt M, *et al.* Metallothionein reduces central nervous system inflammation, neurodegeneration, and cell death following kainic acid-induced epileptic seizures. *J Neurosci Res* 2005; **79**:522–534.

Penkowa M, Giralt M, Camats J, Hidalgo J. Metallothionein 1 + 2 protect the CNS during neuroglial degeneration induced by 6-aminonicotinamide. *J Comp Neurol* 2002a; **444**:174–189.

Penkowa M, Giralt M, Moos T, *et al.* Impaired inflammatory response to glial cell death in genetically metallothionein-I- and -II–deficient mice. *Exp Neurol* 1999b; **156**:149–164.

Penkowa M, Hidalgo J. Metallothionein I+II expression and their role in experimental autoimune encephalomyelitis. *Glia* 2000; **32**:247–263.

Penkowa M, Hidalgo J. Metallothionein treatment reduces proinflammatory cytokines IL-6 and TNF-alpha and apoptotic cell death during experimental autoimmune encephalomyelitis (EAE). *Exp Neurol* 2001; **170**:1–14.

Penkowa M, Hidalgo J. Treatment with metallothionein prevents demyelination and axonal damage and increases oligodendrocyte precursors and tissue repair during experimental autoimmune encephalomyelitis. *J Neurosci Res* 2003; **72**:574–586.

Penkowa M, Nielsen H, Hidalgo J, *et al.* Distribution of metallothionein I+II and vesicular zinc in the developing central nervous system: Correlative study in the rat. *J Comp Neurol* 1999c; **412**:303–318.

Penkowa M, Poulsen C, Carrasco J, Hidalgo J. M-CSF deficiency leads to reduced metallothioneins I and II expression and increased tissue damage in the brain stem after 6-aminonicotinamide treatment. *Exp Neurol* 2002b; **176**:308–321.

Penkowa M, Quintana A, Carrasco J, *et al.* Metallothionein prevents neurodegeneration and central nervous system cell death after treatment with gliotoxin 6-aminonicotinamide. *J Neurosci Res* 2004; **77**:35–53.

Penkowa M, Tio L, Giralt M, *et al.* Specificity and divergence in the neurobiologic effects of different metallothioneins after brain injury. *J Neurosci Res* 2006b; **83**:974–984.

Potter E, Cheng Y, Brandon Knight J, *et al.* Metallothionein I and II attenuate the thalamic microglial response following traumatic axotomy in the immature brain. *J Neurotrauma* 2007; **24**:28–42.

Prudovsky I, Mandinova A, Soldi R, *et al.* The non-classical export routes: FGF1 and IL-alpha point the way. *J Cell Sci* 2003; **116**:4871–4881.

Puttaparthi K, Gitomer W, Krishnan U, *et al.* Disease progression in a transgenic model of familial amyotrophic lateral sclerosis is dependent on both neuronal and non-neuronal zinc binding proteins. *J Neurosci* 2002; **22**:8790–8796.

Qiu S, Korwek K, Weeber E. A fresh look at an ancient receptor family: Emerging roles for low density lipoprotein receptors in synaptic plasticity and memory formation. *Neurobiol Learn Mem* 2006; **85**:16–29.

Qiu Z, Hyman B, Rebeck G. Apolipoprotein E receptors mediate neurite outgrowth through activation of p44/42 mitogen-activated protein kinase in primary neurons. *J Biol Chem* 2004; **279**:34948–34956.

Quaife C, Findley S, Erickson J, *et al.* Induction of a new metallothionein isoform (MT-IV) occurs during differentiation of stratified squamous epithelia. *Biochemistry* 1994; **33**:7250–7259.

Saito A, Pietromonaco S, Loo AK-C, Farquhar M. Complete cloning and sequencing of rat gp330/"megalin," a distinctive member of the low density lipoprotein receptor gene family. *Proc Natl Acad Sci USA* 1994; **91**:9725–9729.

Sakamoto T, Kawazoe Y, Uchida Y, *et al.* Growth inhibitory factor prevents degeneration of injured adult rat motoneurons. *Neuroreport* 2003; **14**:2147–2151.

Sato M, Bremner I. Biliary excretion of metallothionein and a possible degradation product in rats injected with copper and zinc. *Biochem J* 1984; **223**:475–479.

Sato M, Sasaki M, Hojo H. Tissue specific induction of metallothionein synthesis by tumor necrosis factor-α. *Res Commun Chem Pathol Pharmacol* 1992; **75**:159–172.

Schroeder J, Cousins R. Interleukin 6 regulates metallothionein gene expression and zinc metabolism in hepatocyte monolayer cultures. *Proc Natl Acad Sci USA* 1990; **87**: 3137–3141.

Skabo S, Holloway A, West A, Chuah M. Metallothioneins 1 and 2 are expressed in the olfactory mucosa of mice in untreated animals and during the regeneration of the epithelial layer. *Biochem Biophys Res Commun* 1997; **232**:136–142.

Smirnova I, Bittel D, Ravindra R, *et al.* Zinc and cadmium can promote rapid nuclear translocation of metal response element-binding transcription factor-1. *J Biol Chem* 2000; **275**:9377–9384.

Sogawa CA, Asanuma M, Sogawa N, *et al.* Localization, regulation, and function of metallothionein-III/growth inhibitory factor in the brain. *Acta Med Okayama* 2001; **55**:1–9.

Suttles J, Giri J, Mizel S. IL-1 secretion and processing by calcium ionophores. *J Immunol* 1990; **144**:175–182.

Tanji K, Irie Y, Uchida Y, *et al.* Expression of metallothionein-III induced by hypoxia attenuates hypoxia-induced cell death *in vitro*. *Brain Res* 2003; **976**:125–129.

Taverna S, Ghersi G, Ginestra A, *et al.* Shedding of membrane vesicles mediates fibroblast growth factor-2 release from cells. *J Biol Chem* 2003; **278**:51911–51919.

Trayhurn P, Duncan J, Wood A, Beattie J. Metallothionein gene expression and secretion in white adipose tissue. *Am J Physiol Regul Integr Comp Physiol* 2000; **279**:R2329–R2335.

Trendelenberg G, Prass K, Priller J, *et al.* Serial analysis of gene expression identifies metallothionein-II as major neuroprotective gene in mouse focal cerebral ischemia. *J Neurosci* 2002; **22**:5879–5888.

Tsuji S, Kobayashi H, Uchida Y, *et al.* Molecular cloning of human growth inhibitory factor cDNA and its down-regulation in Alzheimer's disease. *EMBO J* 1992; **11**:4843–4850.

Uchida Y. Growth-inhibitory factor, metallothionein-like protein, and neurodegenerative diseases. *Biol Signals* 1994; **3**:211–215.

Uchida Y, Takio K, Titani K, *et al.* The growth inhibitory factor that is deficient in Alzheimer's disease brain is a 68 amino acid metallothionein-like protein. *Neuron* 1991; **7**: 337–347.

Vela J, Hidalgo J, Gonzalez B, Castellano B. Induction of metallothionein in astrocytes and microglia in the spinal cord from the myelin-deficient jimpy mouse. *Brain Res* 1997; **767**:345–355.

West AK, Chuah MI, Vickers JC, Chung RS. Protective role of metallothioneins in the injured mammalian brain. *Rev Neurosci* 2004; **15**:157–166.

Wicher G, Larsson M, Svenningsen AF, *et al.* Low density lipoprotein receptor-related protein-2/megalin is expressed in oligodendrocytes in the mouse spinal cord white matter. *J Neurosci Res* 2006; **83**:864–873.

Wolff N, Abouhamed M, Verroust P, Thevenod F. Megalin-dependent internalization of cadmium-metallothionein and cytotoxicity in cultured retinal proximal tubule cells. *J Pharmacol Exp Ther* 2006; **318**:782–791.

Yagle M, Palmiter R. Coordinate regulation of mouse metallothionein I and III genes by heavy metals and glucocorticoids. *Mol Cell Biol* 1985; **5**:291–294.

Yamada M, Hayashi S, Hozumi I, *et al*. Subcellular localization of growth inhibitory factor in rat brain: Light and electron microscopic immunohistochemical studies. *Brain Res* 1996; **735**:257–264.

Yanagitani S, Miyazaki H, Nakahashi Y, *et al*. Ischemia induces metallothionein III expression in neurons of rat brain. *Life Sci* 1999; **64**:707–715.

Yin X, Knecht D, Lynes M. Metallothionein mediates leukocyte chemotaxis. *BMC Immunol* 2005; **6**:21–32.

Youn J, Borghesi L, Olson E, Lynes M. Immunomodulatory activities of extracellular metallothionein. II. Effects on macrophage functions. *J Toxicol Environ Health* 1995; **45**:397–413.

Yuguchi T, Kohmura E, Yamada K, *et al*. Expression of growth inhibitory factor mRNA following cortical injury in rat. *J Neurotrauma* 1995a; **2**:299–306.

Yuguchi T, Kohmura E, Yamada K, *et al*. Changes in growth inhibitory factor mRNA expression compared with those in c-jun mRNA expression following facial nerve transection. *Brain Res Mol Brain Res* 1995b; **28**:181–185.

Zou Z, Chung B, Nguyen T, *et al*. Linking receptor-mediated endocytosis and cell signaling evidence for regulated intramembrane proteolysis of megalin in proximal tubule. *J Biol Chem* 2004; **279**:34302–34310.

PART II

METALLOTHIONEIN IN ONCOLOGY

Chapter 7

METALLOTHIONEINS AND ONCOLOGY

Stamatios E. Theocharis

The metallothionein (MT) family is a class of low-molecular-weight, cysteine-rich proteins (MT-1, MT-2, MT-3, and MT-4) with high affinity for metal ions. Apart from their involvement in metal ion homeostasis and detoxification, protection against oxidative damage, and cell proliferation and apoptosis, they are also implicated in drug and radiotherapy resistance and several aspects of the carcinogenic process. Variable MT expression has been observed in different cancer types, reaching statistically significant correlation with clinicopathological parameters in some cases; nevertheless, MT expression as a marker of prognosis or as a predictor for the response to either chemotherapy or radiotherapy remains unclear. The present review examines the expression of MT in different human tumors in correlation with resistance to radiation therapy or chemotherapy and patients' final outcome. Detailed studies focused on the expression of MT isoforms and isotypes in different tumor types could elucidate the role of this group of proteins in patients' prognosis and resistance to treatment strategies.

Keywords: Metallothioneins; cancer; prognosis; drug resistance; radiotherapy resistance.

1. Introduction

Metallothioneins (MTs) are cysteine-rich proteins with a molecular weight of approximately 6000 Da, and have a specific binding capacity for metal ions (Margoshes and Vallee, 1957). In humans, MT proteins are encoded by a family of genes consisting of at least 10 functional MT isoforms, and are subdivided into four groups: MT-1, MT-2, MT-3, and MT-4. Human MT isoforms present tissue-specific expression patterns (Uchida *et al.*, 1991;

Imagawa *et al.*, 1995; Theocharis *et al.*, 2003). Regarding the functional metal-binding capacity, two forms of MT proteins exist: the metal-bound and the metal-free (apo-MT) ones (Pattanaik *et al.*, 1994). Although not all primary functions of the MTs have been defined with certainty, these proteins participate in different biochemical reactions occurring *in vivo* by virtue of their high cysteine content (Theocharis *et al.*, 2003).

MTs, due to their ability to bind metal ions, were initially considered as the mediators of cellular detoxification of metals (Moffatt and Denizeau, 1997; Jiang *et al.*, 1998). By binding biologically essential metals such as Zn and Cu, MTs also act as a reservoir for them, thus facilitating the reversible transfer of these ions to cellular macromolecules. Many cellular proteins, enzymes, DNA and RNA polymerases, and transcription factors require Zn for their biological activity, so MTs control their function by acting as either a Zn donor (MT) or a Zn acceptor (apo-MT). Apo-MT is known to inactivate the DNA-binding capacity of Zn finger transcription factors *in vitro* (Zeng *et al.*, 1991a), whereas Zn bound to MT restores their activity; this reversible activation of transcription factors by MTs is specific for Zn finger proteins (Zeng *et al.*, 1991b). It has also been shown that apo-MTs may activate certain metalloenzymes by abstracting the metal ions bound to a noncatalytic inhibitory site (Maret *et al.*, 1999).

Due to their high cysteine content, MTs can function as potent scavengers of reactive oxygen species, such as hydrogen peroxide (H_2O_2), superoxide (O_2^-), nitric oxide (NO), and most potent hydroxyl (OH) radicals, protecting cellular macromolecules from these highly reactive compounds which are constantly generated *in vivo* by various metabolic processes. Mechanisms underlying this function may include direct free radical interception, complexation of redox-sensitive transition metals, alteration of Zn homeostasis, or interaction with glutathione (GSH) (Zangger *et al.*, 2001). Thus, MTs are stress-inducible proteins with antioxidant attributes that may participate independently of or in conjunction with GSH to protect cells against injurious agents and stress stimuli (Lazo *et al.*, 1998; Kondoh *et al.*, 2001).

2. MTs and Cancer

Several lines of evidence suggest a role for MTs in cancer development, treatment resistance, and prognosis (Theocharis *et al.*, 2003; Theocharis *et al.*, 2004). The induction of MT synthesis by cytokines, hormones, and other cytotoxic agents is indicative of its involvement in cell proliferation and differentiation as well as in cellular defense mechanisms. The clinical significance of MT expression in various human tumors, associated with either processes related to carcinogenesis, histopathological variables, or resistance against radiation and chemotherapy, has been extensively reported (Theocharis *et al.*, 2004; Jasani and Schmid, 1997).

MT expression in tumor tissues has been mainly correlated with tumor cells' proliferative capacity or with the induction of apoptosis (Abdel-Magge and Agrawal, 1997; Kondo *et al.*, 1997). MT expression was considered as cell-cycle–dependent and was used as a marker of cell proliferation in certain tumor types, presenting preferential expression in cells in the S-phase (Cherian *et al.*, 1994). It is also known that perinuclear MT mRNA localization is important for the protective function of MT against DNA damage and apoptosis induced by external stress stimuli (Levadoux-Martin *et al.*, 2001).

One of the most important problems in cancer chemotherapy is acquired drug resistance, and a mechanism that possibly contributes to this phenomenon is the sequestration of alkylating agents by MT *in vivo*. However, data obtained from multiple sources indicate that the resistance to chemotherapy exhibited by different human tumor types cannot be explained by a single mechanism. Recent findings suggested that subcellular MT localization may dictate functionality and that MT may help to determine the threshold for apoptosis. In addition, modulation of MT expression might provide a strategy for altering cellular resistance to chemotherapeutic compounds and apoptosis (Shimoda *et al.*, 2003).

The protective effect of MT in radiation-induced injury has been documented both *in vivo* and *in vitro* (i.e. in cells after pre-induction of MT synthesis by various agents). This radioprotective effect may be due either to MT itself or to other cellular factors modulated by MT action (Cai *et al.*, 1999). Repeated doses of radiation exposure proved to be more effective than a large single one for inducing MT

Table 1 MT expression in relation to prognosis (survival rate) in patients with different types of malignancy.

Type of malignancy	Prognosis (survival)	Reference
Head and neck cancer		
Esophageal SCC	Yes	Hishikawa *et al.* (1997)
Gastrointestinal cancer		
Gastric cancer	Controversial[a]	Tuccari *et al.* (2000b); Theocharis *et al.* (2001); Monden *et al.* (1997); Janssen *et al.* (2002); Theocharis *et al.* (1997)
Colon cancer	Yes	Ofner (1994); Hishikawa *et al.* (2001); Sutoh *et al.* (2000)
Pancreatic cancer	Yes	Ohshio *et al.* (1996)
Lung cancer	Yes	Joseph *et al.* (2001)
Renal cancer	Yes	Tuzel *et al.* (2001); Mitropoulos *et al.* (2005)
Urinary bladder cancer	Controversial [a]	Bahnson *et al.* (1994); Siu *et al.* (1998)
Breast cancer	Controversial[a]	Goulding *et al.* (1995); Bay *et al.* (2006); Schmid *et al.* (1993); Fresno *et al.* (1993); Oyama *et al.* (1996); Haerslev *et al.* (1995); Zhang *et al.* (2000); Haerslev *et al.* (1994); Sens *et al.* (2001)
Ovarian cancer	Controversial[a]	Hengstler *et al.* (2001); Wrigley *et al.* (2000); Surowiak *et al.* (2005)
Neurological neoplasia		
Ependymoma	Yes	Korshunov *et al.* (1999)
Soft tissue and bone neoplasia		
Osteosarcoma	No	Shnyder *et al.* (1998); Uozaki *et al.* (1997)
Melanoma	Yes	Sugita *et al.* (2001); Goldmann *et al.* (2001)
Hematological malignancies		
Childhood ANLL	Yes	Sauerbrey *et al.* (1998)

[a] Controversial: value not always found, depending on the series.
Note: SCC, small cell carcinoma; ANLL, acute nonlymphocytic leukemia.

Table 2 MT expression in relation to chemotherapy or radiotherapy resistance in patients with different types of malignancy.

Type of malignancy	Chemotherapy resistance	Radiotherapy resistance	Reference
Head and neck cancer			
Esophageal SCC	Yes	Yes	Kishi *et al.* (2002)
Gastrointestinal cancer			
Gastric cancer	Yes		Suganuma *et al.* (2003)
Lung cancer	Yes		Volm *et al.* (2002); Matsumoto *et al.* (1997)
Urinary bladder cancer	Yes		Katoh *et al.* (1994); Wood *et al.* (1993); Wülfing *et al.* (2007); Bahnson *et al.* (1994); Siu *et al.* (1998); Siegsmund *et al.* (1999)
Breast cancer	No		Haerslev *et al.* (1994)
Ovarian cancer	Yes		Surowiak *et al.* (2005); Surowiak *et al.* (2007); Cheng *et al.* (2006)
Testicular cancer	Contradictory[a]		Chin *et al.* (1993); Koropatnick *et al.* (1995); Meijer *et al.* (2000); Eid *et al.* (1998)

[a]Contradictory: favorable or unfavorable value, depending on the series.
Note: SCC, small cell carcinoma.

synthesis. It is unclear whether radiation itself can induce MT synthesis or whether MT induction represents a secondary event due to the formation of free radicals and/or release of cytokines. The role of MT in radiation exposure may be important not only in radioprotection, but also in the treatment of certain cancer types with drugs and/or radiation. The presence of MT may be one of several factors involved in the radioresistance of tumor tissue and the adaptive response in low-dose ionizing radiation.

A great number of studies have examined the expression of MT in certain human malignancies in relation to clinicopathological parameters, tumor proliferative capacity, apoptosis, drug and radiation resistance, and patients' survival. In the present study, the available clinical data regarding the predictive and prognostic value of MT expression in human neoplasia will be referred (Tables 1 and 2). Additionally, evidence regarding MT implication in chemotherapy or radiation resistance obtained from *in vitro* experiments will also be included.

3. MT Expression in Human Neoplasia

3.1. *Head and Neck Cancer*

MT was overexpressed in the majority of small cell carcinomas (SCCs) of the esophagus resected from patients with a curative intent. MT positivity was more frequent in stage III and IV tumors, although no difference in clinicopathological variables was noted between MT-positive and MT-negative cases. Patients with MT-positive tumors who were treated with cisplatin presented a worse 5-year survival rate than those with MT-negative ones, implying that MT positivity is a poor prognostic factor (Hishikawa *et al.*, 1997). It was recently shown that patients with MT- and p53-positive esophageal SCCs presented poor response to treatment, suggesting that the combined evaluation of p53 and MT expression might be helpful to identify patients with advanced esophageal SCC who will benefit from pre-operative chemoradiation therapy (Kishi *et al.*, 2002).

In normal epithelium of the nasopharynx, only cells at the basal layer were positively stained for MT. MT immunoreactivity was also prominent in nasopharyngeal cancer (NPC) cases, presenting a nuclear pattern of staining, being inversely correlated with the apoptotic index. According to this study, it was suggested that MT overexpression in NPC may protect tumor cells from entering the apoptotic process and thereby from contributing to tumor expansion. The nuclear MT localization in NPC cells might also enhance radioresistance, since radiotherapy is known to eradicate tumor cells by free radical-induced apoptosis (Jayasura *et al.*, 2000).

3.2. *Gastrointestinal Cancer*

In gastric cancer, although MT immunoreactivity was noted in more than 50% of cases, no relationship was found between

MT immunoreactivity, tumor type, tumor grade, tumor stage, and patients' survival (Tuccari *et al.*, 2000b; Theocharis *et al.*, 2001; Monden *et al.*, 1997). On the other hand, according to another study, MT levels, measured by radioimmunoassay, in gastric tumors were associated with poorer patients' prognosis (Janssen *et al.*, 2002). Additionally, it was shown using cDNA microarray technology that MT gene upregulation was correlated with chemoresistance in gastric cancer cases (Suganuma *et al.*, 2003).

Variable MT immunostaining was noted in primary colon adenocarcinomas, lymph node metastases, and distant hepatic and omental metastases, although it was not always concordant with that reported in the corresponding primary tumor (Giuffre *et al.*, 1996). MT protein was expressed in all normal colorectal mucosa biopsies and in 74% of adenomatous polyps, while only 29% of colorectal adenocarcinoma cases presented focal MT positivity, suggesting that during the transformation of normal colorectal tissue to adenomatous polyps and adenocarcinoma, a progressive decrease of MT expression occurs (Theocharis *et al.*, 1997). MT expression was statistically significantly associated with the tumor stage and the lymph node involvement at the time of operation. Additionally, MT positivity in colon cancer was associated with a favorable clinical outcome, possibly indicating different biological behavior. Different survival rates were noted between patients with MT-positive tumors and those having MT-negative ones (Ofner *et al.*, 1994). It was also proposed that MT overexpression may occur as a direct consequence of somatic mutation, either *cis*-activating mutation(s) of the MT gene itself or *trans*-activating mutation(s) of other genes involved in controlling MT expression, implying an important mechanism of MT overexpression in neoplasia. Such mutation-induced aberrant MT expression may be involved in the acquisition of selective cellular growth or survival advantage during tumor progression (Jasani *et al.*, 1998). Very low or absence of MT positivity was noted in liver metastases of colorectal carcinoma (Stenram *et al.*, 1999; Hishikawa *et al.*, 2001). When MT was examined immunohistochemically in cases of primary and metastatic sites of colorectal adenocarcinoma, MT was detected in 21% of primary tumors, but none in metastatic tumors in the liver. It was also shown that MT positivity in primary tumors was statistically significantly correlated with lymph node involvement. Patients with MT-positive tumors presented worse survival rates when compared with MT-negative ones. MT expression in colorectal cancer could be a potential marker affecting lymph node metastasis and might be a predictor of poor prognosis, particularly in patients with synchronous liver metastasis (Hishikawa *et al.*, 2001). Patients with concurrent expression of drug-resistant-related proteins, including MT, presented worse prognosis, suggesting the potential use of MT in selecting adjuvant chemotherapy and in predicting prognosis more accurately (Sutoh *et al.*, 2000). In another study, it was shown that cytoplasmic and nuclear MT expression in rectal cancer cases presented similar expressions after tumor irradiation. MT was expressed more frequently in tumors from responders to radiotherapy than in those from nonresponders; while there was no correlation between MT expression and tumor stage, histology after irradiation, or survival. These findings contradict the hypothesis that MT overexpression at the end of chemotherapy is a marker related with radiation resistance (Bouzourene *et al.*, 2002). It is not yet clear if MT expression seems to indicate aggressive biological behavior in colorectal adenocarcinomas (Ioachim *et al.*, 1999b), although some studies imply that high MT expression in colorectal cancer patients is significantly associated with poor overall survival, independently of their clinicopathological features (Monden *et al.*, 1997; Theocharis *et al.*, 1997).

Although different studies referred MT expression in cases of liver and gallbladder neoplasia, no reference exists regarding its predictive and prognostic value (Theocharis *et al.*, 2004). In pancreatic adenocarcinoma, most of the examined cases were MT-negative. Nevertheless, in MT-positive cases, MT expression was related to histological grade, presence of metastasis, and poor prognosis (Ohshio *et al.*, 1996). Significant correlation between MT expression and *in vitro* resistance to cisplatin was also not noted (Ohshio *et al.*, 1996).

3.3. *Lung Cancer*

Immunohistochemically detected MT expression was enhanced in tumoral compared to microscopically normal lung tissue. In the normal lung, MT expression was affected by smoking habits (Koomagi *et al.*, 1996). It was also shown that MT expression was positively correlated with tumor volume and grade of differentiation (Mattern *et al.*, 2002). In another study, elevated MT expression was also evident in untreated squamous cell lung carcinoma (SCLC) and adenocarcinoma cases, but absent in small cell lung carcinoma ones (Theocharis *et al.*, 2002). The MT intensity of staining was statistically significantly different between SCLC and adenocarcinoma cases, while the pattern and extent of MT positivity in tumoral cells were not correlated with histopathological parameters such as tumor type and grade (Theocharis *et al.*, 2002). In another study, it was shown that MT expression was the only marker independently predicting short-term survival in patients with small cell lung carcinoma undergoing chemotherapy (Joseph *et al.*, 2001).

Drug-resistance proteins, proliferative factors, apoptotic factors, angiogenic factors, proto-oncogenes, and suppressor genes' products were evaluated mainly by immunohistochemistry in specimens of primary non-SCLC (NSCLC) and compared with the response of tumors to doxorubicin *in vitro*. The cluster analysis revealed three different resistance profiles, with upregulation of MT in the most frequent one (Volm *et al.*, 2002). Following chemotherapy, MT expression was increased conferring resistance in lung cancer, especially in NSCLC (Matsumoto *et al.*, 1997). No single mechanism could explain the drug resistance and the poor prognosis of patients with lung cancer. Different resistance-related proteins (including MT), cell cycle-related proteins, angiogenic factors, proto-oncogenes, and tumor suppressor genes have been expressed in lung carcinomas, but each one could not explain the tumor-resistant phenotype. A key future challenge involves determining the relative quantitative contributions of each of these mechanisms to overall resistance (Volm and Mattern, 1996; Rosell and Felip, 2001).

A short-term *in vitro* test that measures changes in the rate of radioactive nucleic acid precursors incorporating into tumor cells after the addition of doxorubicin was used to determine the response of NSCLC cases to doxorubicin, in correlation with various cellular parameters including MT immunohistochemical expression which was statistically significantly correlated with the results obtained by the short-term *in vitro* test (Volm and Rittgen, 2000).

3.4. *Renal Cancer*

In renal cell carcinoma (RCC), sarcomatoid tumors presented significantly higher MT expression than that found in clear cell, papillary, granular, or chromophobe tumors, suggesting that MT overexpression seems to be associated with malignant behavior and poor prognosis in RCC (Tuzel *et al.*, 2001). An inverse-correlation MT positivity (Tuzel *et al.*, 2001) or MT overexpression (Mitropoulos *et al.*, 2005) with patients' survival was also noted.

3.5. *Urinary Bladder Cancer*

The relationship between MT expression and chemotherapy with certain anticancer drugs was studied in cisplatin- and adriamycin-resistant cell lines. MT expression in drug-resistant cell lines was increased, contrary to other proteins involved in drug resistance that remained unchanged (Saika *et al.*, 1994). Additionally, MT-overexpressing tumors presented resistance to cisplatin-based chemotherapy, suggesting that MT immunostaining might be a useful predictor of tumors' response to cisplatin therapy (Katoh *et al.*, 1994; Wood *et al.*, 1993; Wülfing *et al.*, 2007). Prechemotherapy bladder tumor tissues were retrospectively analyzed in patients treated with neoadjuvant cisplatin, methotrexate, and vinblastine. Patients with nondetectable MT staining were statistically more likely to sustain a complete pathological response postchemotherapy than those with MT-positive tumors. Nevertheless, no relationship between MT positivity and patients' survival was apparent (Bahnson *et al.*, 1994).

MT overexpression in patients with metastatic urothelial transitional cell carcinoma (TCC) was associated with a shorter survival postchemotherapy, possibly due to cisplatin resistance (Siu *et al.*, 1998). In more recent studies on cisplatin-resistant urinary bladder cell lines, an enhanced MT-2 but not MT-1 expression was noted, implying MT-2 for cisplatin resistance (Siegsmund *et al.*, 1999).

The effect of MT synthesis inhibition on drug resistance has been previously examined in nude mice inoculated with human bladder tumor. Tumor-bearing mice pretreated with Zn salts presented increased MT content (in both normal and tumoral tissues) and reduced cisplatin, adriamycin, and melphalan antitumor activity. Injection of the cystathionase inhibitor, propargylglycine, decreased MT induction by Zn in the tumor, resulting in diminished drug resistance (Satoh *et al.*, 1994).

3.6. *Breast Cancer*

A significant association was observed between MT expression and either tumor type or local recurrence, as decreased MT expression in breast tumors was proved to be a good prognostic factor (Goulding *et al.*, 1995; Bay *et al.*, 2006). MT immunohistochemical expression was variably found in the invasive components of pT_1 and pT_2 ductal carcinoma cases, whereas all invasive lobular carcinomas examined presented weak or negative staining. MT-positive invasive ductal carcinomas developed more frequently metastases during follow-up with poorer prognostic outcomes than the MT-negative ones, particularly the pT_2 invasive ductal breast carcinoma cases (Schmid *et al.*, 1993). MT positivity proved to be predictive of worse prognosis in the subgroup of lymph node and estrogen receptor (ER)-negative patients. Consequently, according to those data, MT staining may be a useful marker of a less differentiated and more agressive breast cancer phenotype (Fresno *et al.*, 1993). Vazquez-Ramirez *et al.* (2000) reported that MT positivity in breast cancer cases was associated with metastatic potential, but not with patients' survival or recurrences. The overall survival rate in MT-positive cases was worse than that found in MT-negative ones, being nonstatistically significant.

It has been proposed that cancer cell dedifferentiation may play a role in MT protein induction (Oyama *et al.*, 1996). MT expression in primary breast carcinoma cases was a poor predictor of overall patients' survival by univariate analysis; on the contrary, when multivariate analysis was applied, MT positivity failed to be of any prognostic significance (Haerslev *et al.*, 1995). Low and high MT expression affected the overall and disease-free survival in breast cancer patients. The assessment of MT expression, according to these studies, appeared to provide prognostic information and might have important implications for understanding breast cancer development (Satoh *et al.*, 1994; Zhang *et al.*, 2000). MT expression was examined immunohistochemically in breast carcinoma primary sites, and their concomitant lymph node metastases showed statistically significant correlation with poor overall survival. In one fourth of the patients, the lymph node metastases presented had increased MT expression compared to that found at the primary site. These patients had a poorer, but not statistically significantly different, survival. Additionally, MT expression was not correlated to chemotherapy or radiation therapy (Haerslev *et al.*, 1994). More recently, it was also shown using immunohistochemistry that the MT-3 protein was overexpressed in the cytoplasm in the majority of breast cancer cases, being associated with poor prognosis (Sens *et al.*, 2001).

According to the studies mentioned above, the correlation of MT expression with prognosis cannot be postulated with certainty, although it has been proposed that MT expression could be characteristic of the early phase of breast carcinogenesis (Ioachim *et al.*, 1999a; Head *et al.*, 2002).

3.7. *Ovarian Cancer*

MT content in ovarian tumors was considered higher than that found in either normal ovaries or benign ovarian surface epithelial tumors, suggesting a possible role of MT in ovarian tumorigenesis. The MT content of tumors obtained from patients who had completed chemotherapy was not different to that from untreated ones. Additionally, the levels of MT were not influenced by age, stage, histology, tumor differentiation state, or response to therapy in ovarian carcinoma patients (Murphy *et al.*, 1991; McGown *et al.*, 1994; Germain *et al.*, 1996). Germain *et al.* (1996) referred a weak concordance between MT expression in the tumor before chemotherapy and postchemotherapy. The percentage of MT positively stained cells was significantly higher in borderline than in benign tumors, and was more elevated in malignant mucinous ovarian ones (Tan *et al.*, 1999b). McCluggage *et al.* (2002) reported a similar MT positivity in different types of ovarian malignant tumors (endometrioid, mucinous, serous neoplasms). In mucinous and serous ovarian tumors, an increasing percentage of MT expression was observed during the progression of malignancy (Tan *et al.*, 1999a; McCluggage *et al.*, 2002; Tan *et al.*, 1999b; Hengstler *et al.*, 2001). MT expression either prior to chemotherapy or postchemotherapy was not associated with the survival of ovarian epithelial carcinoma patients, although an increasing trend for MT expression was found postchemotherapy (Wrigley *et al.*, 2000).

More recent studies verified that MT expression is correlated with unfavorable prognosis and drug resistance in cisplatin-treated ovarian cancer patients (Surowiak *et al.*, 2005; Surowiak *et al.*, 2007). Hengstel *et al.* (2002) referred that MT expression was negatively associated with the survival rate when all patients with primary ovarian carcinoma were analyzed. It was firstly reported that MT elevation may be one of the mechanisms responsible for cisplatin resistance in ovarian carcinoma, although *in vitro* selection of resistant-to-cisplatin ovarian carcinoma cell clones did not verify this effect (Andrews *et al.*, 1987). More recently, it was shown using cDNA microarray technology that MT gene expression could predict chemotherapy resistance in ovarian cancer patients (Cheng *et al.*, 2006). After this report, many *in vitro* and clinical studies examined the participation of MT in ovarian cancer drug resistance, supporting evidence that MT expression is a rather weak candidate in ovarian cancer chemoresistance *in vivo*. On the other hand, ovarian cancer represents a highly heterogeneous neoplasm, and the expression of markers of chemoresistance might reflect this heterogeneity (Cheng *et al.*, 2006; Helleman *et al.*, 2006). In human ovarian cancer cell lines, the acquisition of cisplatin resistance appeared to be related mainly to gamma-glutamylcysteine synthetase (gamma-GCS), MT, and topoisomerase-II genes, and partly to that of topoisomerase-I and GST-pi ones (Kikuchi *et al.*, 1997).

3.8. *Endometrial and Cervical Cancer*

In endometrial adenocarcinoma, including the papillary serous type which presents aggressive behavior, the elevated MT content observed was associated with high tumor grade and stage, but failed to be an independent prognostic factor (McCluggage *et al.*, 1999). MT expression was also examined in the progression of cervical intraepithelial neoplasia (CIN) to cervical carcinoma. MT positivity was prominent in the basal cervical layer cells of normal and CIN 1 patients; while during the progression of CIN 2 to CIN 3, MT positivity was enhanced (McCluggage *et al.*, 1998; Theocharis *et al.*, 1999). In invasive squamous cell carcinoma cases, intense cytoplasmic and nuclear MT expression was observed in all cells (McCluggage *et al.*, 1998) or in the vast majority of them (Theocharis *et al.*, 1999).

3.9. *Prostate Cancer*

In prostatic glandular epithelia, cytoplasmic and nuclear MT expression strong in the peripheral prostate zones and weak in the central prostate zones was referred (Suzuki *et al.*, 1991). In prostate cancer, the positive correlation of MT expression with the Gleason score suggested its participation in prostate oncogenesis (Moussa *et al.*, 1997). In prostate cancer and prostatic intraepithelial neoplasia

(PIN) lesions, MT-3 was overexpressed, being highly correlated with the Gleason score.

In PC-3 prostate cancer cells, MT-3 overexpression proved to influence the growth rate and the chemotherapeutic drug resistance (Dutta *et al.*, 2002). In four human hormone-independent prostatic carcinoma cell lines, the amount of MT, its isotype expression, and its subcellular distribution were examined. Both PC-3 and DU-145 cells were thiol-rich with similar MT and GSH levels, while HPC36M and PC-3 MA2 cells were thiol-poor with lower MT and GSH levels. All of these prostatic cell lines expressed the MT-2A isoform at a basal level, and DU-145 cells also expressed MT-1E mRNA constitutively. It was shown that the cytoplasmic and nuclear MT subcellular distribution was cell type-specific. Resistance to cisplatin treatment was correlated with nuclear MT content, suggesting that the subcellular localization of MT was functionally important in cellular protection against cisplatin in this type of tumoral cell (Kondo *et al.*, 1995). In a more recent study, it was shown that Zn treatment induced MT expression in LNCaP and C4-2 prostate cancer cells that was associated with resistance to cisplatin chemotherapy and radiotherapy in these cells (Smith *et al.*, 2006).

3.10. *Testicular Cancer*

Seminomas were either weakly or negatively stained for MT, regardless of the clinical stage, while most nonseminomas were MT intensively stained. In the more advanced-stage nonseminoma cases, increased MT expression was noted, suggesting a possible role of MT in cisplatin resistance observed in germ cell tumors (Chin *et al.*, 1993). In a variety of human germ tumor cell lines, the increased MT expression observed was concomitant with increased cisplatin resistance (Koropatnick *et al.*, 1995; Meijer *et al.*, 2000). On the contrary, in patients with germ cell testicular tumors, high MT immunoreactivity was found to predict a better response rate to chemotherapy, opposing the hypothesis that MT overexpression contributes to cisplatin resistance in this tumor type (Eid *et al.*, 1998). Additionally, it is known that metastatic testicular germ cell tumors were cured in most cases using cisplatin-based combination chemotherapy (Masters *et al.*, 1996). No difference in MT expression was found between responding and nonresponding patients, so MT could not be used to predict response to chemotherapy (Maijer *et al.*, 2000).

3.11. *Neurological Neoplasia*

MT expression was found in a large percentage of meningiomas, especially in atypical and malignant ones. Additionally, MTs and other drug resistance-related proteins were expressed in normal and neoplastic vessels, which may confer to impaired penetration of therapeutic agents through the blood-brain barrier and blood-tumor barrier. No difference was found in MT expression between primary, recurrent, or irradiated tumors (Tews *et al.*, 2001).

Malignant gliomas proved to be largely resistant to current chemotherapeutic strategies, often displaying a multidrug-resistant phenotype. Nevertheless, MT immunohistochemical expression did not differ between primary and secondary glioblastomas or reveal any correlation to precursor or recurrent tumors (Tews *et al.*, 2000). In glioblastomas, the MT extent of positivity presented a statistically significant inverse relationship with p53 expression (Maier *et al.*, 1997). It has been referred that MT might play a significant role in astrocytic neoplasm growth, with an acquired enhanced ability of MT expression as the malignant tumor potential increases (Hiura *et al.*, 1998). Intracranial ependymomas are the third most common primary brain tumor in children. Increased MT expression was noted in low-grade ependymomas, while the progression-free survival time was found to be significantly shorter in MT-immunoreactive tumors of both low and high grades (Korshunov *et al.*, 1999). It was also noted from experimental data obtained by the use of the cisplatin-resistant human neuroblastoma cell line BM1R2 that MT could exert crucial roles on cisplatin resistance (Yasuno *et al.*, 1999).

3.12. *Soft Tissue and Bone Neoplasia*

In osteosarcoma patients, no significant difference was found in the survival rate between MT-expressing and MT-nonexpressing tumors, suggesting that lack of MT expression did not confer any advantage in terms of patients' survival (Shnyder *et al.*, 1998; Uozaki *et al.*, 1997).

3.13. *Dermal Cancer and Melanoma*

The differential pattern of MT expression in different dermal entities, including melanoma and nonmelanoma malignancies, suggested participation of this protein in the pathogenesis of these lesions (Aroni *et al.*, 1998). MT expression was correlated with neoplastic transformation in malignant melanoma cases, being significantly associated with progressive disease and representing a useful prognostic indicator (Zelger *et al.*, 1993; Weinlich *et al.*, 2007; Weinlich and Zelger, 2007). MT expression was also increased in metastatic compared with nonmetastatic states of melanoma cases (Goldmann *et al.*, 1998). In more recent studies, MT expression was strongly correlated with the level of tumor invasion and thickness in melanoma cases. Additionally, MT-positive melanoma cases presented worse survival rates than MT-negative ones (Sugita *et al.*, 2001; Goldmann *et al.*, 2001). MT immunoreactivity was found in most ocular melanoma cases. No differences in MT immunostaining were found in relation to age, sex, tumor size, histotype, or amount of pigment. Univariate analysis of survival data showed no prognostic significance of MT expression in this type of neoplasia (Tuccari *et al.*, 2000a).

3.14. *Hematological Malignancies*

Alterations of MT genes have been reported in hematologic malignancies. Somatic cell hybridization has shown that most of the MT genes, MT-1A, MT-1B, and MT-2A, were localized to band q22 of human chromosome 16. This chromosomal band was also a breakpoint in two specific rearrangements, the inv(16)(p13q22) and t(16;16)(p13;q22) ones, found in a subgroup of patients with acute myelomonocytic leukemia (AMML) (Le Beau *et al.*, 1985). MT expression was independent of clinical prognostic factors such as age, sex, immunological subtype, and initial blast cell count in patients with initial or relapsed acute lymphoblastic leukemia (Sauerbrey *et al.*, 1994). The expression of the resistance-related proteins Pgp, GST-pi, topoisomerase-II, TS, and MT was investigated in leukemic cells of 19 children with newly diagnosed acute nonlymphoblastic leukemia (ANLL). Different percentages of positivity for the examined proteins were noted. MT was expressed in leukemic cells of 68% of cases with newly diagnosted ANLL. Patients who developed relapse, exerting poorer prognosis, expressed frequently more than two resistance-related proteins (including MT) compared to patients who remained in remission (Sauerbrey *et al.*, 1998). In children with acute leukemia at diagnosis and during treatment, MT expression *in vivo* by tumoral cells constitutes a cellular protective mechanism preventing chemotherapy-induced apoptosis (Tsangaris *et al.*, 2000).

4. Perspectives

Alterations of basal MT levels in organs and tissues have been observed in different stages of human development. In tumor cell pathobiology, the role of MT becomes evident, as certain tumors express high levels of MT while others are essentially devoid of this protein. The knowledge accumulated from the literature on MT expression in relation to histopathological parameters, as well as drug or radiotherapy resistance, patients' survival, and prognosis, is contradictory among different cancer types. This is mainly due either to the specific tissue or cell type involved in neoplasia, or to the methods applied using antibodies which are unable to distinguish between specific MT isoforms (i.e. metal-bound and metal-free forms of the protein). Consequently, the functional significance of MT expression in human tumors remains to be determined, in relation to specific MT isoforms and/or apo-MT, which are capable of differentially influencing the growth and survival pattern of tumor cells in specific tissue types.

The purpose of the expression of different MT isoforms in certain tissues is not clearly evident. It would be challenging to identify and characterize the tissue-specific factors responsible for the expression of genes encoding MT isoforms. Analysis of the expression of different MT isoforms in human tumors in relation to their metal ion composition, cellular antioxidant defense mechanisms, and genetic alterations in specific genes related to MT, using appropriately sensitive and discriminant methods, should elucidate the participation of this group of proteins in carcinogenesis. Further studies on differential MT isoform and isotype expression in different tumor types could clarify the clinical significance of this family of proteins in patients' prognosis and monitoring of treatment.

References

Abdel-Mageed A, Agrawal KC. Antisense down-regulation of metallothionein induces growth arrest and apoptosis in human breast carcinoma cells. *Cancer Gene Ther* 1997; **4**:199–207.

Andrews PA, Murphy MP, Howell SB. Metallothionein-mediated cisplatin resistance in human ovarian carcinoma cells. *Cancer Chemother Pharmacol* 1987; **19**:149–154.

Aroni K, Theocharis S, Tsagroni E, Davaris P. Immunohistochemical expression of metallothionein in different dermal lesions. *Arch Anat Cytol Pathol Clin Exp Pathol* 1998; **46**:344.

Bahnson RR, Becich M, Ernstoff MS, *et al.* Absence of immunohistochemical metallothionein staining in bladder tumor specimens predicts response to neoadjuvant cisplatin, methotrexate and vinblastine chemotherapy. *J Urol* 1994; **152**:2272–2275.

Bay B-H, Jin R, Huang J, Tan P-H. Metallothionein as a prognostic biomarker in breast cancer. *Exp Biol Med* 2006; **231**:1516–1521.

Bouzourene H, Chaubert P, Gebhard S, *et al.* Role of metallothionein in irradiated human rectal carcinoma. *Cancer* 2002; **95**:1003–1008.

Cai L, Satoh M, Tohyama C, Cherian MG. Metallothionein in radiation exposure: Its induction and protective role. *Toxicology* 1999; **132**:85–98.

Cheng TC, Manorek G, Samimi G, *et al.* Identification of genes whose expression is associated with cisplatin resistance in human ovarian carcinoma cells. *Cancer Chemother Pharmacol* 2006; **58**:384–385.

Cherian G, Howell SB, Imura N, *et al.* Role of metallothionein in carcinogenesis. *Toxicol Appl Pharmacol* 1994; **126**:1–5.

Chin JL, Banerjee D, Kadhim SA, *et al.* Metallothionein in testicular germ cell tumors and drug resistance. *Cancer* 1993; **72**:3029–3035.

Dutta R, Sens DA, Somji S, *et al.* Metallothionein isoform 3 expression inhibits cell growth and increases drug resistance of PC-3 prostate cancer cells. *Prostate* 2002; **52**:89–97.

Eid H, Geczi L, Bodrogi I, *et al.* Do metallothioneins affect the response to treatment in testis cancers? *J Cancer Res Clin Oncol* 1998; **124**:31–36.

Fresno M, Wu W, Rodriguez JM, Nadji M. Localization of metallothionein in breast carcinomas. An immunohistochemical study. *Virchows Arch A Pathol Anat Histopathol* 1993; **423**:215–219.

Germain I, Tetu B, Brisson J, *et al.* Markers of chemoresistance in ovarian carcinomas: An immunohistochemical study of 86 cases. *Int J Gynecol Pathol* 1996; **15**:54–62.

Giuffre G, Barresi G, Sturniolo GC, *et al.* Immunohistochemical expression of metallothionein in normal human colorectal mucosa, in adenomas and in adenocarcinomas and their associated metastases. *Histopathology* 1996; **29**:347–354.

Goldmann T, Moorkamp A, Wiedorn KH, *et al.* The prognostic value of the expression of collagenase IV, cathepsin D and metallothionein in squamous cell carcinomas of the skin determined by immunohistochemistry. *Arch Dermatol Res* 2001; **293**:115–120.

Goldmann T, Ribbert D, Suter L, *et al.* Tumor characteristics involved in the metastatic behaviour as an improvement in primary cutaneous melanoma prognostics. *J Exp Clin Cancer Res* 1998; **17**:483–489.

Goulding H, Jasani B, Pereira H, *et al.* Metallothionein expression in human breast cancer. *Br J Cancer* 1995; **72**:968–972.

Haerslev T, Jacobsen GK, Zedeler K. The prognostic significance of immunohistochemically detectable metallothionein in primary breast carcinomas. *APMIS* 1995; **103**:279–285.

Haerslev T, Jacobsen K, Nedergaard L, Zedeler K. Immunohistochemical detection of metallothionein in primary breast carcinomas and their axillary lymph node metastases. *Pathol Res Pract* 1994; **190**:675–681.

Head JF, Elliott RL, Yang DC. Gene targets of antisense therapies in breast cancer. *Expert Opin Ther Targets* 2002; **6**:375–385.

Helleman J, Burger H, Hamelers IH, *et al.* Impaired cisplatin influx in an A2780 mutant cell line: Evidence for a putative, *cis*-configuration-specific platinum influx transporter. *Cancer Biol Ther* 2006; **5**:943–949.

Hengstler JG, Pilch H, Schmidt M, *et al.* Metallothionein expression in ovarian cancer in relation to histopathological parameters and molecular markers of prognosis. *Int J Cancer* 2001; **95**:121–127.

Hishikawa Y, Abe S, Kinugasa S, *et al.* Overexpression of metallothionein correlates with chemoresistance to cisplatin and prognosis in esophageal cancer. *Oncology* 1997; **54**:342–347.

Hishikawa Y, Kohno H, Ueda S, *et al.* Expression of metallothionein in colorectal cancers and synchronous liver metastases. *Oncology* 2001; **61**:162–167.

Hiura T, Khalid H, Yamashita H, *et al.* Immunohistochemical analysis of metallothionein in astrocytic tumors in relation to tumor grade, proliferative potential, and survival. *Cancer* 1998; **83**:2361–2369.

Imagawa M, Ishikawa Y, Shimano H, *et al.* CTG triplet repeat in mouse growth inhibitory factor/metallothionein III gene promoter represses the transcriptional activity of the heterologous promoters. *J Biol Chem* 1995; **270**:20898–20900.

Ioachim E, Kamina S, Demou A, *et al.* Immunohistochemical localization of metallothionein in human breast cancer in comparison with cathepsin D, stromelysin-1, CD44, extracellular matrix components, p53, Rb, C-erbB-2, EGFR, steroid receptor content and proliferation. *Anticancer Res* 1999a; **19**:2133–2139.

Ioachim EE, Goussia AC, Agnantis NJ, *et al.* Prognostic evaluation of metallothionein expression in human colorectal neoplasms. *J Clin Pathol* 1999b; **52**:876–879.

Janssen AM, van Duijn W, Kubben FJ, *et al.* Prognostic significance of metallothionein in human gastrointestinal cancer. *Clin Cancer Res* 2002; **8**:1889–1896.

Jasani B, Campbell F, Navabi H, *et al.* Clonal overexpression of metallothionein is induced by somatic mutation in morphologically normal colonic mucosa. *J Pathol* 1998; **184**:144–147.

Jasani B, Schmid KW. Significance of metallothionein overexpression in human tumours. *Histopathology* 1997; **31**:211–214.

Jayasura A, Bay BH, Yap WM, Tan NG. Correlation of metallothionein expression with apoptosis in nasopharyngeal carcinoma. *Br J Cancer* 2000; **82**:1198–1203.

Jiang LJ, Maret W, Vallee BL. The glutathione redox couple modulates zinc transfer from metallothionein to zinc-depleted sorbitol dehydrogenase. *Proc Natl Acad Sci USA* 1998; **95**:3483–3488.

Joseph MG, Banerjee D, Kocha W, *et al.* Metallothionein expression in patients with small cell carcinoma of the lung: Correlation with other molecular markers and clinical outcome. *Cancer* 2001; **92**:836–842.

Katoh S, Naito S, Sakamoto N, *et al.* Metallothionein expression is correlated with cisplatin resistance in transitional cell carcinoma of the urinary tract. *J Urol* 1994; **152**:1267–1270.

Kikuchi Y, Hirata J, Yamamoto K, *et al.* Altered expression of gamma-glutamylcysteine synthetase, metallothionein and topoisomerase I or II during acquisition of drug resistance to cisplatin in human ovarian cancer cells. *Jpn J Cancer Res* 1997; **88**:213–217.

Kishi K, Doki Y, Miyata H, *et al.* Prediction of response to chemoradiation and prognosis in oesophageal squamous cancer. *Br J Surg* 2002; **89**:597–603.

Kondo Y, Kuo SM, Watkins SC, Lazo JS. Metallothionein localization and cisplatin resistance in human hormone-independent prostatic tumor cell lines. *Cancer Res* 1995; **55**:474–477.

Kondo Y, Rusnak JM, Hoyt DG, *et al.* Enhanced apoptosis in metallothionein null cells. *Mol Pharmacol* 1997; **52**:195–201.

Kondoh M, Inoue Y, Atagi S, *et al.* Specific induction of metallothionein synthesis by mitochondrial oxidative stress. *Life Sci* 2001; **69**:2137–2146.

Koomagi R, Stammler G, Manegold C, *et al.* Expression of resistance related proteins in tumoral and peritumoral tissues of patients with lung cancer. *Cancer Lett* 1996; **110**:129–136.

Koropatnick J, Kloth DM, Kadhim S, *et al.* Metallothionein expression and resistance to cisplatin in a human germ cell tumor cell line. *J Pharmacol Exp Ther* 1995; **275**:1681–1687.

Korshunov A, Sycheva R, Timirgaz V, Golanov A. Prognostic value of immunoexpression of the chemoresistance-related proteins in ependymomas: An analysis of 76 cases. *J Neurooncol* 1999; **45**:219–227.

Lazo JS, Kuo SM, Woo ES, Pitt BR. The protein thiol metallothionein as an antioxidant and protectant against antineoplastic drugs. *Chem Biol Interact* 1998; **111–112**:255–262.

Le Beau MM, Diaz MO, Karin M, Rowley JD. Metallothionein gene cluster is split by chromosome 16 rearrangement in myelomonocytic leukaemia. *Nature* 1985; **313**:709–711.

Levadoux-Martin M, Hesketh JE, Beattie JH, Wallace HM. Influence of metallothionein-1 localization on its function. *Biochem J* 2001; **355**:473–479.

Maier H, Jones C, Jasani B, *et al.* Metallothionein overexpression in human brain tumours. *Acta Neuropathol* 1997; **94**:599–604.

Maret W, Jacob C, Vallee BL, Fischer EH. Inhibitory sites in enzymes: Zinc removal and reactivation by thionein. *Proc Natl Acad Sci USA* 1999; **96**:1936–1940.

Margoshes M, Vallee BC. A cadmium binding protein for equine kidney cortex. *J Am Chem Soc* 1957; **79**:4813–4814.

Masters JR, Thomas R, Hall AG, *et al.* Sensitivity of testis tumour cells to chemotherapeutic drugs: Role of detoxifying pathways. *Eur J Cancer* 1996; **32A**:1248–1253.

Matsumoto Y, Oka M, Sakamoto A, *et al.* Enhanced expression of metallothionein in human non-small-cell lung carcinomas following chemotherapy. *Anticancer Res* 1997; **17**:3777–3780.

Mattern J, Koomagi R, Volm M. Expression of drug resistance gene products during progression of lung carcinomas. *Oncol Rep* 2002; **9**:1181–1184.

McCluggage WG, Maxwell P, Bharucha H. Immunohistochemical detection of metallothionein and MIB1 in uterine cervical squamous lesions. *Int J Gynecol Pathol* 1998; **17**:29–35.

McCluggage WG, Maxwell P, Hamilton PW, Jasani B. High metallothionein expression is associated with features predictive of aggressive behaviour in endometrial carcinoma. *Histopathology* 1999; **34**:51–55.

McCluggage WG, Strand K, Abdulkadir A. Immunohistochemical localization of metallothionein in benign and malignant epithelial ovarian tumors. *Int J Gynecol Cancer* 2002; **12**:62–65.

McGown AT, Murphy D, Swindell R, *et al.* Correlation of drug resistance-associated parameters in ovarian tumor biopsies. *Int J Gynecol Cancer* 1994; **4**:241–249.

Meijer C, Timmer A, De Vries EG, *et al.* Role of metallothionein in cisplatin sensitivity of germ-cell tumours. *Int J Cancer* 2000; **85**:777–781.

Mitropoulos D, Kyroudi-Voulgari A, Theocharis S, *et al.* Prognostic significance of metallothionein expression in renal cell carcinoma. *World J Surg Oncol* 2005; **3**:5.

Moffatt P, Denizeau F. Metallothionein in physiological and physiopathological processes. *Drug Metab Rev* 1997; **29**:261–307.

Monden N, Abe S, Sutoh I, *et al.* Prognostic significance of the expression of metallothionein, glutathione-S-transferase-pi, and P-glycoprotein in curatively resected gastric cancer. *Oncology* 1997; **54**:391–399.

Moussa M, Kloth D, Peers G, *et al.* Metallothionein expression in prostatic carcinoma: Correlation with Gleason grade, pathologic stage, DNA content and serum levels of prostate-specific antigen. *Clin Invest Med* 1997; **20**:371–380.

Murphy D, McGown AT, Crowther D, *et al.* Metallothionein levels in ovarian tumors before and after chemotherapy. *Br J Cancer* 1991; **63**:711–714.

Ofner D, Maier H, Riedmann B, *et al.* Immunohistochemical metallothionein expression in colorectal adenocarcinoma: Correlation with tumour stage and patient survival. *Virchows Arch* 1994; **425**:491–497.

Ohshio G, Imamura T, Okada N, *et al.* Immunohistochemical study of metallothionein in pancreatic carcinomas. *J Cancer Res Clin Oncol* 1996; **122**:351–355.

Oyama T, Take H, Hikino T, *et al.* Immunohistochemical expression of metallothionein in invasive breast cancer in relation to proliferative activity, histology and prognosis. *Oncology* 1996; **53**:12–17.

Pattanaik A, Shaw CF 3rd, Petering DH, *et al.* Basal metallothionein in tumors: Widespread presence of apoprotein. *J Inorg Biochem* 1994; **54**:91–105.

Rosell R, Felip E. Predicting response to paclitaxel/carboplatin-based therapy in non-small cell lung cancer. *Semin Oncol* 2001; **28**:37–44.

Saika T, Tsushima T, Ochi J, *et al.* Over-expression of metallothionein and drug-resistance in bladder cancer. *Int J Urol* 1994; **1**:135–139.

Satoh M, Cherian MG, Imura N, Shimizu H. Modulation of resistance to anti-cancer drugs by inhibition of metallothionein synthesis. *Cancer Res* 1994; **54**: 5255–5257.

Sauerbrey A, Zintl F, Hermann J, Volm M. Multiple resistance mechanisms in acute non-lymphoblastic leukemia (ANLL). *Anticancer Res* 1998; **18**:1231–1236.

Sauerbrey A, Zintl F, Volm M. Expression of metallothionein in initial and relapsed childhood acute lymphoblastic leukemia. *Ann Hematol* 1994; **69**:111–115.

Schmid KW, Ellis IO, Gee JM, *et al.* Presence and possible significance of immuno-histochemically demonstrable metallothionein over-expression in primary invasive ductal carcinoma of the breast. *Virchows Arch A Pathol Anat Histopathol* 1993; **422**:153–159.

Sens MA, Somji S, Garrett SH, *et al.* Metallothionein isoform 3 overexpression is associated with breast cancers having a poor prognosis. *Am J Pathol* 2001; **159**:21–26.

Shimoda R, Achanzar WE, Qu W, *et al.* Metallothionein is a potential negative regulator of apoptosis. *Toxicol Sci* 2003; **73**:294–300.

Shnyder SD, Hayes AJ, Pringle J, Archer CW. P-glycoprotein and metallothionein expression and resistance to chemotherapy in osteosarcoma. *Br J Cancer* 1998; **78**:757–759.

Siegsmund MJ, Marx C, Seemann O, *et al.* Cisplatin-resistant bladder carcinoma cells: Enhanced expression of metallothioneins. *Urol Res* 1999; **27**:157–163.

Siu LL, Banerjee D, Khurana RJ, *et al.* The prognostic role of p53, metallothionein, P-glycoprotein, and MIB-1 in muscle-invasive urothelial transitional cell carcinoma. *Clin Cancer Res* 1998; **4**:559–565.

Smith DJ, Jaggi M, Zhang W, *et al.* Metallothionein and resistance to cisplatin and radiation in prostate cancer. *Urology* 2006; **67**:1341–1347.

Stenram U, Ohlsson B, Tranberg KG. Immunohistochemical expression of metallothionein in resected hepatic primary tumors and colorectal carcinoma metastases. *APMIS* 1999; **107**:420–424.

Suganuma K, Kubota T, Saikawa Y, *et al.* Possible chemoresistance-related genes for gastric cancer detected by cDNA microarray. *Cancer Sci* 2003; **94**:355–359.

Sugita K, Yamamoto O, Asahi M. Immunohistochemical analysis of metallothionein expression in malignant melanoma in Japanese patients. *Am J Dermatopathol* 2001; **23**:29–35.

Surowiak P, Materna V, Kaplenko I, *et al.* Augmented expression of metallothionein and glutathione-S-transferase pi as unfavourable prognostic factors in cisplatin-treated ovarian cancer patients. *Virchows Arch* 2005; **447**:626–633.

Surowiak P, Materna V, Maciejczyk A, *et al.* Nuclear metallothionein expression correlates with cisplatin resistance of ovarian cancer cells and poor clinical outcome. *Virchows Arch* 2007; **450**:279–285.

Sutoh I, Kohno H, Nakashima Y, *et al.* Concurrent expressions of metallothionein, glutathione S-transferase-pi, and P-glycoprotein in colorectal cancers. *Dis Colon Rectum* 2000; **43**:221–232.

Suzuki T, Umeyama T, Ohma C, *et al.* Immunohistochemical study of metallothionein in normal and benign prostatic hyperplasia of human prostate. *Prostate* 1991; **19**:35–42.

Tan Y, Sinniah R, Bay BH, Singh G. Expression of metallothionein and nuclear size in discrimination of malignancy in mucinous ovarian tumors. *Int J Gynecol Pathol* 1999a; **18**:344–350.

Tan Y, Sinniah R, Bay BH, Singh G. Metallothionein expression and nuclear size in benign, borderline, and malignant serous ovarian tumours. *J Pathol* 1999b; **189**:60–65.

Tews DS, Fleissner C, Tiziani B, Gaumann AK. Intrinsic expression of drug resistance-associated factors in meningiomas. *Appl Immunohistochem Mol Morphol* 2001; **9**:242–249.

Tews DS, Nissen A, Kulgen C, Gaumann AK. Drug resistance-associated factors in primary and secondary glioblastomas and their precursor tumors. *J Neurooncol* 2000; **50**:227–237.

Theocharis S, Karkantaris C, Philipides T, *et al.* Expression of metallothionein in lung carcinoma: Correlation with histological type and grade. *Histopathology* 2002; **40**:143–151.

Theocharis S, Kounelis S, Agapitos E, *et al.* Expression of metallothionein in malignant transformation of cervical neoplasia: An immunohistochemical study. *Virchows Arch* 1999; **435**:234.

Theocharis S, Liatsos C, Rokkas T, *et al.* Immunohistochemical expression of metallothionein in colorectal adenomatous polyps. *Gastroenterology* 1997; **112**:A668.

Theocharis S, Margeli A, Koutselinis A. Metallothionein: A multifunctional protein from toxicity to cancer. *Int J Biol Markers* 2003; **18**:162–169.

Theocharis SE, Kouraklis GP, Kakisis JD, *et al.* Metallothionein expression in human gastric cancer. *J Eur Surg Res* 2001; **33**:137.

Theocharis SE, Margeli AP, Klijanienko JT, Kouraklis GP. Metallothionein expression in human neoplasia. *Histopathology* 2004; **45**:103–118.

Tsangaris GT, Vamvoukakis J, Politis I, *et al.* Metallothionein expression prevents apoptosis. II: Evaluation of the role of metallothionein expression on the chemotherapy-induced apoptosis during the treatment of acute leukemia. *Anticancer Res* 2000; **20**:4407–4411.

Tuccari G, Fedele F, Giuffre G, *et al.* Immunohistochemical demonstration of metallothionein in eyes with choroidal melanomas. *Eur J Ophthalmol* 2000a; **10**:312–317.

Tuccari G, Giuffre G, Arena F, Barresi G. Immunohistochemical detection of metallothionein in carcinomatous and normal human gastric mucosa. *Histol Histopathol* 2000b; **15**:1035–1041.

Tuzel E, Kirkali Z, Yorukoglu K, *et al.* Metallothionein expression in renal cell carcinoma: Subcellular localization and prognostic significance. *J Urol* 2001; **165**:1710–1713.

Uchida Y, Takio K, Titani K, *et al.* The growth inhibitory factor that is deficient in the Alzheimer's disease brain is a 68 amino acid metallothionein-like protein. *Neuron* 1991; **7**:337–347.

Uozaki H, Horiuchi H, Ishida T, *et al.* Overexpression of resistance-related proteins (metallothioneins, glutathione-S-transferase pi, heat shock protein 27, and lung resistance-related protein) in osteosarcoma. Relationship with poor prognosis. *Cancer* 1997; **79**:2336–2344.

Vazquez-Ramirez FJ, Gonzalez-Campora JJ, Hevia-Alvarez E, *et al.* P-glycoprotein, metallothionein and NM23 protein expression in breast carcinoma. *Pathol Res Pract* 2000; **196**:553–559.

Volm M, Koomagi R, Mattern J, Efferth T. Protein expression profiles indicative for drug resistance of non-small cell lung cancer. *Br J Cancer* 2002; **87**:251–257.

Volm M, Mattern J. Resistance mechanisms and their regulation in lung cancer. *Crit Rev Oncol* 1996; **7**:227–244.

Volm M, Rittgen W. Cellular predictive factors for the drug response of lung cancer. *Anticancer Res* 2000; **20**:3449–3458.

Weinlich G, Topar G, Eisendle K, *et al.* Comparison of metallothionein overexpression with sentinel lymph node biopsy as prognostic factors in melanoma. *J Eur Acad Dermatol Venereol* 2007; **21**:669–677.

Weinlich G, Zelger B. Metallothionein overexpression, a highly significant prognostic factor in thin melanoma. *Histopathology* 2007; **51**:280–283.

Wood DP Jr, Klein E, Fair WR, Chaganti RS. Metallothionein expression in bladder cancer exposed to cisplatin. *Mod Pathol* 1993; **6**:33–35.

Wrigley E, Verspaget HW, Jayson GC, McGown AT. Metallothionein expression in epithelial ovarian cancer: Effect of chemotherapy and prognostic significance. *J Cancer Res Clin Oncol* 2000; **126**:717–721.

Wülfing C, van Ahlen H, Eltze E, *et al.* Metallothionein in bladder cancer: Correlation of overexpression with poor outcome after chemotherapy. *World J Urol* 2007; **25**:199–205.

Yasuno T, Matsumura T, Shikata T, *et al.* Establishment and characterization of a cisplatin-resistant human neuroblastoma cell line. *Anticancer Res* 1999; **19**:4049–4057.

Zangger K, Oz G, Haslinger E, *et al.* Nitric oxide selectively releases metals from the amino-terminal domain of metallothioneins: Potential role at inflammatory sites. *FASEB J* 2001; **15**:1303–1305.

Zelger B, Hittmairr A, Schir M, *et al.* Immunohistochemically demonstrated metallothionein expression in malignant melanoma. *Histopathology* 1993; **23**:257–263.

Zeng J, Heuchel R, Schaffner W, Kagi JHR. Thionein (apoprotein) can modulate DNA binding and transcription activation by zinc finger containing factor SP1. *FEBS Lett* 1991a; **279**:310–312.

Zeng J, Vallee BL, Kagi JHR. Zinc transfer from transcription factor IIIA fingers to thionein clusters. *Proc Natl Acad Sci USA* 1991b; **88**:9984–9988.

Zhang R, Zhang H, Wei H, Luo X. Expression of metallothionein in invasive ductal breast cancer in relation to prognosis. *J Environ Pathol Toxicol Oncol* 2000; **19**:95–97.

Chapter 8

METALLOTHIONEIN AND MELANOMA

Georg Weinlich and Bernhard Zelger

Melanoma is one of the most aggressive human neoplasms, and its incidence is still increasing. To estimate the risk of possible progression and overall survival, Breslow tumor thickness and the invasion level (Clark level) are the most established markers for melanomas at the time of primary diagnosis. In the last decades, overexpression of immunohistochemically labeled metallothioneins (MTs) on paraffin-embedded tissues has turned out to be a highly significant prognostic marker in different tumors. We report the results of a large prospective study on melanoma patients in which MT overexpression was a highly significant marker for progression and survival. In contrast to most other markers, MT overexpression was independent from tumor thickness, and was highly specific even in thin (low-risk) melanoma patients. In high-risk melanoma patients, sentinel lymph node (SLN) biopsy — a surgical technique with predictive value for progression — was performed. The benefit of this procedure for the individual patient's overall survival remains unclear. Our results corroborate the validity of MT overexpression in primary melanoma as a useful prognostic marker: its accuracy is comparable, and to some degree supplementary to, the results of SLN biopsy.

Keywords: Melanoma; metallothionein (MT); progression; survival; sentinel lymph node (SLN) biopsy.

1. Introduction

Approximately, 150 000 people per year worldwide develop a melanoma, and the incidence of this neoplasm is increasing. Once a melanoma has spread to different organs, the cancer is considered incurable. The treatment options are limited, and the 5-year survival rate for patients with distant metastases is less than 10%.

In 1989, it was observed that 50%–60% of the elevated copper concentrations in melanoma tissue was associated with metallothioneins (MTs), although an explanation for this was still missing (Krauter *et al.*, 1989). In mouse melanoma cell lines with a higher expression of MTs, resistance against anticancer agents such as cisplatin was found (Koropatnick and Pearson, 1990). In the following years, several reports documented the role of MTs for the chemoresistance in different human malignancies (Okozaki *et al.*, 1998; Hishikawa *et al.*, 1997; Chin *et al.*, 1993; Sunada *et al.*, 2005), as well as their function to protect cells against ultraviolet (UV)/ionic radiation (Hanada *et al.*, 1998; Hansen *et al.*, 1997; Reeve *et al.*, 2000). Although MTs participate in the carcinogenic process, their use as a potential marker for tumor differentiation or cell proliferation and as a marker of poor prognosis remains controversial (Cherian *et al.*, 2003; Theocharis *et al.*, 2004).

In the last decade, several reports disclosed MT overexpression as a useful prognostic factor for tumor progression in different human cancers. Similar results could be found in small retrospective studies on melanoma and nonmelanoma skin cancers with at most 150 patients (Zelger *et al.*, 1993; Goldmann *et al.*, 1998; Sugita *et al.*, 2001; Zelger *et al.*, 1994; Rossen *et al.*, 1997). In 1993, we carried out a prospective study on melanoma patients to investigate the role of MT overexpression as a prognostic marker for progression and survival. This trial included the investigation of most well-known progression markers in primary melanoma, their comparison with MT overexpression, and their comparison with the predictive value of the surgical procedure of sentinel lymph node (SLN) biopsy in high-risk melanomas.

2. Experimental Procedures

2.1. *Patient Selection*

Between 1993 and 2004, a total of 3386 patients with histological diagnosis of cutaneous melanoma were enrolled in this study, of which 1270 patients were evaluated for further statistical analysis. Besides Breslow's tumor thickness and Clark's level of invasion, ulceration, location of the primary lesion, age, and gender were added for statistical analysis. Endpoint measurements were taken when a progression (lymph node and/or distant metastasis) was detected and when a patient died due to widespread disease. Of the statistically eligible patients, 158 had SLN dissection between 1998 and 2004, due to high-risk melanomas thicker than 1 mm and/or a Clark level of IV. This group was investigated in an additional statistical analysis.

2.2. *Immunohistopathological Analysis*

Formalin-fixed and paraffin-embedded tissues from melanoma patients were routinely stained with hematoxylin and eosin. For immunohistochemical investigations, 4-μm sections were mounted on chrome gel-coated glass slides, deparaffinized in xylene, rehydrated in a graded series of alcohol, and rinsed in Tris buffer. Endogenous peroxidase was blocked by means of sodium azide, glucose, and glucose oxidase. The primary monoclonal MT mouse IgG1 antibody E9 (clone E9; Dako, Denmark), which reacts with both human epitopes of MT-I and MT-II isoforms, was applied, followed by a peroxidase-conjugated rabbit anti-mouse antibody. The enzyme reaction was developed in a freshly prepared solution containing 3-amino-9-ethylcarbazole (AEC) and 0.01% H_2O_2. Then, the sections were counterstained with hematoxylin, dehydrated, cleared in xylene, and mounted with Entellan (Evering *et al.*, 1990; Jasani and Elmes, 1991).

The stained sections were independently assessed by two dermatohistopathological observers by an eyeball estimate without prior knowledge of the clinical data. Overall, good correlation and reproducibility were obtained by the two observers. Only tumor cells staining well above the background level (faint labeling was occasionally observed around strongly positive cells, most likely due to antigen diffusion) were considered to be positive. The reactivity of basal keratocytes as well as the proliferating epithelium of the follicular bulb and ductal epithelium of eccrine and apocrine glands served as positive internal controls.

The specimens were scored as negative if there was a complete lack of MT-positive cells, i.e. only a faint hue of reactivity compared with positive internal controls or only a minor component of up to 10% of positive tumor cells. A reactivity of more than 10% was interpreted as MT overexpression. A limiting value of 10% was chosen to prevent false-positive results. In most of the MT-positive cases,

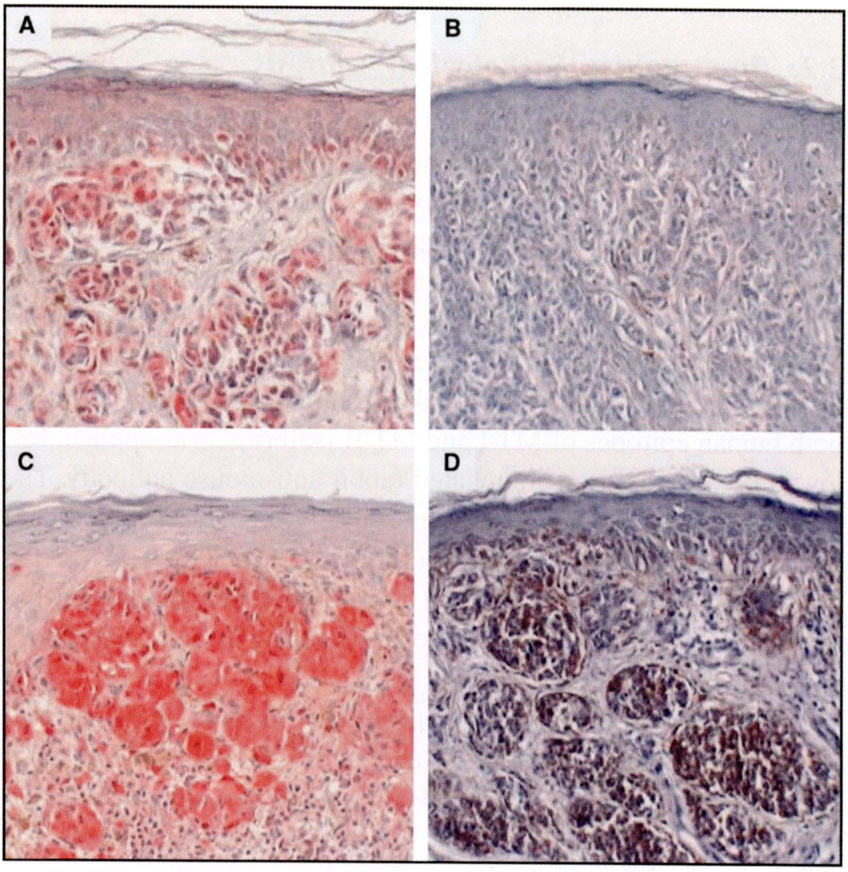

Fig. 1 MT-positive (A, C; red color of AEC indicates positivity) and MT-negative (B, D; brown color in D is melanin) melanomas with comparable tumor thickness (A and B: 0.6 mm, Clark's level III; C and D: 1.0 mm, Clark's level IV).

the MT monoclonal antibody diffusely labeled the majority of tumor cells (Fig. 1).

2.3. *Statistical Analysis*

The ascertained data were statistically analyzed using SPSS statistical software (SPSS for Windows, version 12.0; SPSS Inc., Chicago, IL). The correlation between the immunohistochemical data (absence or presence of MT overexpression) was compared with the clinicopathological data (the occurrence or not of metastasis and/or death due to melanoma) using the two-tailed Fisher's exact test and the Mantel–Haenszel χ^2 test. Kaplan–Meier curves were used to estimate the progression and survival (Kaplan and Meier, 1958). Odds ratios (ORs) and 95% confidence intervals (CIs) were calculated for all known prognostic factors for progression and survival using Cox regression analysis, assuming proportional hazards in a univariate and multivariate approach (Spiegel, 1990). A p-value of less than 0.05 was considered statistically significant in every statistical analysis.

3. Results

Between 1993 and 2004, a total of 3386 patients were recruited for this study, of which 1270 were evaluated for statistical analysis. As only patients with at least 3 months of observation could enter this study, 1226 patients had to be dropped out because they were lost to follow-up. The majority of the lost patients had low-risk melanomas of <1.0 mm or *in situ* melanomas ($n = 1167$ or 95%). Another 890 eligible patients with *in situ* melanomas had to be dropped out for the statistical analysis, as these lesions were not able to metastasize.

Males and females were evenly balanced in our cohort (51%/49%), and the median age at the time of excision was 54 years (range, 7–95 years). Breslow tumor thickness varied between 0.12 mm and 30 mm (median value, 0.7 mm; mean, 1.3 mm). In 11 years of recruitment (median observation time, 32 months), 167 out of 1270 patients (13.1%) showed a relapse of their disease and the median time to progression was 18.0 months (mean value, 26.0 months). Another 110 of these patients (8.7%) died due to metastatic melanoma. The median time from detection of the first metastasis to death was 9.0 months (mean, 15.3 months), and the median time from the primary diagnosis to death was 26.5 months (mean, 35.1 months). None of our eligible patients with a tumor thickness of <0.5 mm developed metastases.

The majority of patients who showed progression, i.e. 117 out of 167 (70.1%), and of those who died due to metastasis, i.e. 80 out of 110 (72.7%), showed MT overexpression of their primary melanoma ($p < 0.001$, χ^2 test; Fisher's exact test). Even in patients with low-risk melanomas of <1.0 mm (846 out of 1270), the majority of those with progression, i.e. 21 out of 32 (65.6%; $\chi^2 = 63.9$, df $= 1$, $p < 0.001$), and of those who died, i.e. 11 out of 15 (73.3%; $\chi^2 = 39.0$, df $= 1$, $p < 0.001$), showed a statistically significant MT overexpression in their primary tumor (Table 1).

Estimated by the Kaplan–Meier method, 55.9% of the MT-positive group, but only 10.9% of the MT-negative group, developed metastasis after 10 years of observation ($p < 0.0001$). About 44.1% of MT-positive patients died due to progression of melanoma,

Table 1 Characteristics of metallothionein overexpression in primary melanoma.

		MT-positive	MT-negative
Number of melanomas	1270	310 (24.40%)	960 (75.60%)
Tumor thickness			
<1.0 mm	846	131 (15.50%)	715 (84.50%)
1.01–2.0 mm	236	78 (33.10%)	158 (66.90%)
2.01–4.0 mm	104	53 (51.00%)	51 (49.00%)
>4.01 mm	84	48 (57.10%)	36 (42.90%)
Progression ($n = 167$, 13.1%) ($\chi 2 = 217.2$, df $= 1$, $p < 0.001$)		117 (70.10%)	50 (29.90%)
Death ($n = 110$, 8.7%) ($\chi 2 = 152.4$, df $= 1$, $p < 0.001$)		80 (72.70%)	30 (27.30%)

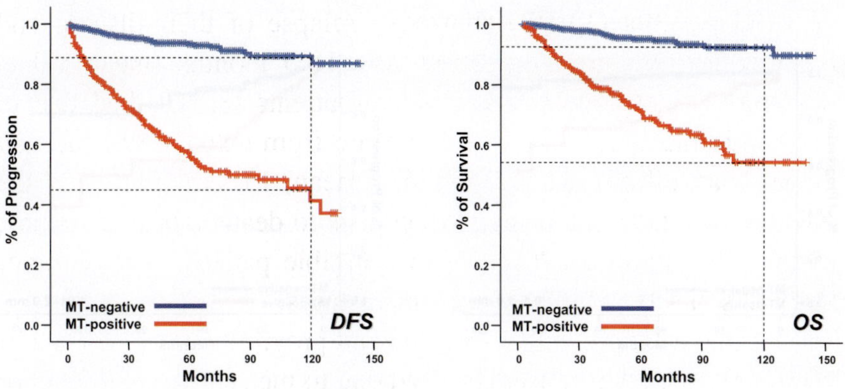

Fig. 2 Kaplan–Meier estimates for the disease-free survival (DFS) and overall survival (OS).

compared to only 6.8% of patients in the MT-negative group ($p < 0.0001$) (Fig. 2).

According to the TNM classification of the American Joint Committee on Cancer 2001, we subdivided our patients into four groups — (<1.0 mm, 1.01–2.0 mm, 2.01–4.0 mm, and >4.01 mm) — with similar results. Even in the group of low-risk melanomas thinner than 1.0 mm, the prognosis for the MT-positive group was worse in comparison with the MT-negative group. Figure 3 illustrates the Kaplan–Meier curves for the progress-free interval (analogous data for survival not shown).

As percentages of MT-positive melanomas were increasing with a higher Clark level and Breslow tumor thickness, we tried to prove the independence of MT overexpression as a prognostic marker using Cox regression models. In a univariate Cox proportional hazard model for disease progression, we found a high significance of MT overexpression with an expected risk of 7.4 (95% CI, 5.3–10.2). MT overexpression as well as all groups of Breslow tumor thickness had p-values <0.001 in the univariate analysis (Table 2). The patient's data were also calculated according to survival with similar results (expected risk, 7.2; 95% CI, 4.7–10.9; $p < 0.001$).

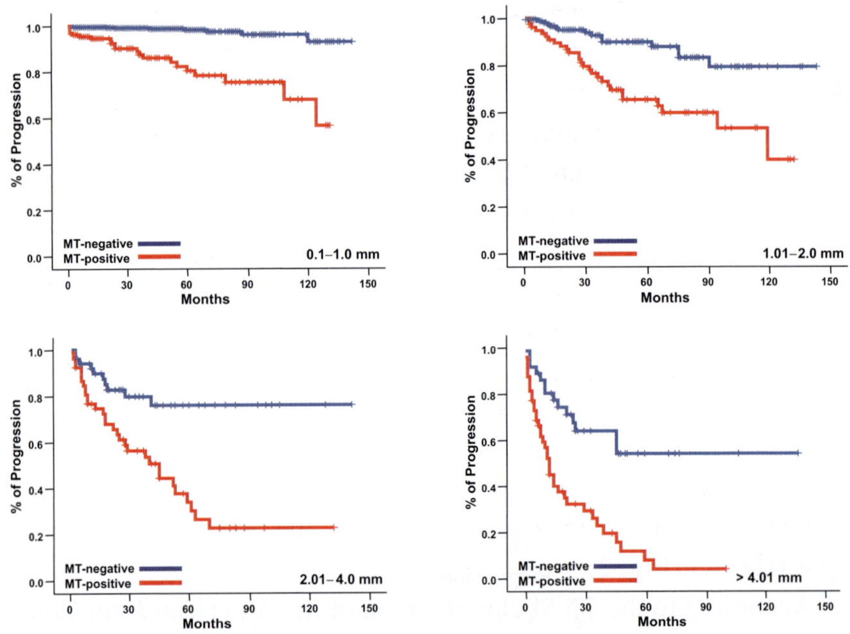

Fig. 3 Kaplan–Meier curves for the progression in melanoma patients in different groups of Breslow tumor thickness.

Table 2 Univariate Cox proportional hazard model for disease progression and survival (metallothionein overexpression/tumor thickness) ($n = 1270$).

	Progression			Survival		
	Relative risk	95% CI	*p*-value	Relative risk	95% CI	*p*-value
MT overexpression	7.36	5.28–10.25	<0.001	7.16	4.71–10.9	<0.001
Tumor thickness						
<1.0 mm	1			1		
1.01–2.0 mm	4.24	2.68–6.72	<0.001	5.44	2.9–10.19	<0.001
2.01–4.0 mm	11.1	6.99–17.62	<0.001	11.56	6.03–22.16	<0.001
>4.01 mm	25.25	16.2–39.34	<0.001	35.06	19.49–63.07	<0.001

In the next step, the data of our patients were adjusted according to the other prognostic factors tested, such as Breslow tumor thickness, Clark level, ulceration, tumor localization, median age, gender, and MT overexpression in a multivariate Cox regression analysis. The strongest significant and independent parameters for progression were tumor thickness and MT overexpression; only a Clark level of V reached a significance of $p < 0.05$, and ulceration was not significant in our cohort (Table 3). In the Cox regression model for the survival, we found similar results with two highly significant and independent parameters: MT overexpression and Breslow tumor thickness. Even when we adapted the Cox regression analyses with Breslow tumor

Table 3 Multivariate Cox regression analysis results for disease progression and survival in melanoma patients ($n = 1270$).

	Progression			Survival		
	Relative risk	95% CI	*p*-value	Relative risk	95% CI	*p*-value
MT overexpression	3.94	2.77–5.6	<0.001	3.49	2.25–5.43	<0.001
Tumour thickness						
<1.0 mm	1			1		
1.01–2.0 mm	1.97	1.14–3.41	0.015	2.55	1.24–5.25	0.011
2.01–4.0 mm	3.46	1.91–6.28	<0.001	3.64	1.63–8.13	0.002
>4.01 mm	6.58	3.49–12.43	<0.001	8.83	3.93–19.83	<0.001
Clark level						
CL II	1			1		
CL III	2.11	0.64–6.95	0.22	3.09	0.40–23.57	0.277
CL IV	3.29	0.94–11.50	0.062	4.94	0.61–39.79	0.133
CL V	5.75	1.51–21.90	0.01	8.36	0.98–71.55	0.053
Ulceration						
Present	1.39	0.96–2.03	0.079	1.46	0.93–2.28	0.094
Age (years)						
>54	1.62	1.15–2.27	0.05	1.52	1.00–2.30	0.05
Gender						
Female	0.84	0.60–1.16	0.279	1.04	0.69–1.54	0.862
Location						
TANS[a]	0.88	0.64–1.22	0.443	1.48	0.99–2.22	0.059

[a]TANS: thorax, upper arm, neck, scalp.

thickness and age as continuous variables, we found similar results (table not shown).

In a second analysis, we investigated the validity of MT overexpression as a progression marker in contrast to SLN biopsy. Between 1998 and 2004, a total of 158 patients from our study cohort were referred to our department to perform SLN biopsies due to high-risk melanomas. In 57 out of the 158 patients (36.1%), a positive SLN could be detected; and in 12 patients, further micrometastases could be found in the following complete lymph node dissection of the accordant regional basin. Four of these 12 patients showed progression in the following months, and two of them died by disseminated metastasis; the other eight patients did not show progression over a mean observation time of 37.9 months (median, 36.0 months). In total, 28 out of 158 patients (17.7%) showed progression of their disease. The majority of all patients who developed metastasis (17 out of 28 patients or 60.7%), and of those who died due to metastasis (9 out of 17 patients or 52.9%), showed MT overexpression of their primary melanoma ($p = 0.002$ for progression, $p = 0.11$ for survival, Pearsons χ^2-test).

In contrast, only 13 patients (46.4%) of those who relapsed and 9 (52.9%) of those who died had a positive SLN, which did not give significant results in the χ^2-test ($p = 0.21$ for progression, $p = 0.13$ for survival). SLN-positive patients had a significantly shorter time to progression than SLN-negative ones (mean, 17 vs. 30 months) and to death after primary diagnosis (mean, 27 vs. 38 months). These differences in the mean time between primary diagnosis and progression, as well as the time between primary diagnosis and death, could not be detected in the MT-positive vs. MT-negative group of primary melanomas.

Figure 4 illustrates the progression-free and overall survival of the melanoma patients over a time period of more than 70 months estimated by the Kaplan–Meier method. After 6 years of observation, 50% of MT-positive patients, but only 35% of SLN-positive ones, showed a progression of their disease. Almost 40% of MT-positive

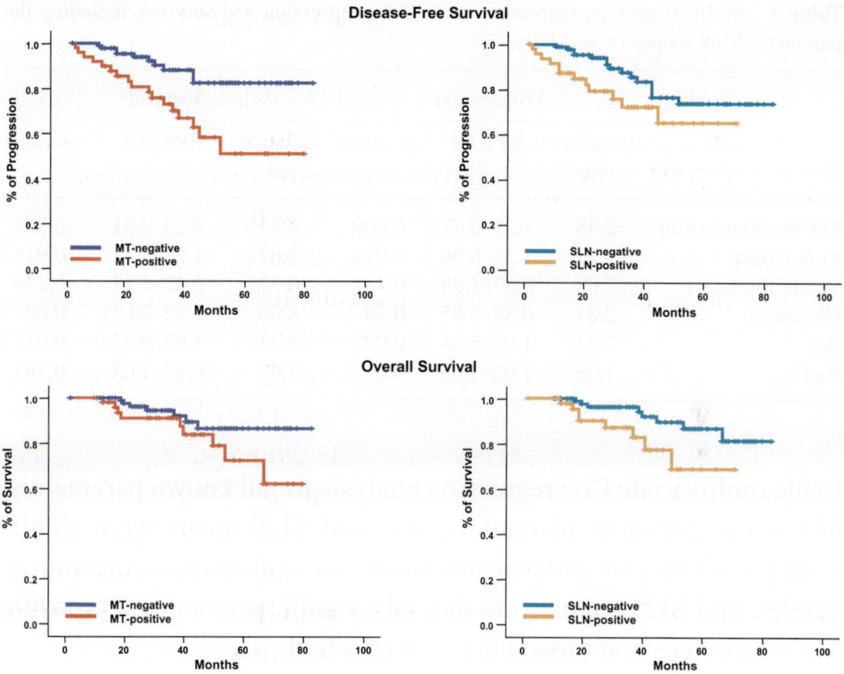

Fig. 4 Kaplan–Meier estimates for the progression and survival in melanoma patients (left side, MT overexpression; right side, SLN positivity).

patients, but only about 30% of SLN-positive ones, died due to metastatic melanoma.

To see whether MT overexpression and SLN positivity turned out as significant and independent risk factors for progression and overall survival, univariate and multivariate Cox proportional hazard models were used again. Univariate analysis for disease progression found a high significance of MT overexpression with an expected risk of 2.861 (95% CI, 1.3–6.1; $p = 0.007$). When the data were calculated according to overall survival, the p-value lost significance (expected risk, 1.91; 95% CI, 0.7–5.0; $p = 0.183$). In contrast, SLN biopsy could not reach a significant p-value for progression in this model (expected risk, 1.88; 95% CI, 0.9–4.0; $p = 0.097$); for survival, the p-value was 0.03 (expected risk, 2.99; 95% CI, 1.1–8.0).

Table 4 Multivariate Cox regression model for progression and survival, including the parameter SLN biopsy ($n = 1270$).

	Progression			Survival		
	Relative risk	95% CI	p-value	Relative risk	95% CI	p-value
MT overexpression	2.98	1.31–6.77	0.009	0.85	0.29–2.51	0.765
SLN biopsy	2.79	1.23–6.36	0.014	3.62	1.28–10.27	0.015
Tumor thickness	1.08	0.87–1.34	0.481	0.71	0.45–1.12	0.139
Ulceration	2.27	0.88–5.85	0.09	8.66	2.52–29.71	0.001
Sex	2.33	1.08–5.04	0.032	3.17	1.07–9.37	0.037
Age	1.05	1.02–1.08	0.01	1.02	0.99–1.05	0.191

In the multivariate Cox regression analysis for all known parameters, MT overexpression, ulceration, age, and SLN status were significant parameters for progression. In the survival analysis, ulceration, gender, and SLN status were the only significant parameters; while MT overexpression missed a $p < 0.05$ (Table 4).

4. Conclusion

Melanoma is a highly malignant neoplasm with still-increasing incidence. As there is still no effective curative treatment for patients with disseminated melanoma, it is important to estimate the risk of progression in the individual patient preferably at the time of primary diagnosis. To date, Breslow tumor thickness is still the best prognostic marker in primary melanoma. Over the preceding years, a wide variety of different prognostic factors have been investigated: Clark level, tumor stage, growth phase, ulceration, mitotic counts, tumor-infiltrating lymphocytes, and others. Most of these prognostic markers often derive their predictive value from a direct or secondary correlation with the leading predictor, Breslow tumor thickness.

The results of this large prospective study with a long follow-up time corroborate previously published data which stress MT overexpression as an important prognostic parameter in melanoma patients. In all statistical analyses performed (Kaplan–Meier curves, two-tailed Fisher's exact test, Pearson's χ^2-test, Cox regression analysis), MT overexpression turned out as a highly significant and independent factor for progression and survival. It only failed to reach significance in the Cox regression analysis for overall survival in comparison to SLN biopsy, presumably due to the small number of patients who died of disseminated disease and/or the relative short time of observation in this group.

In the study cohort, none of the patients with tumors <0.5 mm (591 patients) developed metastasis in the subsequent years. On the other hand, MT-positive melanoma thinner than 1.0 mm had a significantly higher risk for developing metastasis. Approximately 5.3% of the patients in this "low-risk" group (9 out of 170 MT-positive melanomas <1.0 mm) showed a progression of their disease; their relative risk was roughly comparable to MT-negative melanomas with a thickness of 2.1–4.0 mm. This may be used to more carefully follow-up these MT-positive patients and/or probably even serve as a tool to indicate and perform SLN biopsy, and this group of patients could probably profit from adjuvant treatment.

MT overexpression probably has an additional value. In stage IV melanoma patients, anticancer drugs and irradiation therapy are known to often show only a frustrating rate of clinical responses. These therapeutic failures may be partially related to an enhanced MT overexpression in the tumor cells, although the involvement of MTs in conferring resistance to chemotherapeutics remains under discussion (Cherian *et al.*, 2003; Okazaki *et al.*, 1998; Hishikawa *et al.*, 1997; Chin *et al.*, 1993). As a variety of endogenous factors (e.g. glucocorticosteroids, interleukins, interferon γ, tumor necrosis factor α) are involved in the induction of the synthesis of intracellular MTs, one may suggest that this may on the one hand lead to an overprotection of tumor cells against apoptosis, and on the other hand support the metastatic behavior of the tumor (Nath *et al*, 1988; Miles *et al.*, 2000; Tsangaris and Tzortzatou-Stathopoulou, 1998; Karin *et al.*, 1985; Nishimura *et al.*, 2000; Schroeder and Cousins, 1990; Sato and Sasaki, 1992; Karasawa *et al.*, 1987).

So far, the data of all studies performed in melanoma patients were able to prove that MT overexpression is a useful and elegant tool for prognostication (Zelger *et al*, 1993; Goldmann *et al.*, 1998; Sugita *et al.*, 2001; Weinlich *et al.*, 2003; Weinlich *et al.*, 2006; Weinlich *et al.*, 2007; Weinlich and Zelger, 2007). This marker is highly significant and independent from tumor thickness, and is already predictive in low-risk melanomas thinner than 1.0 mm. In high-risk melanomas >1.0 mm where SLN biopsy is performed, the immunohistochemical marker MT overexpression might have supplementary information to the result of the SLN biopsy. One could argue that MT overexpression is the indicator for the principal tendency of the neoplasm to spread (step 1 in the metastasizing cascade), whereas positive SLN is one possible first sign of an already-occurred progression (step 2 in the cascade).

References

Cherian MG, Jayasurya A, Bay BH. Metallothioneins in human tumors and potential roles in carcinogenesis. *Mutat Res* 2003; **533**:201–209.

Chin JL, Banerjee D, Kadhim SA, *et al.* Metallothionein in testicular germ cell tumors and drug resistance. Clinical correlation. *Cancer* 1993; **72**:3029–3035.

Evering WE, Haywood S, Elmes ME, *et al.* Histochemical and immunocytochemical evaluation of copper and metallothionein in the liver and kidney of copper-loaded rats. *J Pathol* 1990; **160**:305–312.

Goldmann T, Ribbert D, Suter L, *et al.* Tumour characteristics involved in the metastatic behaviour as an improvement in primary cutaneous melanoma prognostics. *J Exp Clin Cancer Res* 1998; **17**:483–489.

Hanada K, Sawamura D, Tamai K, *et al.* Novel function of metallothionein in photoprotection: Metallothionein-null mouse exhibits reduced tolerance against ultraviolet B injury in the skin. *J Invest Dermatol* 1998; **111**:582–585.

Hansen C, Ablett E, Green A, *et al.* Biphasic response of the metallothionein promoter to ultraviolet radiation in human melanoma cells. *Photochem Photobiol* 1997; **65**:550–555.

Hishikawa Y, Abe S, Kinugasa S, *et al.* Overexpression of metallothionein correlates with chemoresistance to cisplatin and prognosis in oesophageal cancer. *Oncology* 1997; **54**:342–347.

Jasani B, Elmes ME. Immunohistochemical detection of metallothionein. *Methods Enzymol* 1991; **205**:95–107.

Kaplan EL, Meier P. Nonparametric estimations from incomplete observations. *J Am Stat Assoc* 1958; **53**:457–481.

Karasawa M, Hosoi J, Hashiba H, *et al.* Regulation of metallothionein gene expression by 1 alpha,25-dihydroxyvitamin D3 in cultured cells and in mice. *Proc Natl Acad Sci USA* 1987; **84**:8810–8813.

Karin M, Imbra RJ, Heguy A, Wong G. Interleukin 1 regulates human metallothionein gene expression. *Mol Cell Biol* 1985; **5**:2866–2869.

Koropatnick J, Pearson J. Zinc treatment, metallothionein expression, and resistance to cisplatin in mouse melanoma cells. *Somat Cell Mol Genet* 1990; **16**(6):529–537.

Krauter B, Nagel W, Hartmann HJ, Weser U. Copper-thionein in melanoma. *Biochim Biophys Acta* 1989; **1013**(3):212–217.

Miles AT, Hawksworth GM, Beattie JH, Rodilla V. Induction, regulation, degradation, and biological significance of mammalian metallothioneins. *Crit Rev Biochem Mol Biol* 2000; **35**:35–70.

Nath R, Kambadur R, Gulati S, *et al.* Molecular aspects, physiological functions, and clinical significance of metallothioneins. *Crit Rev Food Sci Nutr* 1988; **27**:41–85.

Nishimura N, Reeve VE, Nishimura H, *et al.* Cutaneous metallothionein induction by ultraviolet B irradiation in interleukin-6 null mice. *J Invest Dermatol* 2000; **114**: 343–348.

Okazaki Y, Miura N, Satoh M, *et al.* Metallothionein-mediated resistance to multiple drugs can be induced by several anticancer drugs in mice. *Biochem Biophys Res Commun* 1998; **245**:815–818.

Reeve VE, Nishimura N, Bosnic M, *et al.* Lack of metallothionein-I and -II exacerbates the immunosuppressive effect of ultraviolet B radiation and *cis*-urocanic acid in mice. *Immunology* 2000; **100**:399–404.

Rossen K, Haerslev T, Hou-Jensen K, Jacobsen GK. Metallothionein expression in basaloid proliferations overlying dermatofibromas and in basal cell carcinomas. *Br J Dermatol* 1997; **136**:30–34.

Sato M, Sasaki M. Tissue-specific induction of metallothionein synthesis by tumor necrosis factor α. *Res Commun Chem Pathol Pharmacol* 1992; **75**:159–172.

Schroeder JJ, Cousins RJ. Interleukin-6 regulates metallothionein gene expression and zinc metabolism in hepatocyte monolayer cultures. *Proc Natl Acad Sci USA* 1990; **87**:3137–3141.

Spiegel MR. *Statistik* (2nd ed.). McGraw-Hill, Hamburg, 1990.

Sugita K, Yamamoto O, Asahi M. Immunohistochemical analysis of metallothionein expression in malignant melanoma in Japanese patients. *Am J Dermatopathol* 2001; **23**:29–35.

Sunada F, Itabashi M, Ohkura H, Okumura T. p53 negativity, CDC25B positivity, and metallothionein negativity are predictors of a response of esophageal squamous cell carcinoma to chemoradiotherapy. *World J Gastroenterol* 2005; **11**:5696–5700.

Theocharis SE, Margeli AP, Klijanienko JT, Kouraklis GP. Metallothionein expression in human neoplasia. *Histopathology* 2004; **45**:103–118.

Tsangaris GT, Tzortzatou-Stathopoulou F. Metallothionein expression prevents apoptosis: A study with antisense phosphorothioate oligodeoxynucleotides in a human T cell line. *Anticancer Res* 1998; **18**:2423–2434.

Weinlich G, Bitterlich W, Mayr V, *et al.* Metallothionein-overexpression as a prognostic factor for progression and survival in melanoma. A prospective study on 520 patients. *Br J Dermatol* 2003; **149**:535–541.

Weinlich G, Eisendle K, Hassler E, *et al.* Metallothionein-overexpression as a highly significant prognostic factor in melanoma. A prospective study on 1270 patients. *Br J Cancer* 2006; **94**(6):835–841.

Weinlich G, Topar G, Eisendle K, *et al.* Comparison of metallothionein-overexpression with sentinel lymph node biopsy as prognostic factors in melanoma. *J Eur Acad Dermatol* 2007; **21**(5):669–677.

Weinlich G, Zelger B. Metallothionein overexpression, a highly significant prognostic factor in thin melanoma. *Histopathology* 2007; **51**(2):280–283.

Zelger B, Hittmair A, Schir M, *et al.* Immunohistochemically demonstrated metallothionein expression in malignant melanoma. *Histopathology* 1993; **23**:257–263.

Zelger B, Sidoroff A, Höpfl R, *et al.* Metallothionein expression in nonmelanoma skin cancer. An immunohistochemical study. *Appl Immunohistochem* 1994; **2**:254–260.

Chapter 9

METALLOTHIONEIN AND BREAST CANCER

Yiyang Lai, George Wai-Cheong Yip, Puay-Hoon Tan, Srinivasan Dinesh Kumar and Boon-Huat Bay

Mammoth efforts have been made in an attempt to understand the etiology of breast neoplasia. Metallothioneins (MTs) are ubiquitous, low-molecular-weight, metal-binding proteins. In addition to their essential roles in metal homeostasis, MTs have been reported to be overexpressed in breast cancers. Generally, MTs are involved in many processes that support tumorigenesis, encompassing the promotion of cell proliferation and metastasis as well as the disruption of apoptotic mechanisms in breast cancers. There is also increasing evidence which shows the association of MT with tumor grade and chemoresistance, supporting the notion that MT could potentially be a biomarker of prognosis. In this chapter, we shall focus on the contributions of individual MT isoforms in relation to breast carcinogenesis and its eventual pathological outcome.

Keywords: Breast carcinoma; functional metallothionein isoforms; biomarkers; breast carcinogenesis; prognosis.

1. Introduction

Breast cancer is a worldwide disease that plagues many women. Although not exclusive to women, gender difference does contribute to breast cancer susceptibility (American Cancer Society, 2007). In Singapore, it is the most prevalent type of cancer reported in the female population (Seow *et al.*, 2004). Most cases of breast cancer originate from epithelial cells (Fig. 1), from which they can be categorized into ductal, lobular, medullary and papillary, or other carcinoma subtypes. Amongst them, ductal and lobular subtypes account for the majority of documented breast cancer cases. Mutation of the tumor suppressor *BRCA* gene, age, hormone replacement therapy, family history, and usage of oral contraceptives are some of the risk factors listed by the American Cancer Society that are associated with breast carcinogenesis. Since cancer is ultimately a disease of altered gene expression, researchers have also turned their attention to screening the entire human genome to identify candidate oncogenes that are responsible for neoplasm of the mammary gland.

Metallothioneins (MTs) are ubiquitous, low-molecular-weight (6–7 kDa), metal-binding proteins. In humans, 10 functional MT isoforms are coded by genes located on chromosome 16q13, namely MT-1A, MT-1B, MT-1E, MT-1F, MT-1G, MT-1H, MT-1X, MT-2A, MT-3, and MT-4 (West *et al.*, 1990; Stennard *et al.*, 1994;

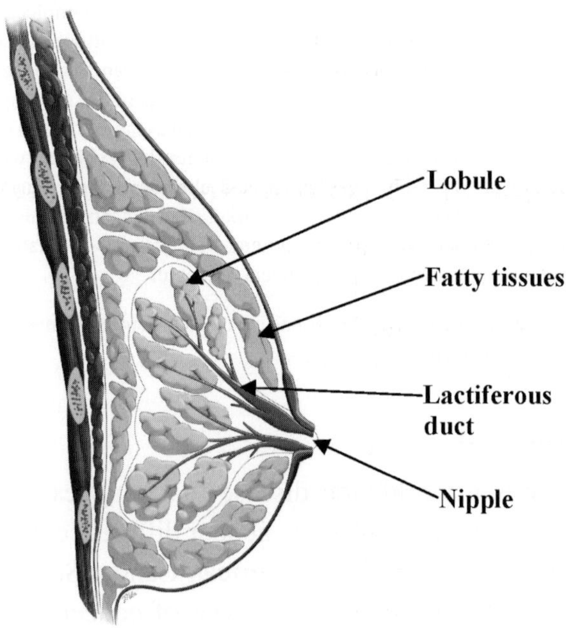

Fig. 1 Gross structure of the human mammary gland.

Mididoddi *et al.*, 1996). Averaging 61–68 amino acids in length, this group of highly conserved polypeptides binds to heavy metals such as zinc, copper, and cadmium by virtue of the unique clusters of cysteine residues located within their equally sized globular domains (Kägi, 1991). Given their affinity for divalent metals, MTs have been implicated in mediating heavy metal detoxification (Ebadi *et al.*, 1996; Roesijadi, 2000). Before 1987, most research was focused on the metalloregulatory roles of MTs, until Nartey *et al.* (1987) reported the expression of MT in thyroid cancer. Since then, there has been extensive interest in the contribution of MT to processes linked to carcinogenesis, such as cell proliferation, differentiation, apoptosis, and radioresistance/chemoresistance (Kägi, 1991; Satoh *et al.*, 1994; Nagel and Vallee, 1995; Kelly *et al.*, 1988; Jayasurya *et al.*, 2000a; Jayasurya *et al.*, 2000b; Kling and Olsson, 2000; Cai and Cherian, 2003). To date, overexpression of MTs has been well documented in tumors of the breast, prostate, and colon (Nagel and Vallee, 1995; Jin *et al.*, 1999; Sens *et al.*, 2000).

2. Expression of MT Isoforms in Breast Tumors

The majority of breast cancer research has centered on luminal epithelial cells, as these are believed to be the progenitors of most carcinomas of the breast (Fig. 2) (Ronnov-Jessen *et al.*, 1996). In healthy and nonneoplastic human breast tissue, MT expression is chiefly limited to myoepithelial cells that border their luminal neighbors within the bilayered milk duct (Fresno *et al.*, 1993; Bier *et al.*, 1994; Jin *et al.*, 2001b). Conversely, MT positivity and upregulation were annotated in invasive ductal carcinomas and their stromal components, as opposed to other classes of breast malignancies where either weak or negative MT staining was recorded (Fresno *et al.*, 1993; Schmid *et al.*, 1993; Bier *et al.*, 1994; Dutsch-Wicherek *et al.*, 2005; Gurel *et al.*, 2005).

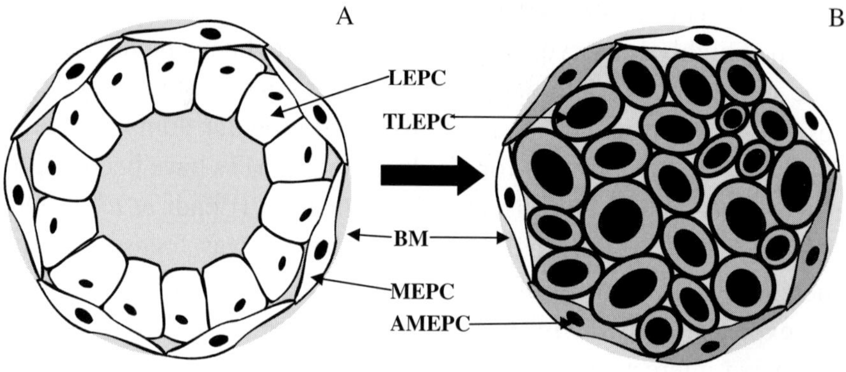

Fig. 2 Oncogenic transformation of epithelium of the milk duct to ductal carcinoma *in situ* (DCIS). (A) Cross-sectional representation of normal milk duct showing a bilayered orientation of luminal and myoepitheial cells. (B) Transformed luminal epithelial cells shed milk production duties and occupy the lumen. In DCIS, myoepithelial cells have altered expression of proto-oncogenes, which in turn propagate tumor cells. MEPC, myoepithelial cells; AMEPC, altered myoepithelial cells; LEPC, luminal epithelial cells; TLEPC, transformed luminal epithelial cells; BM, basement membrane.

In the laboratory, the E9 antibody has been used to study the expression of MT-1 and MT-2 in breast cancer via immunohistochemistry. Examination of breast invasive ductal cancer cell lines and surgically resected tissue samples showed that MTs display nuclear and cytoplasmic expression (Fig. 3). However, as MT-1 and MT-2 isoforms share a conserved epitope, the antibody is unable to differentiate between both subtypes.

Studies pertaining to specific MT isoforms are usually carried out by quantifying the respective mRNA level via reverse transcription–polymerase chain reaction (RT-PCR) or *in situ* hybridization using gene-specific primers and probes. Studies employing these strategies showed that various MT isoforms were diversely and variably expressed in invasive ductal carcinoma (Sens *et al.*, 2001; Jin *et al.*, 2002; Tai *et al.*, 2003). Although both MCF-7 and MDA-MB-231 breast cancer cell lines are of adenocarcinoma lineage, they demonstrated distinct expression profiles for various MT isoforms (Fig. 4). This might hint at the possibility of breast cancer cells altering their

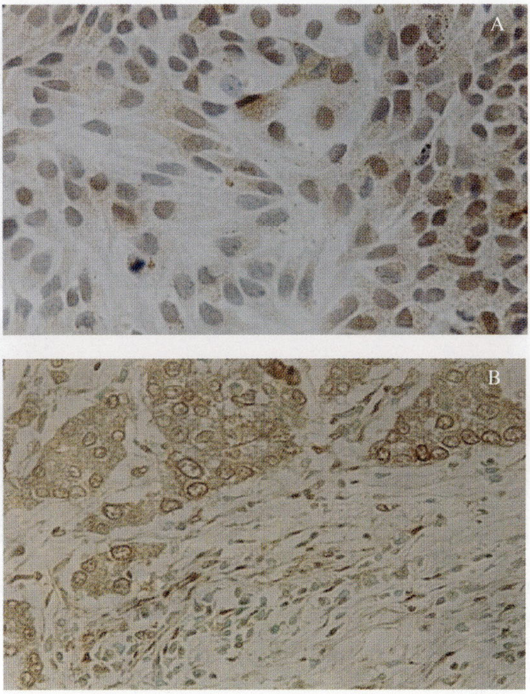

Fig. 3 Positive immunostaining for MT-1/MT-2 proteins in breast cancer. (A) MCF-7 breast cancer cells. Hematoxylin counterstain. (B) Invasive ductal breast cancer tissue section. Methyl-green counterstain. (Magnification, 400×)

expression signatures for various isoforms of MT during tumor development (Cherian *et al.*, 2003) and, in the case of MDA-MB-231, during acquisition of invasive potential.

Tissue-specific isoforms such as MT-1B and MT-4 were notably absent in human breast cancer, while low levels of MT-1A and MT-1H transcripts were detected in breast cancer cell lines and tissue samples. A recent finding by Piotrowski *et al.* (2006) showed that hypermethylation of CpG islets was responsible for MT-1A down-regulation, while the PLU-1/JARID1B transcription repressor negatively regulated MT-1H (Scibetta *et al.*, 2007). Expression of MT-1G was exclusive to cancerous tissue in a study conducted by Tai *et al.* (2003), in which they surmised that its expression was due to the

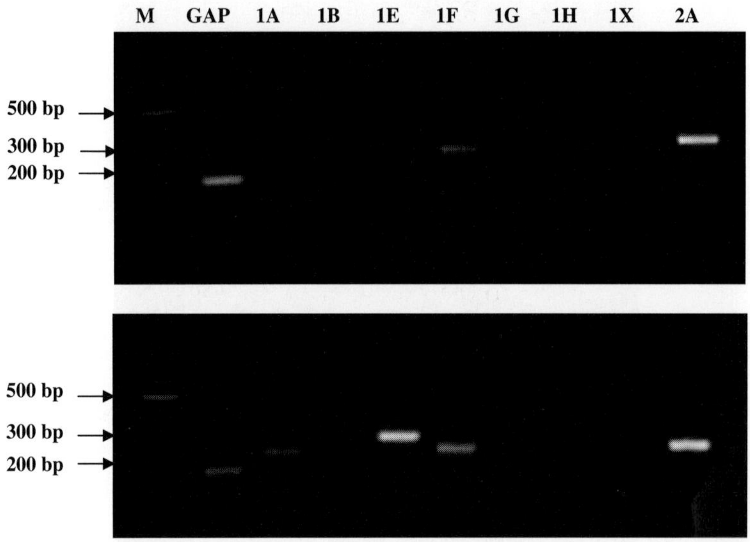

Fig. 4 Gel electrophoresis of PCR products of MT isoforms. Upper panel: MCF-7 breast cancer cells show expression of MT-1F, MT-1X, and MT-2A mRNA transcripts. Lower panel: MDA-MB231 breast cancer cells show expression of MT-1A, MT-1E, MT-1F, and MT-2A mRNA transcripts. The PCR products of G3PDH, MT-1A, MT-1E, MT-1F, MT-1X, and MT-2A are 160 bp, 219 bp, 284 bp, 232 bp, 151 bp, and 259 bp, respectively. M, marker; GAP, G3PDH.

presence of myoepithelial cells in the stroma, the only cell type known to react immunopositively to MT staining in normal breast ductal tissue (Jin *et al.*, 2001b; Gurel *et al.*, 2005). MT-1E was found to be highly expressed in the estrogen receptor (ER)-negative subset of invasive ductal cancer (Friedline *et al.*, 1998; Jin *et al.*, 2000). Interestingly, MT-1F, MT-1X, and MT-2A expressions were detected in breast cancer cell lines and tissues, with MT-2A having the highest expression amongst all MT isoforms. The higher expression level of MT-2A was positively associated with aggressiveness of breast cancer, suggesting a role for MT-2A in breast cancer progression (Jin *et al.*, 2002). MT-1F was upregulated in advanced breast cancer, whilst MT-1X expression was increased in breast cancer compared to prostate cancer (Jasani and Schmid, 1997; Jin *et al.*, 2001a). MT-3 was reported to be overexpressed in ductal carcinoma *in situ* (DCIS), which was confirmed via RT-PCR and immunohistochemistry with an MT-3–specific antibody (Sens *et al.*, 2001).

3. Roles of MT Isoforms in Breast Carcinogenesis

Carcinogenesis is a multi-step process. Hanahan and Weinberg (2000) proposed that transformed somatic cells would require essential modifications to their cellular physiology, coined as "hallmarks of cancers," in order to achieve immortality and malignancy. In that review paper, alterations to pathways governing apoptosis and cell proliferation were described as common targets of most cancers in the quest to fulfill their tumorigenic agenda. MTs are well known to influence tumor growth by affecting key processes such as cell proliferation and apoptosis in cancer development (Kägi, 1991). This section shall address the contributions of proto-oncogenic MTs to the manifestation of cancerous lesions of the breast.

Breast cancer exhibits differential expression of MT isoforms. As previously suggested by Cherian *et al.* (2003), changes in the expression profile of various isoforms may be attributed to alterations in proliferation or differentiation in the event of tumor progression. Based on their findings, Jasani and Schmid (1997) also postulated that the differential expression of specific MT isoforms may hold the key to the functional significance of MT upregulation in tumor tissue. Table 1 summarizes the validated function(s) of individual isoforms in the context of breast cancer.

In 2002, Jin and colleagues observed the association of MT-2A with enhanced cell proliferation. Using double immunolabeling and colocalization studies with Ki-67, a marker of cell proliferative activity, the investigators showed a positive correlation between increased MT-2A mRNA transcripts and cell growth (Jin *et al.*, 2002). Conversely, a reciprocal study employing antisense oligonucleotide-mediated MT-2A silencing performed by Abdel-Mageed and Agrawal (1997) led to growth arrest and apoptosis. The same investigators also proposed that MT-2A interacted

Table 1 Functional significance of MT isoforms in breast cancer.

MT isoforms	Expression level in cell	Expression level in tissue	Roles in carcinogenesis	Prognosis
MT-1A	Low	Low	Downregulated (Piotrowski *et al.*, 2006)	N/A
MT-1B	Absent	Absent	Unknown	N/A
MT-1E	High and only in $ER^{-/-}$ cell line	High	Replace ER function (Jin *et al.*, 2000)	Poor (Jin *et al.*, 2000)
MT-1F	Moderate	Moderate	Tumor differentiation (Jin, 2001)	Poor (Jin, 2001)
MT-1G	Only in myoepithelial cells	Low	Unknown	N/A
MT-1H	Low	Low	Downregulated (Scibetta *et al.*, 2007)	N/A
MT-1X	Low	Low	Unknown	N/A
MT-2A	Highest	Highest	Cell proliferation, interact with NF-κB (Jin *et al.*, 2002); disrupts p53 (Fan and Cherian, 2002)	Poor (Jin *et al.*, 2002)
MT-3	Absent	High	Disrupts p53 (Sens *et al.*, 2001)	Poor (Sens *et al.*, 2001)
MT-4	Absent	Absent	Unknown (Jin, 2001)	N/A

Note: Expression levels of MT isoforms were semiquantitated via RT–PCR. All expression data were derived from Tai *et al.* (2003) unless otherwise stated. ER, estrogen receptor; N/A, not applicable.

with NF-κB, a transcriptional activator, to bring about the cell proliferative effect (Abdel-Mageed and Agrawal, 1998). Other breast cancer-related binding partners of MT include protein kinase Cμ and Rab3A GTPase (Rao *et al.*, 2003; Knipp *et al.*, 2005).

As MT is a metalloprotein, other hypotheses on its cancer-promoting effects revolve around the affinity that MTs have for zinc. MTs could potentially disrupt p53 function by chelating zinc, which is required for p53 structural stability (Fan and Cherian, 2002). Regarded as the "guardian" of the genome, the anticancer apoptotic and transcriptional activities regulated by p53 could be impaired due to the formation of the MT–p53 complex (Ostrakhovitch *et al.*, 2006). A downstream product of p53 regulation, it would seem rather contradictory of MT to destabilize p53, but the overexpression of MT as observed in breast cancers was found to be independent of p53 induction (Ostrakhovitch *et al.*, 2007). More studies are required to understand the mechanism and rationale underlying the dysregulation of MTs, and to provide further insights on the neoplastic transformation of breast tissues. On the other hand, MT–Zn could also function as an intracellular reservoir for Zn proteins which regulate cell proliferation, including ER and transcription factor IIIa (Cano-Gauci and Sarkar, 1996; Jacobs *et al.*, 1998; Huang *et al.*, 2004). There is a strong association between ER status and MT-1E expression in breast carcinoma, with a high proportion of ER-negative cancers displaying higher MT-1E expression (Friedline *et al.*, 1998; Jin *et al.*, 2000). It has been suggested that MT-1E may serve as a replacement to carry out functions of ER in the ER-null state (Mckenzie and Sukumar, 1996; Ysaziji *et al.*, 2000).

Even though the majority of the available literature appears to associate MT overexpression with less favorable outcomes in invasive ductal breast cancers, Gurel *et al.* (2003) reported that overexpression of MT-3 in selective breast cancer cell lines devoid of this isoform showed growth inhibition. This counterproliferative trait is unique to MT-3, despite sharing 70% homology with MT-1/MT-2, and is most probably linked to the growth inhibitory property of MT-3 previously established in murine neural cells (Uchida *et al.*, 1991; Tsuji *et al.*, 1992; Amoureux *et al.*, 1995; Swell *et al.*, 1995).

Preventive measures have also been incorporated within the microenvironment of acinar milk ducts to deter epithelial cells from undergoing oncogenic transformation. Myoepithelial cells, the effector cells responsible for this mechanism, may be the first line of defense to keep luminal cells in check, in addition to their (myoepithelial cells') milk ejection role (Hamperl, 1970). The tumor-suppressing capabilities of these cells, including perturbation to key processes such as angiogenesis and cell invasion as well as induction of G_2/M growth arrest, have been shown *in vitro* (Sternlicht *et al.*, 1996; Shao *et al.*, 1998; Nguyen *et al.*, 2000). During the course of carcinogenesis, microevolution in cancerous breast tissue could result in adenocarcinoma cells assuming a myoepithelial phenotype, evident by anomalous MT expression, and subsequently overwhelming their myoepithelial counterparts (Jin *et al.*, 2001b; Adriance *et al.*, 2005). Loss of myoepithelial cells, particularly their tumor-suppressive functions, coupled to the mimicry to secrete myoepithelial-related oncoproteins may provide a competitive edge for these rogue epithelial cells to thrive and metastasize (Weaver *et al.*, 2002).

4. Association of MT Isoforms with Pathological Parameters and Prognostication

Higher MT expression in breast cancers has generally been shown to predict worse survival for patients (Vazquez-Ramirez *et al.*, 2000; Zhang *et al.*, 2000; Ioachim *et al.*, 2003). This implies that MT may have putative prognostic utility in breast tumors. Early studies conducted in pT_2 invasive ductal breast carcinoma indicated that increased MT expression is associated with poorer prognosis and a shorter disease-free survival period (Schmid *et al.*, 1993; Goulding *et al.*, 1995).

Conventionally, histological grade has always been an essential criterion in the pathological assessment of invasive ductal carcinoma. Using guidelines laid down by Bloom and Richardson (1957), breast cancers are graded based on the extent of tubule formation, nuclear pleomorphism, and mitotic rate as a predictive index of prognosis. Bay *et al.* (2006) have established the association between high expression of MT-1F and MT-2A isoforms with histological grade 3 invasive ductal carcinoma, indicating a poorer survival rate (Jin *et al.*, 2001b; Jin *et al.*, 2002). The bad outcome is not surprising given the ability of MT to promote metastasis and cell proliferation (Schmid *et al.*, 1993; Vazquez-Ramirez *et al.*, 2000; Jin *et al.*, 2002). Similarly, overexpression of the MT-3 isoform was strongly associated with poor prognosis in DCIS (Sens *et al.*, 2000). On the contrary, several investigations have shown that MT expression does not significantly correlate with clinicopathological parameters such as age, tumor size, or lymph node metastasis (Fresno *et al.*, 1993; Goulding *et al.*, 1995; Oyama *et al.*, 1996; Jin *et al.*, 2002).

In breast cancer therapy, the hormonal status of the cancerous tissue is critical in chemotherapeutic assignment. ER-positive cancers receiving adjuvant tamoxifen treatment are less likely to suffer a relapse compared to their ER-negative counterparts (Early Breast Cancer Trialist Collaborative Group, 1998; Duffy, 2001). Since high MT-1E expression is inversely related to ER positivity, screening for MT-1E might strengthen the prediction of tamoxifen responsiveness in breast cancer treatment now that MTs have been proven to mediate tamoxifen resistance (Surowaik *et al.*, 2005).

Radioresistance and chemoresistance are major factors contributing to poor prognosis, and MTs are believed to be key mediators. High levels of MTs were cited to confer resistance to radiation and antineoplastic drugs by sequestration of free radicals, drugs, and their metabolites, thereby reducing the efficacy of these therapeutic regimes (Cherian *et al.*, 2003). Genomic analysis has singled out MT-2A as the culprit mediating resistance towards the novel anticancer drug gallium nitrate in lymphoma (Yang *et al.*, 2007). In a separate study, Yang *et al.* (1994) verified the involvement of MT-2A in cisplatin resistance. Given that its expression is the highest in most invasive ductal carcinomas, the role of MT-2A in chemoresistance remains to be elucidated and reinforces its candidature as a biomarker of prognosis in breast tumor.

5. Conclusion

Breast cancer is a fast-growing disease in many countries. As the level of awareness for this mammary gland disorder is raised globally, more and more women are stepping forward to go for routine mammogram screening, which could translate into a foreseeable surge in the number of breast cancer cases detected. Although the outcome need not be fatal, the physical, psychological, and financial impacts inflicted extend beyond the patient. A better understanding of this illness will equip healthcare workers with knowledge to combat it and will provide a glimmer of hope for breast tumor-stricken females.

Much remains to be learnt about the function of MTs in breast carcinogenesis and chemotherapy, especially with regard to what role each of the isoforms performs in these processes. The lack of an antibody capable of differentiating the various isoforms of MT-1/MT-2 has greatly hampered the advances made in relation to unmasking the function that each isoform performs in cell physiology. With the help of biotechnology, isoform-targeting tools can be incorporated to help solve the mystery underlying aberrant expression of MT isoforms in breast neoplasia. Elevated MT expression in the stromal component, as well as the surrounding healthy tissue, should be assessed given the heterogeneity of cells encompassing the cancerous lesion.

In today's knowledge-driven society, the focus on translational studies will pave the way for better strategies and therapeutics to improve cancer treatment. Tumor-suppressing elements in myoepithelial cells warrant further attention, with the hope of being developed into potential anticancer treatment. Speculative as it remains, information such as the predictive capabilities of MTs to tamoxifen resistance could be adopted by clinicians to strengthen their assessment prior to drug administration and alleviate the patient's suffering. While MTs present as promising prognostic biomarkers, the validation of their usefulness in prospective patient studies for use in clinical settings entails more reaffirmation.

Regardless of any potential that anti-MT therapy may have in the pipeline, it is unlikely to be a "be-all and end-all" solution in our continuous fight against this disease given the multifactorial origin of cancer.

Acknowledgments

Special thanks go out to Song-Lin Bay, Rongxian Jin, and Yong-Chiat Wong for their technical assistance rendered. The authors are grateful to the National University of Singapore and Singapore National Medical Research Council for funding this work.

References

Abdel-Mageed A, Agrawal KC. Antisense down-regulation of metallothionein induces growth arrest and apoptosis in human breast carcinoma cells. *Cancer Gene Ther* 1997; **4**:199–207.

Abdel-Mageed A, Agrawal KC. Activation of nuclear factor kappa B: Potential role in metallothionein-mediated mitogenic response. *Cancer Res* 1998; **58**:2335–2338.

Adriance MC, Inman JL, Petersen OW, Bissell MJ. Myoepithelial cells: Good fences make good neighbors. *Breast Cancer Res* 2005; **7**:190–197.

American Cancer Society. *Breast Cancer Facts and Figures 2007–2008*. American Cancer Society, Atlanta, 2007.

Amoureux M-C, Wurch T, Pauwels PJ. Modulation of metallothionein-III mRNA content and growth rate of rat C6-gial cells by transformation with human 5-HT$_{1D}$ receptor genes. *Biochem Biophys Res Commun* 1995; **214**:639–645.

Bay BH, Jin R, Huang J, Tan PH. Metallothionein as a prognostic biomarker in breast cancer. *Exp Biol Med* 2006; **231**:1516–1521.

Bier B, Douglas-Jones A, Totsch M, *et al.* Immunohistochemical demonstration of metallothionein in normal human breast tissue and benign and malignant lesions. *Breast Cancer Res Treat* 1994; **30**:213–221.

Bloom HJ, Richardson WW. Histological grading and prognosis in breast cancer. A study of 1409 cases of which 359 have been followed for 15 years. *Br J Cancer* 1957; **11**:359–377.

Cai L, Cherian MG. Zinc-metallothionein protects from DNA damage induced by radiation better than gluthathione and copper- or cadmium-metallothionein. *Toxicol Lett* 2003; **136**:193–198.

Cano-Gauci DF, Sarkar B. Reversible zinc exchange between metallothionein and the estrogen receptor zinc finger. *FEBS Lett* 1996; **386**:1–4.

Cherian MG, Jayasurya A, Bay BH. Metallothionein in human tumors and potential roles in carcinogenesis. *Mutat Res* 2003; **533**:201–209.

Duffy MJ. Biochemical markers in breast cancer: Which ones are clinically useful? *Clin Biochem* 2001; **34**:347–352.

Dutsch-Wicherek M, Popiela TJ, Klimek M, *et al.* Metallothionein stroma reaction in tumor adjacent healthy tissue in head and neck squamous cell carcinoma and breast adenocarcinoma. *Neuro Endocrinol Lett* 2005; **26**:567–574.

Early Breast Cancer Trialist Collaborative Group. Tamoxifen for early cancer: An overview of the randomized trials. *Lancet* 1998; **351**:1451–1467.

Ebadi M, Leuschen MP, El Refaey H, *et al.* The antioxidant properties of zinc and metallothionein. *Neurochem Int* 1996; **29**:159–166.

Fan LZ, Cherian MG. Potential role of p53 on metallothionein induction in human epithelial breast cancer cells. *Br J Cancer* 2002; **87**:1019–1026.

Fresno M, Wu W, Rodriguez JM, Nadji M. Localization of metallothionein in breast carcinomas. An immunohistochemical study. *Virchows Arch A Pathol Anat Histopathol* 1993; **423**:215–219.

Friedline JA, Garrett SH, Somji S, *et al.* Differential expression of the MT-1E gene in estrogen-receptor-positive and -negative human breast cancer cell lines. *Am J Pathol* 1998; **152**:23–27.

Goulding H, Jasani H, Pereira A, *et al.* Metallothionein expression in human breast cancer. *Br J Cancer* 1995; **72**:968–972.

Gurel V, Sens DA, Somji S, *et al.* Stable transfection and overexpression of metallothionein isoform 3 inhibits the growth of MCF-7 and Hs578T cells but not that of T-47D or MDA-MB-231 cells. *Breast Cancer Res Treat* 2003; **80**:181–191.

Gurel V, Sens DA, Somji S, *et al.* Post transcriptional regulation of metallothionein isoform 1 and 2 expression in the human breast and the MCF-10A cell line. *Toxicol Sci* 2005; **85**:906–915.

Hamperl H. The myothelia (myoepithelial cells). Normal state; regressive changes; hyperplasia; tumors. *Curr Top Pathol* 1970; **53**:797–818.

Hanahan D, Weinberg RA. The hallmarks of cancers. *Cell* 2000; **100**:57–70.

Huang M, Shaw CF III, Petering DH. Interprotein metal exchange between transcription factor IIIa and apo-metallothionein. *J Inorg Biochem* 2004; **98**:639–648.

Ioachim E, Tsanou E, Briasoulis E, *et al.* Clinicopathological study of the expression of hsp27, pS2, cathepsin D and metallothionein in primary invasive breast cancer. *Breast* 2003; **12**:111–119.

Jacobs C, Maret W, Vallee BL. Control of zinc exchange between metallothionein and the estrogen receptor zinc finger. *Proc Natl Acad Sci USA* 1998; **95**:3489–3494.

Jasani B, Schmid KW. Significance of metallothionein overexpression in human tumours. *Histopathology* 1997; **3**:211–214.

Jayasurya A, Bay BH, Yap WM, Tan NG. Correlation of metallothionein expression with apoptosis in nasopharyngeal carcinoma. *Br J Cancer* 2000a; **82**:1198–1203.

Jayasurya A, Bay BH, Yap WM, *et al.* Proliferative potential in nasopharyngeal carcinoma: Correlations with metallothionein expression and tissue zinc levels. *Carcinogenesis* 2000b; **21**:1809–1812.

Jin R. Metallothionein expression in invasive ductal breast carcinoma. Thesis dissertation, National University of Singapore, Singapore, 2001, pp. 105.

Jin R, Bay BH, Chow VT, Tan PH. Metallothionein 1F mRNA expression correlates with histological grade in breast carcinoma. *Breast Cancer Res Treat* 2001a; **66**:265–272.

Jin R, Bay BH, Chow VT, *et al.* Metallothionein 1E mRNA is highly expressed in oestrogen receptor-negative human invasive ductal breast cancer. *Br J Cancer* 2000; **83**:319–323.

Jin R, Bay BH, Chow VT, *et al.* Significance of metallothionein expression in breast myoepithelial cells. *Cell Tissue Res* 2001b; **303**:221–226.

Jin R, Bay BH, Tan P, Tan BK. Metallothionein expression and zinc levels in invasive ductal carcinoma. *Oncol Rep* 1999; **6**:871–875.

Jin R, Chow VT, Tan PH, *et al.* Metallothionein 2A expression is associated with cell proliferation in breast cancer. *Carcinogenesis* 2002; **23**:81–86.

Kägi JHR. Overview of metallothionein. *Methods Enzymol* 1991; **205**:613–626.

Kelly S, Basu A, Teicher BA, *et al.* Overexpression of metallothionein confers resistance to anticancer drugs. *Science* 1988; **241**:1813–1815.

Kling PG, Olsson P. Involvement of differential metallothionein expression in free radical sensitivity of RTG-2 and CHSE-214 cells. *Free Radic Biol Med* 2000; **28**: 1628–1637.

Knipp M, Meloni G, Roschitzki B, Vasak M. Metallothionein-3 and the synaptic vesicle cycle: Interaction of metallothionein-3 with small GTPase Rab3A. *Biochemistry* 2005; **44**:3159–3165.

McKenzie K, Sukumar S. In: Huff J, Boyd J, Barrett JC (eds.), *Cellular and Molecular Mechanisms of Hormonal Carcinogenesis: Environmental Influences*, Wiley-Liss, New York, 1996, pp. 183–209.

Mididoddi S, McGuirt JP, Sens MA, *et al.* Isoform-specific expression of metallothionein mRNA in developing and adult human kidney. *Toxicol Lett* 1996; **685**:17–27.

Nagel WW, Vallee BL. Cell cycle regulation of metallothionein in human colonic cancer cells. *Proc Natl Acad Sci USA* 1995; **92**:579–583.

Nartey N, Cherian MG, Banerjee D. Immunohistochemical localization of metallothionein in human thyroid tumors. *Am J Pathol* 1987; **129**:177–182.

Nguyen M, Lee MC, Wang JL, *et al.* The human myoepithelial cell displays a multi-faceted anti-angiogenic phenotype. *Oncogene* 2000; **19**:3449–3459.

Ostrakhovitch EA, Olsson P, Jiang S, Cherian MG. Interaction of metallothionein with tumour suppressor p53 protein. *FEBS Lett* 2006; **580**:1235–1238.

Ostrakhovitch EA, Olsson PE, von Hofsten J, Cherian MG. P53 mediated regulation of metallothionein transcription in breast cancer cells. *J Cell Biochem* 2007; **102**: 1571–1583.

Oyama T, Take H, Hikino T, *et al.* Immunohistochemical expression of metallothionein in invasive breast cancer in relation to proliferative activity, histology and prognosis. *Oncology* 1996; **53**:112–117.

Piotrowski A, Benetkiewicz M, Menzel U, *et al.* Microarray-based survey of CpG islands identifies concurrent hyper- and hypomethylation patterns in tissues derived from patients with breast cancer. *Genes Chromosomes Cancer* 2006; **45**:656–667.

Rao PS, Jaggi M, Smith DJ, *et al.* Metallothionein 2A interacts with the kinase domain of PKCmu in prostate cancer. *Biochem Biophys Res Commun* 2003; **310**:1032–1038.

Roesijadi G. Metal transfer as a mechanism for metallothionein-mediated metal detoxification. *Cell Mol Biol* 2000; **46**:393–405.

Ronnov-Jessen L, Petersen OW, Bissell MJ. Cellular changes involved in conversion of normal to malignant breast: Importance of the stromal reaction. *Physiol Rev* 1996; **76**:69–125.

Satoh M, Cherian MG, Imura N, Shimizu H. Modulation of resistance to anticancer drug by inhibition of metallothionein synthesis. *Cancer Res* 1994; **54**:5255–5257.

Schmid KW, Ellis IO, Gee JM, *et al.* Presence and possible significance of immunohistochemically demonstrable metallothionien over-expression in primary invasive ductal carcinoma of the breast. *Virchows Arch A Pathol Anat Histopathol* 1993; **422**:153–159.

Scibetta AG, Santangelo S, Coleman J, *et al.* Functional analysis of the transcription repressor PLU-1/JARID1B. *Mol Cell Biol* 2007; **27**:7220–7235.

Sens MA, Somji S, Garrett SH, *et al.* Metallothionein isoform 3 overexpression is associated with breast cancers having a poor prognosis. *Am J Pathol* 2001; **159**:21–26.

Sens MA, Somji S, Lamm DL, *et al.* Metallothionein isoform 3 as a potential biomarker for human bladder cancer. *Environ Health Perspect* 2000; **108**:413–418.

Seow A, Koh WP, Chia KS, *et al.* Cancer incidence in Singapore 1968–2002. Singapore Cancer Registry Report No. 6, Singapore, 2004, pp. 15–41.

Shao ZM, Nguyen M, Alpaugh ML, *et al.* The human myoepithelial cell exerts antiproliferative effects on breast carcinoma cells characterized by $p21^{WAF1/CIP1}$ induction, G_2/M arrest and apoptosis. *Exp Cell Res* 1998; **241**:394–403.

Stennard FA, Holloway AF, Hamilton J, West AK. Characterization of six additional human metallothionein genes. *Biochim Biophys Acta* 1994; **1218**:357–365.

Sternlicht MD, Safarians S, Rivera SP, Barsky SH. Characterization of the extracellular matrix and the proteinase inhibitor content of human myoepithelial tumors. *Lab Invest* 1996; **74**:781–796.

Surowaik P, Matkowski R, Materna V, *et al.* Elevated metallothionein (MT) in invasive ductal breast cancer predicts tamoxifen resistance. *Histol Histopathol* 2005; **20**:1037–1044.

Swell AK, Jensen LT, Erickson JC, *et al.* The bioactivity of metallothionein-3 correlates with its novel β domain sequence rather than metal binding properties. *Biochemistry* 1995; **34**:4740–4747.

Tai SK, Tan OJ, Chow VT, *et al.* Differential expression of metallothionein 1 and 2 isoforms in breast cancer lines with different invasive potential: Identification of a novel nonsilent metallothionein-1H mutant variant. *Am J Pathol* 2003; **163**:2009–2019.

Theocharis SE, Margeli AP, Klijanienko JT, Kouraklis GP. Metallothionein expression in human neoplasia. *Histopathology* 2004; **45**:103–118.

Tsuji S, Kobayashi H, Uchida Y, *et al.* Molecular cloning of human growth inhibitory factor cDNA and its down-regulation in Alzheimer's disease. *EMBO J* 1992; **11**:4843–4850.

Uchida Y, Takio K, Titani K, *et al.* The growth inhibitory factor that is deficient in Alzheimer's disease is a 68 amino acid metallothionein-like protein. *Neuron* 1991; **7**:337–347.

Vazquez-Ramirez FJ, Gonzalez-Campora JJ, Hevia-Alvarez E. P-glycoprotein, metallothionein and NM23 protein expression in breast carcinoma. *Pathol Res Pract* 2000; **196**:553–559.

Weaver VM, Lelievre S, Lakins JN, *et al.* Beta4 integrin-dependent formation of polarized three-dimensional architecture confers resistance to apoptosis in normal and malignant mammary epithelium. *Cancer Cell* 2002; **2**:205–216.

West AK, Stallings R, Hildebrand CE, *et al.* Human metallothionein genes: Structure of the functional locus at 16q13. *Genomics* 1990; **8**:513–518.

Yang M, Kroft SH, Chitambar CR. Gene expression analysis of gallium-resistant and gallium-sensitive lymphoma cells reveals a role for metal-responsive transcription factor-1, metallothionein-2A and zinc-transporter-1 in modulating the antineoplastic activity of gallium nitrate. *Mol Cancer Ther* 2007; **6**:633–643.

Yang YY, Woo ES, Reese CE, *et al.* Human metallothionein isoform gene expression in cisplatin-sensitive and resistant cells. *Mol Pharmacol* 1994; **45**:453–460.

Ysaziji H, Gown AM, Sneige N. Detection of stromal invasion in breast cancer: The myoepithelial markers. *Adv Anat Pathol* 2000; **7**:100–109.

Zhang R, Zhang H, Wei H. Expression of metallothionein in invasive ductal breast cancer in relation to prognosis. *J Environ Pathol Toxicol Oncol* 2000; **19**:95–97.

Chapter 10

METALLOTHIONEINS AND PROSTATE CANCER

Narassa Narayani and K. C. Balaji

Metallothionein (MT) is an avid metal-binding (metal + thiol/sulphur-binding) protein in the human body. It binds to trace elements like zinc and copper as well as heavy metals like cadmium, and plays an important role in metal detoxification and homeostasis. MT isoforms are expressed differentially in benign and malignant prostate tissue, with increased MT expression noted in higher-Gleason-grade prostate cancer. MT expression in prostate has been shown to be regulated by high Zn concentration and promoter hypermethylation. MT is known to play a role in the resistance to chemotherapeutic agents such as cisplatin and radiation treatment, presumably by trace metal or free radical scavenging. MT expression in the prostate gland is of particular interest because heavy metals such as Zn, which is present at the highest concentration in prostate compared to other human organs, induce MT expression and may be amenable to therapeutic manipulation in order to improve sensitivity to chemotherapy and radiation. MT may prove to be a useful therapeutic target for novel approaches such as local or systemic heavy metal chelation therapy and gene vectors for treating patients with prostate cancer.

Keywords: Metallothioneins; prostate cancer; zinc; biomarker; treatment.

1. Introduction

Metallothioneins (MTs) are small-molecular-weight scavenger proteins that have been shown to play a role in normal development and in disease states such as cancer (Cherian *et al.*, 1993; Kägi and Schaffer, 1988). Among cancers in Western countries, prostate cancer is the most commonly diagnosed noncutaneous cancer in men and is associated with several thousand deaths each year (Hellerstedt and

Pienta, 2002; Jemal *et al.*, 2005). The role of MTs in prostate cancer is of particular interest because a variety of MT isoforms are differentially expressed in benign and malignant prostate tissue, and are associated with normal and disease development of the prostate gland (Theocharis *et al.*, 2004; Theocharis *et al.*, 2003). While the human prostate gland is not critical for survival in humans, it contributes to liquefaction of semen and plays an important role in fertility (Lwaleed *et al.*, 2004). In this chapter, we will discuss the expression of MT isoforms in prostate cancer, their association with prostate cancer, and the potential role of MT in disease development and in resistance to treatments such as radiation therapy and chemotherapy (Lazo *et al.*, 1998).

2. Prostate Cancer

2.1. *Prostate Gland Has Highest Levels of Tissue Zinc Concentration Among Human Organs*

Prostate gland is unique among human organs because of its high levels of tissue Zn concentration compared to other organs (Costello and Franklin, 1998; Mawson and Fischer, 1951; Mawson and Fischer, 1952; Mawson and Fischer, 1953). Because MTs are metal-responsive proteins, the high concentration of Zn in prostate glands provides an important regulatory mechanism that is required for understanding MT expression in prostate gland and for its potential therapeutic manipulation to treat prostate diseases.

Interestingly, Zn concentration varies within prostate gland, with about 10-fold higher concentrations demonstrable in benign prostate compared to malignant prostate (Gyorkey *et al.*, 1967; Ogunlewe and Osegbe, 1989). Similar to Zn, citrate levels have also been demonstrated to decrease in prostate malignant tissue compared to benign glands (Costello and Franklin, 1998). Normally, mitochondrial aconitase, which is capable of oxidizing citrate, is kept inhibited by zinc in prostate gland; however, in prostate cancer cells, the low concentration of zinc limits aconitase inhibition, leading to citric acid oxidation and reduction in tissue citrate levels. The low levels of citrate in malignant prostate have been exploited in magnetic resonance imaging (MRI) spectroscopy, where the tissue choline-to-citrate ratio detects prostate cancer within the gland with a positive predictive value of 90% and excludes the presence of cancer with an 83% negative predictive value (Kurhanewicz *et al.*, 2000; Westphalen *et al.*, 2007). Such an improvement in the clinical imaging of prostate provides the impetus to study additional molecular markers that may contribute to improve the clinical care of patients. MTs hold promise for patients with prostate cancer because their expression can be influenced by trace metal concentration (including Zn), and also because of their evolving role in prostate cancer development and in resistance to commonly used treatment modalities such as radiation therapy and chemotherapy.

2.2. *Expression of MT Isoforms in Prostate*

Depending on the age and the type of tissue, most adult mammalian tissue contains very low basal levels of MT. In the human body, it is synthesized primarily in the liver and kidneys. During early fetal development, MT is known to play the role of a temporary reservoir for essential metals like Zn and Cu. While in the early neonatal period, it is detected in the nucleus and the cytoplasm of the cell, in adult tissues it is mainly a cytoplasmic protein (Cherian *et al.*, 1994). A transient localization of MT into nucleus is seen during cell proliferation and differentiation under certain conditions like embryogenesis, early fetal development, and carcinogenesis. These changes in intracellular localization and expression of the MT gene are suggestive of MT being an oncodevelopmental tumor marker (Cherian *et al.*, 1994; Cherian *et al.*, 1993).

Since the discovery of MT in 1957 (Margoshes and Vallee, 1957), at least 10 isoforms of MT have been identified. They have been broadly subdivided into four major subgroups, MT-1 to

MT-4 (Moffatt and Denizeau, 1997). Of these isoforms, MT-1/MT-2, MT-2A, and more recently MT-3 (which was thought to be confined to neural tissue; see Uchida *et al.*, 1991) have been demonstrated in the human reproductive tissue. In the normal prostate gland, MT was seen to be localized more in the glandular epithelial cells of the peripheral zone than in the central zone of the prostate, indicating that this might be due to the functional difference among the prostatic epithelial cells (Suzuki *et al.*, 1991). Normal human prostate expresses the MT-1A, MT-1E, MT-1X, and MT-2A genes; however, MT-1X was not demonstrable in advanced prostate cancer (Garrett *et al.*, 2000).

Prostate cancer is commonly graded using the Gleason grading system, with an increasing grade correlating with high-risk disease and poorer clinical outcome (Albertsen *et al.*, 1998; Moussa *et al.*, 1997). MT expression has been shown to increase with a worsening Gleason grade, suggesting that MT expression may correlate with disease outcome. El Sharkawy *et al.* (2006) specifically addressed the MT-2 expression in human prostate tissue and identified high MT-2 expression in prostatic intraepithelial neoplasia (PIN), commonly considered to be a precursor of prostate cancer, and also noted that an increasing MT-2 expression correlated with a worsening Gleason grade. Contrary to prostate gland, seminal vesicles located adjacent to prostate gland rarely develop malignancy. While 70% of prostate tissue expressed MT, only 40% of seminal vesicle tissue stained for MT, suggesting that MT expression is correlated with an increased risk of malignancy (Pannek *et al.*, 2001).

2.3. *Role of MT in Prostate Cancer*

MT plays a main role in protecting the body against metal toxicity and from oxidative stress by acting as a free radical scavenger (Klaassen *et al.*, 1999; Thornalley and Vašák, 1985). MT has been demonstrated to have antioxidant, anti-inflammatory, and antiapoptotic properties (Aschner and West, 2005; Penkowa *et al.*, 2006), and to play an important role in cellular processes such as cellular proliferation and growth. MT also seems to be involved in regulation of the tumor suppressor protein p53 (Ostrakhovitch *et al.*, 2006). Malignant cells enriched with MT both *in vivo* and *in vitro* have shown to exhibit greater resistance to chemotherapeutic agents (Kelley *et al.*, 1988).

Studies have shown that MT is mainly a cytoplasmic protein in adult tissues, and is detected in the nucleus of normal cells in the early fetal and neonatal period (Chan and Cherian, 1993). When the level of intracellular MT expression increases, MT tends to localize to the nucleus, and by doing so causes rapid proliferation of the cells (Tohyama *et al.*, 1993), inhibits apoptosis (Kondo *et al.*, 1997), and moves the cells into the S phase (Tsujikawa *et al.*, 1991). Several studies have shown that increased MT expression may protect the cells from carcinogenic effects of cadmium or anticancer drugs (e.g. cisplatin, melphelan, chlorambucil) and from ionizing radiation (Bakka *et al.*, 1981; Kondo *et al.*, 1995b). However, this protection of the cells is not always seen when MT is overexpressed, but is seen to occur only when there is subcellular localization into the nucleus (Kondo *et al.*, 1995a).

MT also plays a major role in the detoxification of heavy metals. Cadmium, a heavy metal, is an environmental pollutant and is a major constituent of tobacco smoke. It has been classified as a toxic metal by the International Agency for Research on Cancer (IARC) as a known human carcinogen (Group 1; Hartwig, 1998). Exposure to this heavy metal, which has no known beneficial physiological role, has been linked to a wide range of detrimental effects on mammalian reproduction. Human studies indicate that nearly 7% of the general population suffers renal dysfunction from cadmium exposure (Klaassen *et al.*, 1999). Although it is not possible to quantify the contribution of cadmium to the incidence of prostate cancer (Waalkes and Rehm, 1994), cadmium and zinc have both been implicated as carcinogens for prostate cancer (Habib, 1980; Palmer, 1984; Waalkes and Rehm, 1994; Waalkes *et al.*, 1992). Cadmium is known to have androgen-like activity in the prostate, and has been implied to be the potential mechanism for its carcinogenic effect (Ye *et al.*, 2000). It is known that African-American men in South Carolina have the highest age-adjusted death rate for prostate cancer in the United States; among the many factors is environmental exposure to cadmium and selenium through groundwater, as the soil in these rural areas has cadmium and selenium concentrations unique to South Carolina (Drake *et al.*, 2006). It is conceivable that there might be a possible interaction between the geological and underlying biological factors such as metal transporter gene expression by race in black men (Blackshear *et al.*, 2003).

In vitro and a few *in vivo* studies have shown that cadmium has the ability to stimulate the expression of the cellular proto-oncogenes c-*jun*, c-*fos*, and c-*myc* (Abshire *et al.*, 1996). Metallothionein null cells are more susceptible to chemical-induced apoptosis, while MT-I and MT-II knockout mice are sensitive to the cadmium-induced mRNA expression of c-jun and p53 (Zheng *et al.*, 1996). These data suggest that MT, a trace metal and heavy metal detoxifier, may be one of the important mechanisms that the body uses to prevent cadmium-induced prostate cancer.

2.4. *Regulation of MT Expression in Prostate Cancer*

Human MT synthesis is induced by metals such as Zn and Cd; by endogenous factors such as vitamin D, interferon, reactive oxygen species, interleukin 1, and glucocorticoids; and also by stress (Bremner, 1987). The level of the response to these inducers depends on the MT gene. The human MT gene is located on chromosome 16 (16q13) (West *et al.*, 1990). At least 10 of the 17 genes identified so far have been found to be functional. MT gene expression, which is metal- and isoform-specific, is controlled primarily at the level of transcription (Palmiter, 1998). MT genes present in their promoter-specific proximal sequences include metal response elements (MREs), glucocorticoid response elements (GREs), and antioxidant response element. The transcriptional regulation of MT by metals depends on the metal-responsive transcription factor MTF-1 (metal transcription factor 1), a zinc finger transcription factor (Heuchel *et al.*, 1994) that is an important zinc sensor. MTF-1 in turn is under the control of a Zn-sensitive inhibitor termed MTI (metallothionein inhibitor) (Palmiter, 1994).

Differential expression of MT isoforms MT-1, MT-2, and MT-3 in response to Zn treatment has been demonstrated in human normal and malignant prostate cells and tissues (Wei *et al.*, 2008). The study provided evidence that there is attenuated MT-1/MT-2 expression with prostate tumor progression, and that the zinc induction of MT-1/MT-2 expression results in response to cellular zinc restoration. MT expression is upregulated under hypoxic conditions in prostate cancer cell lines and promotes cell survival (Yamasaki *et al.*, 2007).

MT expression is also regulated by promoter methylation and protein degradation. The MT-I promoter has been shown to be suppressed in human prostate cancer lines PC3 and DU145, probably by promoter methylation; whereas cadmium-induced MT-I in the human prostate cancer line LNCaP seems to be independent of promoter methylation. Degradation of MT protein is regulated primarily by the cellular Zn content and occurs in both lysosomal and nonlysosomal compartments (Chen and Failla, 1989), each being depleted and replenished at a different rate (Steinebach and Wolterbeek, 1992). Cytosolic MT is degraded by the cystosolic 26S proteosome complex (McKim *et al.*, 1992). The role of MT degradation specifically in prostate cancer remains to be investigated.

2.5. *Clinical Utility of MT in Prostate Cancer*

2.5.1. *MT as a Biomarker of Disease Progression in Prostate Cancer*

Studies using paraffinized human prostate tissue demonstrate both nuclear and cytoplasmic MT stains in benign and malignant prostate tissue. An initial study by Zhang *et al.* (1996) demonstrated a positive correlation between increased MT staining and increasing Gleason grade in prostate cancer. Another study specifically studied MT-II isoform expression in hyperplastic, PIN, and neoplastic prostate human tissue. While epithelial cells in normal and benign prostatic tissues from 8 patients demonstrated patchy metallothionein staining, all 6 cases of PIN and 20 of 30 patients with prostatic carcinoma showed positive staining for metallothionein, which increased considerably from low-grade to high-grade tumors (El Sharkawy *et al.*, 2006). A study using imprint smears of prostate cancer tissue demonstrated a positive correlation between MT staining and nuclear proliferation, increasing Gleason score, pathologic stage, and disease recurrence (Athanassiadou *et al.*, 2007). While most studies demonstrate increased MT staining in prostate cancer, the MT-1G isoform has been reported to be downregulated in prostate cancer. Almost a quarter of 121 prostate cancer human samples had demonstrable hypermethylation of the MT-1G promoter, compared to none of the 13 benign prostate tissues examined.

Overall, the published studies suggest that expression of MT isoforms is altered in prostate cancer, and there is some preliminary evidence to suggest that MT may be useful as a biomarker of prostate cancer progression. However, larger studies specifically addressing the various MT isoforms and long-term clinical follow-up are necessary to establish the clinical utility of MT expression in prostate cancer.

2.5.2. *MT as a Target to Improve Sensitivity to Chemotherapy in Prostate Cancer*

Prostate cancer is generally considered resistant to chemotherapy (Kamradt *et al.*, 1999; Urakami *et al.*, 2005). Recently, two large randomized clinical trials demonstrated a modest improvement of 2 months after the administration of docetaxol, a microtubule-stabilizing agent, in patients with advanced prostate cancer who failed first-line androgen ablative treatment (Petrylak *et al.*, 2004;

Tannock *et al.*, 2004). There continues to exist an urgent need to improve chemotherapy or sensitizing agents to current chemotherapy in order to improve the clinical outcome in patients with prostate cancer. Using prostate cancer cell line models, MTs have been shown to induce resistance to adriamycin in DU145 cells (Webber *et al.*, 1988). The ribozyme-induced loss of MT-2A mRNA induced cell death in prostate cancer PC-3 cell lines and was associated with dose-dependent downregulation of the proto-oncogene c-*myc* and the anti-apoptotic gene *Bcl*-2, suggesting that MT-2A is an important cell survival or antiapoptotic factor for prostate cancer cells (Tekur and Ho, 2002). The established role of MT in causing chemoresistance in other cancers, the upregulation of MT with advancing grade and stage of prostate cancer, and the inherent chemoresistance in prostate cancer strongly suggest a potential role for MT in prostate cancer as well. For example, cells with acquired resistance to cisplatin or chlorambucil overexpress MT, which tends to bind these alkylating agents to a higher extent than the nonresistant cells (Ebadi and Iversen, 1994). In addition to sequestering electrophilic anticancer drugs, MT alters the therapeutic efficacy of antineoplastic agents by regulating the activities of zinc-requiring metalloenzymes or scavenging radical species (Ebadi and Iversen, 1994).

We have previously established an excellent *in vitro* cell line model to study the effect of MT expression in prostate cancer following treatment with Zn (Smith *et al.*, 2006). The results from our microarray analysis confirmed that zinc treatment induces 0.33% of the 12 144 genes studied, most of which are MTs, and were significantly associated with resistance to cisplatin in prostate cancer cells. Others have demonstrated the presence of an MT-like zinc-binding protein in prostate cancer cell lines exhibiting relative resistance to cisplatin, and nuclear localization of MT has been shown to be associated with cisplatin resistance in prostate cancer cell lines (Kondo *et al.*, 1995a; Metcalfe *et al.*, 1986). Although no specific MT inhibitor has yet been described, inhibition of cysteine synthesis by propargylglycine has been shown to significantly reduce MT induction in mice inoculated with human or murine bladder tumor cells, which markedly enhanced the antitumor activity of cisplatin and other drugs (Saga *et al.*, 2004). Because MT may be therapeutically manipulated, further studies are needed to explore MT as a biomarker for the chemosensitivity of prostate cancer to cisplatin treatment and to possibly improve the efficacy of cisplatin as a radiosensitizer.

2.5.3. *MT as a Target to Improve Sensitivity to Radiation Treatment in Prostate Cancer*

Radiation to prostate is a primary treatment option for patients with prostate cancer, and has been shown to produce a satisfactory long-term clinical outcome (Altundag *et al.*, 2005; Bao *et al.*, 1991). However, about a third of patients undergoing radiation treatment eventually fail, emphasizing the need to improve long-term response to radiation treatment (Pollack *et al.*, 2003). MT plays a homeostatic role in the control and detoxification of heavy metals; several evidences indicate that MT has the capacity to scavenge reactive oxygen metabolite (ROM), particularly the hydroxyl radical. These substances — which are produced continuously during normal aerobic metabolism — may become noxious in situations of imbalance with endogenous antioxidants, leading to cellular destruction, chromosomal aberrations, and finally cancer. Paradoxically, by anticancer treatment such as radiotherapy and chemotherapy, tumor cells are killed by generating toxic amounts of ROM.

Prostate is a readily accessible organ that may be amenable to injection of local agents which could improve sensitivity to radiation. Cisplatin is a well-established chemotherapy drug that is known to improve sensitivity to radiation (Ebadi and Iversen, 1994). Alternatively, agents that influence MT expression either directly (such as gene vectors) or indirectly (through heavy metals such as Zn) can be administered either directly into prostate or systemically to improve sensitivity to radiation treatment. EDTA, a heavy metal chelator, has been used in human clinical trials in patients with coronary artery disease with an established safety profile (Knudtson *et al.*, 2002).

Using a general heavy metal chelator (EDTA) or a more Zn-specific chelator (TPEN) (Hashemi *et al.*, 2007) at the time of radiation treatment for prostate cancer (clinically not used yet) may decrease Zn-mediated MT induction and improve sensitivity to radiation. Such novel approaches could potentially improve the long-term clinical outcome to radiation treatment for prostate cancer and decrease disease recurrence. Clearly, further preclinical and clinical studies are needed prior to establishing MT as a useful target in patients undergoing radiation treatment for prostate cancer.

3. Conclusion

Metallothioneins (MTs) are trace metal and free radical scavenging small-molecular-weight proteins that are expressed in several human organs including the prostate, and seem to play an important role in development and cellular homeostasis. While the various isoforms of MT are expressed differentially in benign and malignant prostate tissue, MT expression increases with a worsening grade of prostate cancer. MT expression in prostate is of particular interest because heavy metals such as Zn, which is present at the highest concentration in prostate compared to other human organs, induce MT expression and may be amenable to therapeutic manipulation. In addition to induction by Zn, MT expression may also be regulated through promoter methylation in prostate tissue. MT has been demonstrated to play a role in the resistance to cisplatin chemotherapy and radiation treatment in prostate cancer. Because of the high concentration of Zn in prostate, MT may prove to be a useful therapeutic target in prostate cancer in order to improve the sensitivity to chemotherapy and radiation therapy treatments.

References

Abshire MK, Buzard GS, Shiraishi N, Waalkes MP. Induction of c-myc and c-jun proto-oncogene expression in rat L6 myoblasts by cadmium is inhibited by zinc preinduction of the metallothionein gene. *J Toxicol Environ Health* 1996; **48**:359–377.

Albertsen PC, Hanley JA, Gleason DF, Barry MJ. Competing risk analysis of men aged 55 to 74 years at diagnosis managed conservatively for clinically localized prostate cancer [see comments]. *JAMA* 1998; **280**:975–980.

Altundag O, Altundag K, Morandi P, Hanrahan E. Cisplatin as a radiosensitizer in the treatment of locally advanced head and neck cancer. *Oral Oncol* 2005; **41**:435.

Aschner M, West AK. The role of MT in neurological disorders. *J Alzheimers Dis* 2005; **8**:139–145.

Athanassiadou P, Bantis A, Gonidi M, *et al.* The expression of metallothioneins on imprint smears of prostate carcinoma: Correlation with clinicopathologic parameters and tumor proliferative capacity. *Tumori* 2007; **93**:189–194.

Bakka A, Endresen L, Johnsen ABS, *et al.* Resistance against *cis*-dichlorodiammineplatinum in cultured cells with a high content of metallothionein. *Toxicol Appl Pharmacol* 1981; **61**:215–226.

Bao YH, Yuan XC, Wu JD, *et al.* Cisplatin as a radiosensitizer in clinical practice: A pilot study. *Tumori* 1991; **77**:21–24.

Blackshear PJ, Phillips RS, Vazquez-Matias J, Mohrenweiser H. Polymorphisms in the genes encoding members of the tristetraprolin family of human tandem CCCH zinc finger proteins. *Prog Nucleic Acid Res Mol Biol* 2003; **75**:43–68.

Bremner I. Nutritional and physiological significance of metallothionein. *Experientia Suppl* 1987; **52**:81–107.

Chan HM, Cherian MG. Ontogenic changes in hepatic metallothionein isoforms in prenatal and newborn rats. *Biochem Cell Biol* 1993; **71**:133–140.

Chen M, Failla M. Degradation of zinc-metallothionein in monolayer cultures of rat hepatocytes. *Proc Soc Exp Biol Med* 1989; **191**:130–138.

Cherian MG, Howell SB, Imura N, *et al.* Role of metallothionein in carcinogenesis. *Toxicol Appl Pharmacol* 1994; **126**:1–5.

Cherian MG, Huang PC, Klaassen CD, *et al.* National Cancer Institute workshop on the possible roles of metallothionein in carcinogenesis. *Cancer Res* 1993; **53**:922–925.

Costello LC, Franklin RB. Novel role of zinc in the regulation of prostate citrate metabolism and its implications in prostate cancer. *Prostate* 1998; **35**:285–296.

Drake BE, Keane TE, Mosley CM, *et al.* Prostate cancer disparities in South Carolina: Early detection, special programs, and descriptive epidemiology. *J SC Med Assoc* 2006; **102**:241–249.

Ebadi M, Iversen PL. Metallothionein in carcinogenesis and cancer chemotherapy. *Gen Pharmacol* 1994; **25**:1297–1310.

El Sharkawy SL, Abbas NF, Badawi MA, El Shaer MA. Metallothionein isoform II expression in hyperplastic, dysplastic and neoplastic prostatic lesions. *J Clin Pathol* 2006; **59**:1171–1174.

Garrett SH, Sens MA, Shukla D, *et al.* Metallothionein isoform 1 and 2 gene expression in the human prostate: Downregulation of MT-1X in advanced prostate cancer. *Prostate* 2000; **43**:125–135.

Gyorkey F, Min KW, Huff JA, Gyorkey P. Zinc and magnesium in human prostate gland: Normal, hyperplastic, and neoplastic. *Cancer Res* 1967; **27**:1348–1353.

Habib FK. Evaluation of androgen metabolism studies in human prostate cancer — Correlation with zinc levels. *Prev Med* 1980; **9**:650–656.

Hartwig A. Carcinogenicity of metal compounds: Possible role of DNA repair inhibition. *Toxicol Lett* 1998; **102–103**:235–239.

Hashemi M, Ghavami S, Eshraghi M, *et al.* Cytotoxic effects of intra and extracellular zinc chelation on human breast cancer cells. *Eur J Pharmacol* 2007; **557**:9–19.

Hellerstedt BA, Pienta KJ. The current state of hormonal therapy for prostate cancer. *CA Cancer J Clin* 2002; **52**:154–179.

Heuchel R, Radtke F, Georgiev O, *et al.* The transcription factor MTF-1 is essential for basal and heavy metal-induced metallothionein gene expression. *EMBO J* 1994; **13**:2870–2875.

Jemal A, Murray T, Ward E, *et al.* Cancer statistics, 2005. *CA Cancer J Clin* 2005; **55**:10–30.

Kägi JH, Schaffer A. Biochemistry of metallothionein. *Biochemistry* 1988; **27**:8509–8515.

Kamradt JM, Klein EA, Pienta KJ. Rational use of chemotherapy. It is not just rat poison. *Urol Clin North Am* 1999; **26**:275–279.

Kelley SL, Basu A, Teicher BA, *et al.* Overexpression of metallothionein confers resistance to anticancer drugs. *Science* 1988; **241**:1813–1815.

Klaassen CD, Liu J, Choudhuri S. Metallothionein: An intracellular protein to protect against cadmium toxicity. *Annu Rev Pharmacol Toxicol* 1999; **39**:267–294.

Knudtson ML, Wyse DG, Galbraith PD, *et al.* Program to Assess Alternative Treatment Strategies to Achieve Cardiac Health Investigators. Chelation therapy for ischemic heart disease: A randomized controlled trial. *JAMA* 2002; **287**:481–486.

Kondo Y, Kuo SM, Watkins SC, Lazo JS. Metallothionein localization and cisplatin resistance in human hormone-independent prostatic tumor cell lines. *Cancer Res* 1995a; **55**:474–477.

Kondo Y, Rusnak JM, Hoyt DG, *et al.* Enhanced apoptosis in metallothionein null cells. *Mol Pharmacol* 1997; **52**:195–201.

Kondo Y, Woo ES, Michalska AE, *et al.* Metallothionein null cells have increased sensitivity to anticancer drugs. *Cancer Res* 1995b; **55**:2021–2023.

Kurhanewicz J, Males RG, Swanson MG, *et al.* The prostate: MR imaging and spectroscopy. Present and future. *Radiol Clin North Am* 2000; **38**:115–138.

Lazo JS, Kuo SM, Woo ES, Pitt BR. The protein thiol metallothionein as an antioxidant and protectant against antineoplastic drugs. *Chem Biol Interact* 1998; **111–112**:255–262.

Lwaleed BA, Greenfield R, Stewart A, *et al.* Seminal clotting and fibrinolytic balance: A possible physiological role in the male reproductive system. *Thromb Haemost* 2004; **92**:752–766.

Margoshes M, Vallee BL. A cadmium protein from equine kidney cortex. *J Am Chem Soc* 1957; **79**:4813–4814.

Mawson CA, Fischer MI. Zinc content of the genital organs of the rat. *Nature* 1951; **167**:859.

Mawson CA, Fischer MI. The occurrence of zinc in the human prostate gland. *Can J Med Sci* 1952; **30**:336–339.

Mawson CA, Fischer MI. Zinc and carbonic anhydrase in human semen. *Biochem J* 1953; **55**:696–700.

McKim JM Jr, Choudhuri S, Klaassen CD. *In vitro* degradation of apo-, zinc-, and cadmium-metallothionein by cathepsins B, C, D. *Toxicol Appl Pharmacol* 1992; **116**:117–124.

Metcalfe SA, Cain K, Hill BT. Possible mechanism for differences in sensitivity to *cis*-platinum in human prostate tumor cell lines. *Cancer Lett* 1986; **31**:163–169.

Moffatt P, Denizeau F. Metallothionein in physiological and physiopathological processes. *Drug Metab Rev* 1997; **29**:261–307.

Moussa M, Kloth D, Peers G, *et al.* Metallothionein expression in prostatic carcinoma: Correlation with Gleason grade, pathologic stage, DNA content and serum level of prostate-specific antigen. *Clin Invest Med* 1997; **20**:371–380.

Ogunlewe JO, Osegbe DN. Zinc and cadmium concentrations in indigenous blacks with normal, hypertrophic, and malignant prostate. *Cancer* 1989; **63**:1388–1392.

Ostrakhovitch EA, Olsson PE, Jiang S, Cherian MG. Interaction of metallothionein with tumor suppressor p53 protein. *FEBS Lett* 2006; **580**:1235–1238.

Palmer S. Public health policy on diet, nutrition, and cancer. *Nutr Cancer* 1984; **6**:274–283.

Palmiter R. Regulation of metallothionein genes by heavy metals appears to be mediated by a zinc-sensitive inhibitor that interacts with a constitutively active transcription factor, MTF-1. *Proc Natl Acad Sci USA* 1994; **91**:1219–1223.

Palmiter RD. The elusive function of metallothioneins. *Proc Natl Acad Sci USA* 1998; **95**: 8428–8430.

Pannek J, Lecksell KL, Partin AW. Expression of metallothionein in seminal vesicles — An immunohistochemical study. *Scand J Urol Nephrol* 2001; **35**:11–14.

Penkowa M, Tio L, Giralt M, *et al.* Specificity and divergence in the neurobiologic effects of different metallothioneins after brain injury. *J Neurosci Res* 2006; **83**:974–984.

Petrylak DP, Tangen CM, Hussain MH, *et al.* Docetaxel and estramustine compared with mitoxantrone and prednisone for advanced refractory prostate cancer. *N Engl J Med* 2004; **351**:1513–1520.

Pollack A, Hanlon AL, Movsas B, *et al.* Biochemical failure as a determinant of distant metastasis and death in prostate cancer treated with radiotherapy. *Int J Radiat Oncol Biol Phys* 2003; **57**:19–23.

Saga Y, Hashimoto H, Yachiku S, *et al.* Reversal of acquired cisplatin resistance by modulation of metallothionein in transplanted murine tumors. *Int J Urol* 2004; **11**:407–415.

Smith DJ, Jaggi M, Zhang W, *et al.* Metallothioneins and resistance to cisplatin and radiation in prostate cancer. *Urology* 2006; **67**:1341–1347.

Steinebach OM, Wolterbeek BT. Metallothionein biodegradation in rat hepatoma cells: A compartmental analysis aided 35S-radiotracer study. *Biochim Biophys Acta* 1992; **1116**:155–165.

Suzuki T, Umeyama T, Ohma C, *et al.* Immunohistochemical study of metallothionein in normal and benign prostatic hyperplasia of human prostate. *Prostate* 1991; **19**:35–42.

Tannock IF, de Wit R, Berry WR, *et al.* Docetaxel plus prednisone or mitoxantrone plus prednisone for advanced prostate cancer. *N Engl J Med* 2004; **351**:1502–1512.

Tekur S, Ho SM. Ribozyme-mediated downregulation of human metallothionein II (a) induces apoptosis in human prostate and ovarian cancer cell lines. *Mol Carcinog* 2002; **33**:44–55.

Theocharis SE, Margeli AP, Klijanienko JT, Kouraklis GP. Metallothionein expression in human neoplasia. *Histopathology* 2004; **45**:103–118.

Theocharis SE, Margeli AP, Koutselinis A. Metallothionein: A multifunctional protein from toxicity to cancer. *Int J Biol Markers* 2003; **18**:162–169.

Thornalley PJ, Vašák M. Possible role for metallothionein in protection against radiation-induced oxidative stress. Kinetics and mechanism of its reaction with superoxide and hydroxyl radicals. *Biochim Biophys Acta* 1985; **827**:36–44.

Tohyama C, Suzuki JS, Hemelraad J, *et al.* Induction of metallothionein and its localization in the nucleus of rat hepatocytes after partial hepatectomy. *Hepatology* 1993; **185**:1193–1201.

Tsujikawa K, Imai T, Kakutani M, *et al.* Localization of metallothionein in nuclei of growing primary cultured adult rat hepatocytes. *FEBS Lett* 1991; **283**:239–242.

Uchida Y, Takio K, Titani K, *et al.* The growth inhibitory factor that is deficient in the Alzheimer's disease brain is a 68 amino acid metallothionein-like protein. *Neuron* 1991; **7**:337–347.

Urakami S, Yoshino T, Kikuno N, *et al.* Docetaxel-based chemotherapy as second-line treatment for paclitaxel-based chemotherapy-resistant hormone-refractory prostate cancer: A pilot study. *Urology* 2005; **65**:543–548.

Waalkes MP, Rehm S. Cadmium and prostate cancer. [Review]. *J Toxicol Environ Health* 1994; **43**:251–269.

Waalkes MP, Rehm S, Perantoni AO, Coogan TP. Cadmium exposure in rats and tumours of the prostate. *IARC Sci Publ* 1992; **118**:391–400.

Webber MM, Rehman SM, James GT. Metallothionein induction and deinduction in human prostatic carcinoma cells: Relationship with resistance and sensitivity to adriamycin. *Cancer Res* 1988; **48**:4503–4508.

Wei HDM, Lin S, Xiao D, *et al.* Differential expression of metallothioneins (MTs) 1, 2, and 3 in response to zinc treatment in human prostate normal and malignant cells and tissues. *Mol Cancer* 2008; **7**:7.

West AK, Stallings R, Hildebrand CE, *et al.* Human metallothionein genes: Structure of the functional locus at 16q13. *Genomics* 1990; **8**:513–518.

Westphalen AC, Coakley FV, Qayyum A, *et al.* Peripheral zone prostate cancer: Accuracy of different interpretative approaches with MR and MR spectroscopic imaging. *Radiology* 2007; **246**:177–184.

Yamasaki M, Nomura T, Sato F, Mimata H. Metallothionein is up-regulated under hypoxia and promotes the survival of human prostate cancer cells. *Oncol Rep* 2007; **18**:1145–1153.

Ye J, Wang S, Barger M, *et al.* Activation of androgen response element by cadmium: A potential mechanism for a carcinogenic effect of cadmium in the prostate. *J Environ Pathol Toxicol Oncol* 2000; **19**:275–280.

Zhang XH, Jin L, Sakamoto H, Takenaka I. Immunohistochemical localization of metallothionein in human prostate cancer. *J Urol* 1996; **156**:1679–1681.

Zheng H, Liu J, Choo KH, *et al.* Metallothionein-I and -II knock-out mice are sensitive to cadmium-induced liver mRNA expression of c-jun and p53. *Toxicol Appl Pharmacol* 1996; **136**:229–235.

Chapter 11

METALLOTHIONEIN AND COLORECTAL TUMORS

Daisuke Shirasaka

There have been some results reported concerning the expression of metallothionein (MT) in colorectal tumors. This paper reviews four aspects: the contribution of MT in colorectal carcinogenesis, the prognostic significance of MT, MT expression in colorectal liver metastases, and the clinical significance of MT as a drug-resistance protein.

Keywords: Carcinogenesis; prognosis; liver metastases.

1. Contribution of Metallothionein in Colorectal Carcinogenesis

Ofner *et al.* (1994) found considerable metallothionein (MT) staining (i.e. 81%) of transitional and/or normal mucosa, as opposed to only 31% of colorectal carcinomas; in addition, 47% of the tumors almost completely lacked MT staining. Giuffrè *et al.* (1996) also reported a significant decrease in MT content in carcinomas compared with normal colonic mucosa (Table 1). Janssen *et al.* (2002) reported that more than 74% of the carcinomas were found to have a lower MT level than their corresponding normal mucosas.

In our study, MT expression was observed in all portions of normal colorectal mucosa adjacent to the tumor (Fig. 1), and there was a significant decrease in MT expression as the tumor progressed (Kuroda *et al.*, 2002) (Table 1). In adenomas, MT expression was positive in 58 (67%) cases; however, in carcinomas, MT expression was positive

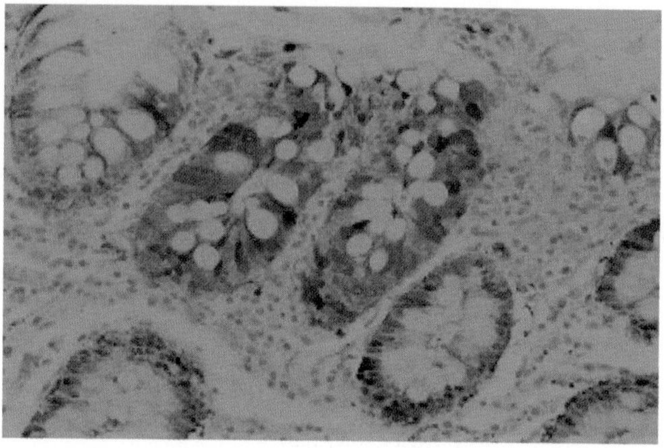

Fig. 1 MT immunostaining in normal colonic glands.

Table 1 MT positivity and colorectal tumors.

Reference	Normal tissue	Adenoma tissue	Cancer tissue
Ofner *et al.* (1994)	81% (63/78)	—	31% (34/109)
Giuffré *et al.* (1996)	10% (20/20)	100% (20/20)	62% (53/85)
Kuroda *et al.* (2002)	—	67% (58/87)	49% (63/128) T1 60% (52/86) T2 26% (11/42)

in 63 (49%) cases. Concerning the depth of invasion, MT expression was positive in 52 (60%) cases of T1 carcinomas, but in only 11 (26%) cases of T2 carcinomas. These results suggest that decreasing MT expression may be an early event in colorectal tumor progression and reflect local invasion.

Moreover, in our study, the positive rate of MT expression in the depressed-type colorectal tumors was significantly lower than that in the polypoid-type colorectal tumors. The depressed-type colorectal cancers have been increasingly reported with recent advances in endoscopic instruments, and have a more malignant potential than the polypoid-type carcinomas. Therefore, the decrease of MT expression is associated with high malignant potential in colorectal cancer.

Ioachim *et al.* (1999) reported that MT expression was inversely correlated with CD44 in the series of adenocarcinomas. CD44 expression has been shown to be associated with metastasis and poor prognosis in colorectal cancer (Tanabe *et al.*, 1993). The inverse correlation with CD44 could suggest that decreased MT expression may contribute to the metastatic spread of the lymph node involvement in colorectal cancer. Although MT and p53 were observed in both benign and malignant lesions, no correlation was detected between them in either group. These results indicate that MT and p53 overexpression may arise from independent mechanisms in early neoplasia.

Brüwer *et al.* (2001) reported MT expression in up to 40% of patients with ulcerative colitis, and a correlation between the degree of MT expression and the severity of inflammation. Brüwer *et al.* (2002) also revealed that MT expression was found in most ulcerative colitis and low-grade dysplasia cases, but in only a small percentage of high-grade dysplasia and ulcerative colitis-associated colorectal carcinoma, and concluded that MT overexpression may present an important early step in the development of colorectal carcinoma.

2. Prognostic Significance of Metallothionein

The prognostic role of MT in human cancer survival has been evaluated previously in several immunohistochemical studies in different types of tumors. Ofner *et al.* (1994) reported that immunohistochemical MT expression of colorectal carcinomas was inversely related to tumor stage and appearance of metastatic spread (Table 2). In addition, MT positivity was significantly associated with a favorable clinical outcome in a univariate analysis, but this significance was lost in a multivariate analysis with Dukes' stage as a stratification factor. However, Janssen *et al.* (2002) reported that the MT level was correlated with Dukes' stage of colorectal carcinomas. In colorectal cancer patients, a high MT level in both the carcinomas and normal mucosas was significantly associated with a poor overall survival (Janssen *et al.*, 2000; Janssen *et al.* 2002). Hishikawa *et al.*

Table 2 MT expression and prognosis.

Reference	Association with prognosis
Ofner *et al.* (1994)	Associated with better prognosis
Janssen *et al.* (2000); Janssen *et al.* (2002)	Associated with worse prognosis
Hishikawa *et al.* (2001)	Associated with worse prognosis
Ioachim *et al.* (1999)	Not associated with prognosis

(2001) also found that strong MT expression was associated with a worse prognosis (Table 2).

This discrepancy may be a consequence of the different immunoreactivities of the MT antibodies used in studies. It may also be due to the inclusion of a heterogeneous group of patients at different disease stages. Since MT expression was stronger in colorectal cancers at early stage than in advanced carcinomas, the unfavorable prognosis might be the result of prevalence in patients with advanced-stage cancer. Ioachim *et al.* (1999) found that the expression of MT in colorectal cancer might not be of prognostic relevance and does not seem to indicate aggressive biological behavior (Table 2).

3. Metallothionein and Colorectal Liver Metastases

Janssen *et al.* (2000) reported that the MT concentration of colorectal liver metastases was six times less than that of normal-appearing liver, but comparable to primary colorectal carcinomas. Interestingly, the liver metastases seemed to contain less MT than Dukes' D primary tumors. In addition, they reported that the heterogeneous MT expression patterns of liver metastases and primary colorectal carcinomas were very similar.

Giuffrè *et al.* (1996) also found that in 83% of patients, liver metastases were immunohistochemically negative, whereas the corresponding primary colorectal carcinomas were MT-positive (Table 3). Moreover, Deng *et al.* (1998) and Stenram *et al.* (1999) revealed

Table 3 MT expression in colorectal cancer and liver metastases.

Reference	Colorectal cancer	Liver metastases
Giffreè *et al.* (1996)	62% (53/85)	17% (5/29)
Hishikawa *et al.* (2001)	20% (7/34)	0% (0/34)
Stenram *et al.* (1999)		12% (14/117)

that liver metastases have a very low expression of MT (Table 3). Deng *et al.* (1998) reported that primary hepatocellular carcinomas showed moderate MT staining with a small number of apoptotic cells, while liver metastases showed no MT staining with a large number of apoptotic cells. Hishikawa *et al.* (2001) revealed that MT expression was found in 7 out of 34 (20.6%) patients with primary colorectal cancer, but MT was not detected in any of the cases of liver metastases (Table 3). Therefore, Hishikawa *et al.* (2001) concluded that altered MT expression in the liver metastases seems to be produced by some kind of phenotypic change due to a change in the tumor environment, or that the MT-negative metastatic tumor cells are derived from a pool of heterogeneous cells with variable MT expression in the primary tumor.

4. Clinical Significance of MT as a Drug-Resistance Protein

Since Bakka *et al.* (1981) first provided evidence, using human and mouse cell lines, that MT contributes to the resistance to cisplatin, it has been found that MT seems to bind and detoxify heavy metals like Zn, Cu, Cd, and Hg, thus contributing to the resistance against electrophilic anticancer drugs (Bakka *et al.*, 1981). Sutoh *et al.* (2000) reported that a tendency toward unfavorable prognosis after resection was obtained for 76 patients with MT-positive colorectal cancers, compared with 34 patients with MT-negative cancers. In particular, the concurrent expression of drug-resistant proteins — MT, glutathione *S*-transferases, and P-glycoprotein — had worse prognoses in colorectal cancers. MT has a connection with drug resistance to CDDP, but not 5-FU or MMC. Examining drug-resistant proteins in colorectal cancers may be useful in selecting adjuvant chemotherapy. When MT and other drug-resistant proteins are detected, it is better to avoid the application of CDDP.

In colorectal carcinomas, the liver metastasis seems to contain less MT than the primary tumors. The fact that the expression of MT is downregulated in liver metastasis may suggest that metal-bound chemotherapeutic drugs, such as CDDP, are more effective in liver metastases than in primary sites. In esophageal squamous cell carcinoma patients, MT expression is correlated with the patient's prognosis treated with CDDP. Therefore, it may be possible that hepatic arterial infusion therapy using CDDP can improve the prognosis of patients with liver metastases from colorectal carcinomas.

References

Bakka A, Endresen L, Johnsen AB, *et al.* Resistance against *cis*-dichlorodiammineplatinum in cultured cells with a high content of metallothionein. *Toxicol Appl Pharmacol* 1981; **61**:215–226.

Brüwer M, Schmid KW, Krieglstein CF, *et al.* Metallothionein: Early marker in the carcinogenesis of ulcerative colitis-associated colorectal carcinoma. *World J Surg* 2002; **26**:726–731.

Brüwer M, Schmid KW, Metz KA, *et al.* Increased expression of metallothionein in inflammatory bowel disease. *Inflamm Res* 2001; **50**:289–293.

Deng DX, Chakrabarti S, Waalkes MP, Cherian MG. Metallothionein and apoptosis in primary human hepatocellular carcinoma and metastatic adenocarcinoma. *Histopathology* 1998; **32**:340–347.

Giuffrè G, Barresi G, Sturniolo GC, *et al.* Immunohistochemical expression of metallothionein in normal human colorectal mucosa, in adenomas and in adenocarcinomas and their associated metastases. *Histopathology* 1996; **29**:347–354.

Hishikawa Y, Kohno H, Ueda S, *et al.* Expression of metallothionein in colorectal cancers and synchronous liver metastases. *Oncology* 2001; **61**:162–167.

Ioachim EE, Goussia AC, Agnantis NJ, *et al.* Prognostic evaluation of metallothionein expression in human colorectal neoplasms. *J Clin Pathol* 1999; **52**:876–879.

Janssen AM, van Duijn W, Kubben FJ, *et al.* Prognostic significance of metallothionein in human gastrointestinal cancer. *Clin Cancer Res* 2002; **8**:1889–1896.

Janssen AM, van Duijn W, Oostendorp-Van De Ruit MM, *et al.* Metallothionein in human gastrointestinal cancer. *J Pathol* 2000; **192**:293–300.

Kuroda K, Aoyama N, Tamura T, *et al.* Variation in MT expression in early-stage depressed-type and polypoid-type colorectal tumours. *Eur J Cancer* 2002; **38**:1879–1887.

Ofner D, Maier H, Riedmann B, *et al.* Immunohistochemical metallothionein expression in colorectal adenocarcinoma: Correlation with tumour stage and patient survival. *Virchows Arch* 1994; **425**:491–497.

Stenram U, Ohlsson B, Tranberg KG. Immunohistochemical expression of metallothionein in resected hepatic primary tumors and colorectal carcinoma metastases. *Acta Pathol Microbiol Immunol* 1999; **107**:420–424.

Sutoh I, Kohno H, Nakashima Y, *et al.* Concurrent expressions of metallothionein, glutathione S-transferase-pi, and P-glycoprotein in colorectal cancers. *Dis Colon Rectum* 2000; **43**:221–232.

Tanabe KK, Ellis LM, Saya H. Expression of CD44R1 adhesion molecule in colon carcinomas and metastases. *Lancet* 1993; **341**:725–726.

PART III
METALLOTHIONEIN IN CARDIOPATHY

Chapter 12

METALLOTHIONEIN AND CARDIOMYOPATHY

Lu Cai

Cardiomyopathy, a condition in which the heart muscle does not work properly, can affect children and adults. Unlike heart disease due to heart attack, where there is a problem with adequate blood flow to the heart, cardiomyopathy refers to the disease of heart muscle itself. The common types of cardiomyopathy are dilated cardiomyopathy, hypertrophic cardiomyopathy, restrictive cardiomyopathy, and specific cardiomyopathy. The causes of cardiomyopathy are unknown, and these diseases can be inherited or caused by other diseases or environmental toxicants. Coronary artery disease causes ischemic cardiomyopathy. Alcohol, especially when a person has a poor diet, causes alcoholic cardiomyopathy. Toxicants, including the anticancer drug doxorubicin, cause toxic cardiomyopathy. Chronic diseases such as diabetes cause disease-related cardiomyopathy, for instance, diabetic cardiomyopathy. Although there are distinct mechanisms for these cardiomyopathies with different etiologies, there are also certain common pathogenic mechanisms; exploring these common mechanisms will be of great importance for developing preventive approaches to these cardiomyopathies. Metallothionein as a stress protein (adaptive protein) was found to play a critical role in protecting the heart from a variety of pathogenic and environmental risks. Therefore, this chapter will summarize the protective effects of metallothionein on the development of these cardiomyopathies, focusing on the experimental evidence, possible mechanisms, and potential clinical implications.

Keywords: Cardiomyopathy; ethanol; ischemia and reperfusion; doxorubicin; adriamycin; diabetes; diabetic cardiomyopathy.

1. Introduction

Cardiomyopathy is a serious disease in which the heart muscle becomes damaged and does not work as well as it should (Davies,

2000). Cardiomyopathy can be classified as either primary or secondary. The cause of primary cardiomyopathy is unknown, and primary cardiomyopathy cannot be attributed to a specific cause (e.g. high blood pressure, heart valve disease, artery diseases, or congenital heart defects). However, if a patient has secondary cardiomyopathy, the cause is known and it is related to a specific aforementioned cause. Examples of secondary cardiomyopathy include metabolic illness, coronary artery disease, muscular dystrophy, and an alcohol or drug problem. Therefore, secondary cardiomyopathy is often associated with diseases involving other organs as well as the heart (Davies, 2000; Thiene *et al.*, 2000).

Cardiomyopathies can be classified into five types: dilated cardiomyopathy, hypertrophic cardiomyopathy, restrictive cardiomyopathy, arrhythmogenic right ventricular cardiomyopathy, and specific cardiomyopathy. The term "specific cardiomyopathy" is now used to describe heart muscle diseases that are associated with specific cardiac or system disorders (Davies, 2000; Thiene *et al.*, 2000). Diabetic cardiomyopathy, as one type of specific cardiomyopathy, is a diabetes-related myopathy and is characterized mainly by impaired diastolic function (Cai and Kang, 2001; Cai, 2007).

Several mechanisms responsible for these cardiomyopathies have been proposed (Davies, 2000; Thiene *et al.*, 2000; Cai and Kang, 2001; Lang *et al.*, 2005; Cai, 2007; Takemura and Fujiwara, 2007). These include (1) impaired regulation of intracellular calcium, leading to impaired cardiac contractility; (2) mitochondrial dysfunction, leading to overproduction of reactive oxygen species (ROS) and reactive nitrogen species (RNS) and even to cardiac cell death; (3) accumulation of advanced glycated endproducts (AGEs) in the heart, which causes the extracellular matrix to accumulate, leading to cardiac diastolic dysfunction and eventually heart failure; (4) abnormal cellular metabolism in favor of fatty acid β-oxidation, leading to the accumulation of toxic lipids in the heart (i.e. lipotoxic heart disease); (5) depolymerization of cardiac actin that contributes to impaired cardiac contractility; and (6) impaired essential metal homeostasis such as zinc (Zn) and copper (Cu). Furthermore, recent emerging evidence also shows that these pathogenic causes may all result in or from oxidative stress.

Oxidative stress indicates a serious imbalance between ROS and/or RNS generation and its clearance by antioxidant defense systems (Cai and Kang, 2001). Therefore, an efficient antioxidant system, including superoxide dismutase (SOD), catalase, glutathione peroxidase (GSHpx), glutathione (GSH), and α-tocopherol, would be critical to the health of the heart. However, in experimental animal models, the activities of all these antioxidants in the heart are low when compared to liver under normal (Doroshow *et al.*, 1980) or stress challenging conditions (Chen *et al.*, 1994). The low antioxidant capacity makes the heart more susceptible to oxidative damage. Under the pathogenic conditions of these cardiomyopathies, the heart not only overproduces ROS and RNS, but also has a further impaired antioxidant defense system, as shown in Fig. 1. For instance, decreased expression of HSP60 and HO-1 in the diabetic heart leads to the heart being highly susceptible to oxidative damage (Turko *et al.*, 2001; Chen *et al.*, 2005; Gross *et al.*, 2007).

Therefore, new and efficient antioxidant therapy to prevent or delay the development of these cardiomyopathies has been explored. Metallothionein (MT) as a stress protein was found to be upregulated in the heart in response to various pathogenic risks, suggesting

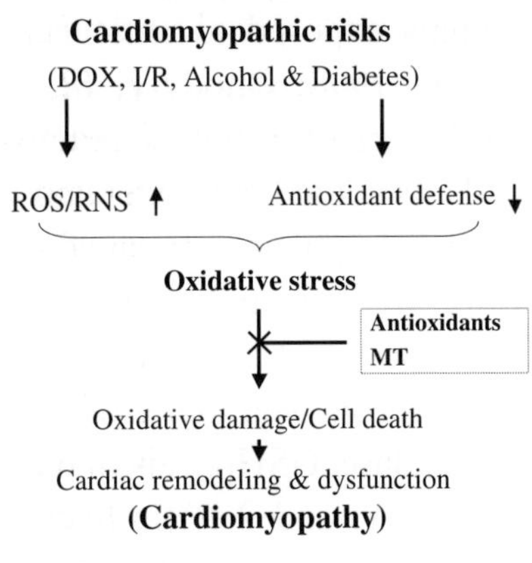

Fig. 1 Outline of the role of oxidative stress in the development of various cardiomyopathies. DOX, doxorubicin; I/R, ischemia/reperfusion.

its possibly protective role as an adaptive mechanism under these pathogenic conditions in preventing heart disease (Fig. 1). This review will briefly introduce the antioxidant feature of MT in the heart and its synthesis in response to toxicants and diabetes; and then will mainly summarize the research achievements for the prevention of these cardiomyopathies by MT, with an emphasis on the experimental evidence, possible mechanisms, and clinical implications.

2. Metallothionein Functions as an Antioxidant in the Heart

MTs are a group of intracellular metal-binding proteins that contain 60–68 amino acids, 20 of which are cysteines (reviewed in Cai *et al.*, 1999; Kang, 1999). Although the antioxidant function of MT has been extensively implicated in *in vitro* and *in vivo* studies (reviewed in Cai *et al.*, 1999; Kang, 1999), MT as an antioxidant in the heart was only recently indicated in a study of Bauman *et al.* (1991). In this study, mice were injected subcutaneously with chemicals that produce oxidative stress (paraquat and diquat to undergo redox cycling, and diethyl maleate to deplete reduced glutathione) (Bauman *et al.*, 1991), suggesting that MT might be involved in protecting against these oxidative stresses.

In support of the early finding, upregulation of cardiac MT was also observed in animals exposed to anticancer drugs, ischemia, and diabetes (Yin *et al.*, 1998; Onody *et al.*, 2003; Song *et al.*, 2005b). For instance, Yin *et al.* (1998) examined whether cardiac MT and other antioxidants could be upregulated when animals were exposed to the anticancer drug doxorubicin (DOX, also called adriamycin or ADR). Mice were treated with DOX and, 4 days after the treatment, cardiac antioxidant activities and mRNA abundances were measured. The results showed that DOX increased the mRNA levels of Cu,Zn-SOD, catalase, GSHpx, and γ-glutamylcysteine synthetase (γ-GCS) [Figs. 2(A) and 2(B)]. On the other hand, DOX increased the activities of catalase and γ-GCS, and slightly decreased

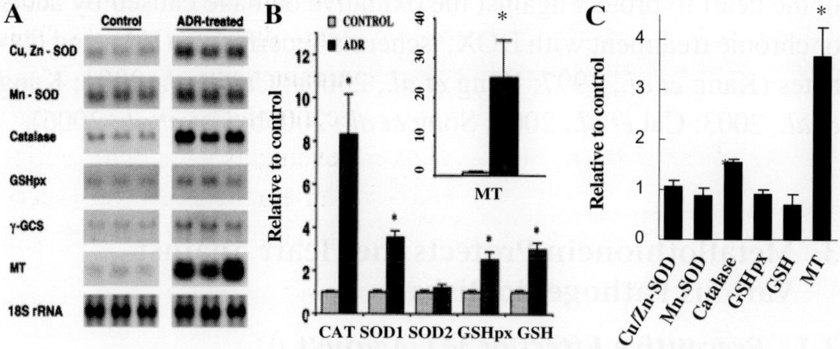

Fig. 2 Northern blot analysis of mRNA from the hearts of mice treated with or without 15 mg/kg of DOX for 4 days. (A) Total heart RNA was isolated, and 10 μg of total RNA from triplicate samples was subjected to Northern blot analysis. Ethidium bromide staining of 18S rRNA was examined to ensure equal loading. (B) The effect of DOX on the abundance of mRNAs for antioxidants was quantitatively analyzed with an autoradiographic image analysis system and expressed as relative to control levels (inset: expression of MT mRNA). Control values are arbitrarily designated as 1. ADR = DOX, SOD1 = Cu,Zn-SOD, and SOD2 = Mn-SOD. The inset shows the MT result. (C) The activity or protein expression of these antioxidants was analyzed as relative to the control levels. *Significantly different from the controls ($P < 0.05$). Panels A and B were extracted from, and panel C was made by the author based on the data from, Yin *et al.* (1998), with permission from Elsevier Limited.

total GSH concentrations in the heart; while Cu,Zn-SOD, Mn-SOD, and GSHpx activities were not changed significantly as compared to the control [Fig. 2(C)]. Importantly, DOX increased the highest expression of both mRNA and protein levels of MT (Fig. 2), suggesting the important role of MT in the cardiac antioxidant defense system (Yin *et al.*, 1998; Kang, 1999).

It is worth mentioning that the most important contribution for directly defining the antioxidant role of MT in the heart is the development of the cardiac-specific, MT-overexpressing transgenic (MT-TG) mice, with overexpression of the human MT-IIA gene in the cardiac myocytes (Kang *et al.*, 1997). As stated in the review of Nath *et al.* (2000), "This opens a new avenue to study the significance of cardiac MTs at the level of molecular biology and pathophysiology." Studies using this unique MT-TG mouse model have provided direct evidence to support the concept that MT functions as an antioxidant in the heart to protect against the oxidative damage caused by acute or chronic treatment with DOX, ischemia/reperfusion (I/R), and diabetes (Kang *et al.*, 1997; Kang *et al.*, 2000a; Cai *et al.*, 2001; Kang *et al.*, 2003; Cai *et al.*, 2005; Song *et al.*, 2005b; Cai *et al.*, 2006).

3. Metallothionein Protects the Heart Against Various Pathogenic Risks

3.1. *Preventive Effect of MT Against DOX-Induced Cardiotoxicity*

Anthracyclines, of which DOX is the leading drug, are highly effective anticancer agents; however, DOX cardiotoxicity is the major obstacle of this drug to be efficiently used in clinics (Takemura and Fujiwara, 2007). DOX-induced cardiomyopathy manifests itself in two ways: acutely, giving rise to temporary and usually benign phenomena; and chronically, giving rise to cardiac failure which is often irreversible. Therefore, the development of preventive approaches to attenuate the risk of DOX-induced cardiotoxicity is of great importance for efficiently using anthracycline, as shown in Fig. 1.

Satoh *et al.* (1988) and Naganuma *et al.* (1988) performed pioneer studies on the protective effect of cardiac MT pre-induction by bismuth subnitrate (BSN) against DOX cardiotoxicity. They demonstrated that pre-induction of MT by oral administration of BSN significantly decreased the cardiotoxicity, along with reduction of the bone marrow toxicity and animal lethality, after mice received a single subcutaneous injection of DOX. A significant increase in the concentration of cardiac MT was observed in mice treated with BSN. The MT level in the heart was significantly correlated with the protective effect of BSN against the cardiotoxicity of DOX. In tumor-bearing mice, pretreatment with BSN did not affect the antitumor activity of DOX, although its cardiotoxicity was significantly depressed. Satoh *et al.* (2000) further demonstrated that cardiac-specific induction of MT by pre-administration of BSN which induced cardiac and bone marrow MT — instead of Zn, which induced cardiac and bone marrow MT as well as tumor MT — could protect the heart without reduction of antitumor efficiency of DOX in the tumor-bearing mouse model. The results described above suggest that DOX toxicity can be attenuated in tissues in which the MT level is elevated, and that the tissue-specific induction of MT synthesis may provide a promising regimen for cancer chemotherapy (Satoh *et al.*, 2000).

Although using MT inducers to pre-induce cardiac MT synthesis is informative, all of the MT inducers are pleitropic and may cause a panoply of biological responses. To provide direct evidence for the role of MT in the cardiac protection of pre-administration of MT inducers with cardiac induction of MT, MT-TG and their wild-type (WT) mice were administered with DOX at a single dose of 20 mg/kg and were killed on the fourth day after treatment (Kang *et al.*, 1997). They demonstrated that MT-TG mice expressing MT activity 10- or 130-fold higher than normal controls exhibited a significant resistance to *in vivo* DOX-induced cardiac morphological changes and an increase in serum creatine phosphokinase activity. Atria isolated from transgenic mice and treated with DOX in tissue bath were also more resistant to functional damage induced by DOX. The results provided direct evidence for the role of MT in cardiac protection against DOX toxicity. This direct evidence was further confirmed by its antiapoptotic effects in the heart against DOX-induced cardiomyopathy (Kang *et al.*, 2000a; Wang *et al.*, 2001a).

3.2. *Preventive Effect of MT Against I/R-Induced Cardiotoxicity*

Chen *et al.* (1997) investigated the potentiality of MT as an antioxidant to protect against cardiac I/R injury. MT content was significantly increased in the heart of rabbits at 24 hours after preconditioning (PC) [Fig. 3(A)] and also in the primary cultures of adult cardiomyocytes at 2 hours and 24 hours after preconditioning (PC) [Fig. 3(B)]. The increased MT content in the hearts of rabbits with

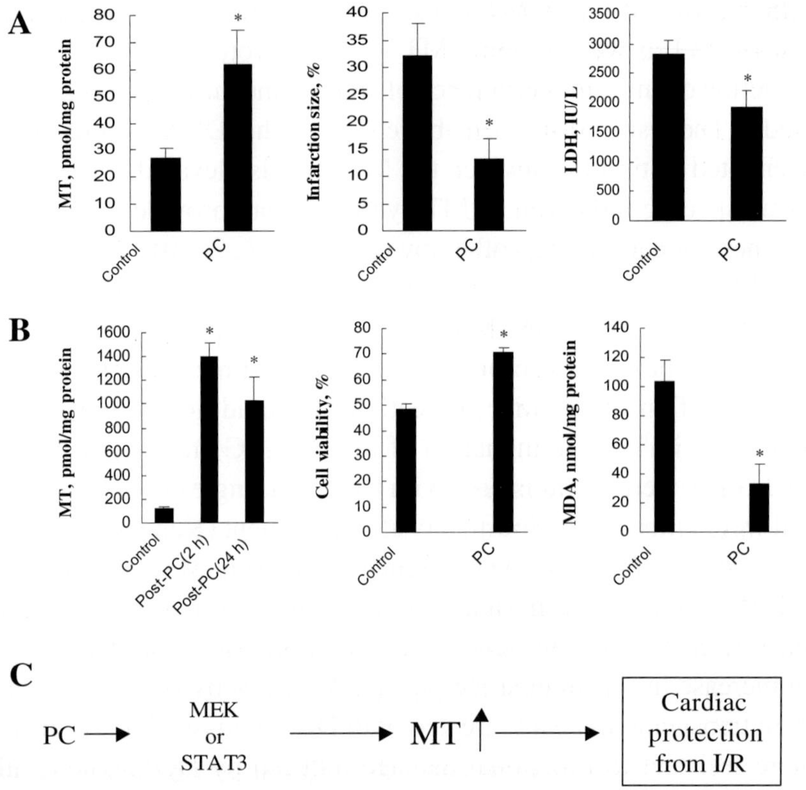

Fig. 3 Preconditioning (PC)-induced MT synthesis in the heart and cardiomyocytes that protected the heart *in vivo* and cardiomyocytes *in vitro*. PC was given to (A) rabbits *in vivo* and (B) cardiomyocytes *in vitro*. At 2 h and 24 h after PC, MT contents were measured in the heart and the cultured cardiomyocytes. Hearts (A) and cardiomyocytes (B) with PC-induced MT synthesis showed high resistance to subsequent I/R-induced cardiac damage (infarction size and serum LDH) and myocyte damage (cell viability and MDA), respectively. Panels A and B were made by the author based on the data from Chen *et al.* (1997). Panel C was made by the author for the potential role of MT in cardiac protection against I/R. * $P < 0.05$ vs. control.

PC treatment was accompanied by significant decreases in cardiac infarct sizes and serum cardiac enzyme lactate dehydrogenase [LDH; Fig. 3(A)]. In the cultured cardiomyocytes, the protective effect of PC with MT induction was also evident for the cell viability and lipid peroxidation, measured by malondialdehyde [MDA; Fig. 3(B)]. All of the delayed protection at 24 hours after PC disappeared completely with the inhibition of MT production with the MAP kinase/ERK kinase (MEK)-specific inhibitor PD098059. Although the specific inhibition of MT induction by MEK inhibitor will be further confirmed and how MEK is involved in MT synthesis remains to be further explored, this study suggests that PC could provide cardiac protection against I/R-induced damage through cardiac MT induction which may be MEK-dependent, as indicated in Fig. 3(C) (Chen *et al.*, 1997).

In the cardiac-specific MT-TG mouse model with a Langendorff heart perfusion preparation, Kang *et al.* (1999) provided direct evidence for the role of MT in cardiac protection against I/R injury. In MT-TG hearts, a significant improvement of the suppressed contractile force postischemia was observed and the efflux of creatine phosphokinase (CPK) from the MT-TG hearts was reduced by more than 50%. In addition, the zone of infarction induced by I/R at the end of reperfusion was suppressed by about 40% in the MT-TG hearts. The same group (Kang *et al.*, 2003) further defined this direct cardiac protection against I/R-induced damage using an open-chest coronary artery occlusion and reperfusion procedure to produce I/R-induced left ventricle infarction. After 30 minutes of ischemia, the left ventricle was reperfused to allow blood flow through the previously occluded coronary artery bed. Cardiac infarction produced after reperfusion for 4 hours was found to be significantly reduced in the MT-TG mice. I/R-induced cardiac lipid peroxidation was also significantly inhibited in the MT-TG heart.

The above two studies suggest that MT in the heart indeed provides a protective action against I/R-induced cardiac damage; therefore, PC-induced MT as observed in the earlier study [Figs. 3(A) and 3(B)] may play a critical role in the cardiac protection against I/R-induced cardiac damage. This hypothesis was confirmed by a recent study from Oshima *et al.* (2005). Since activation of signal transducer and activator of transcription 3 (STAT3) was reported to be correlated with cardiac protection against I/R injury in the ischemic PC model, whether MT is involved in ischemic PC-induced cardiac protection was tested in this study. They demonstrated that cardiac-specific transgenic mice expressing constitutively active STAT3 (ST3-TG) were exposed to I/R, and ST3-TG hearts exhibited infarcts which reduced by 60.3% in size compared with WT mice under the same condition. In parallel, MT was markedly upregulated in ST3-TG hearts, suggesting the involvement of MT in STAT3-mediated cardiac protection against I/R. They further demonstrated that homozygous deletion of the MT gene abrogated the cardiac protection of STAT3 against I/R injury with the cancellation of MT's ROS-scavenging effects (Oshima *et al.*, 2005). Therefore, this study indicated that activation of STAT3 protects the heart from I/R injury *in vivo* at least partially through the MT gene, as outlined in Fig. (3C).

3.3. *Preventive Effect of MT Against Alcohol-Induced Cardiotoxicity*

Moderate consumption of alcohol has been found to provide some degree of protection against ischemic heart disease; however, acute ingestion of large amounts of alcohol can lead to a negative inotropic effect on the heart, together with inhibition of a variety of biochemical reactions in subcellular organelles of the heart. Chronic alcoholism is also associated with the development of a congestive cardiomyopathy (alcoholic cardiomyopathy).

Studies from Dr Ren's laboratory (Li and Ren, 2006a; Li and Ren, 2006b) have demonstrated the protective effect of transgenic overexpression of MT on alcohol-induced cardiac contractile dysfunction. They demonstrated that WT and MT-TG mice were placed on a 4% alcohol or control diet for 12 weeks that significantly depressed the contractibility of cardiomyocytes, along with reduced sarco(endo)plasmic reticulum Ca^{2+}-ATPase (SERCA2a) and Na^{+}-Ca^{2+} exchanger (NCX) abundance in the WT group, but not in the MT-TG group. These results suggested the protective effect of MT against alcoholic cardiomyopathy by measuring the contractibility of isolated cardiomyocytes.

In human, exposure to alcohol typically results in a form of dilated cardiomyopathy that is characterized not only by reduced contractility, ventricular dilatation, and cardiomyocyte apoptosis, but also by increased cardiac fibrosis, often progressing to heart failure. However, studies using experimental animals have not fully duplicated the pathological changes in humans, since animal models of alcoholic cardiac fibrosis are not available. Wang *et al.* (2005) developed a mouse model in which cardiac hypertrophy and fibrosis were produced in MT-knockout (MT-KO) mice fed with an alcohol-containing liquid diet for 8 weeks. The same alcohol feeding did not produce cardiac fibrosis in the WT control mice, although there was no difference in the alcohol-induced heart hypertrophy between the WT controls and the MT-KO mice. Serum CPK activity was significantly higher in the alcohol-administered MT-KO mice than in the WT mice. Therefore, this study also provides direct evidence for the critical role of MT in cardiac protection against alcoholic cardiomyopathy (Wang *et al.*, 2005).

3.4. *Preventive Effect of MT Against Diabetes-Induced Cardiotoxicity*

Diabetic cardiomyopathy, as one type of specific cardiomyopathy, is a diabetes-related cardiomyopathy and is characterized mainly by impaired diastolic function (Cai and Kang, 2001). Studies have shown that diabetes affects cardiac structure and function independent of blood pressure or coronary artery disease, although hypertension can accelerate cardiac damage in diabetic patients. Emerging evidence also shows that, although hyperglycemia, hyperlipidemia, and inflammatory cytokines or peptides cause the pathogenesis of diabetic cardiomyopathy via different mechanisms, all of these pathogenic effects may be related to oxidative stress (Cai and Kang, 2001). Given that MT acts as a potent antioxidant in the heart, we have firstly tested the possible protection of cardiac MT in MT-TG mice against diabetes-induced cardiotoxicity in a study (Cai *et al.*, 2001). We found that a single injection of streptozotocin (STZ) induced type 1 diabetes in both MT-TG and WT mice. There was no difference in hyperglycemia, body-weight gain, and serum lipid peroxide levels between the MT-TG and the WT diabetic mice. However, serum CPK activity as a biomarker of cardiac injury was markedly increased in the WT diabetic mice; significant cardiac injury was also found in the WT diabetic mice as evidenced by morphological changes. In comparison to the WT diabetic mice, cardiac injury was significantly suppressed in the MT-TG diabetic mice. These results thus provided the first evidence indicating the preventive role of MT in the heart against diabetic cardiotoxicity. Since then, the cardiac protection of MT against diabetes-induced cardiomyopathy has been extensively confirmed by studies from several groups including the author's (Liang *et al.*, 2002; Ye *et al.*, 2003; Cai *et al.*, 2005; Fang *et al.*, 2005; Kralik *et al.*, 2005; Song *et al.*, 2005b; Cai *et al.*, 2006b; Ceylan-Isik *et al.*, 2006b; Wang *et al.*, 2006; Wold *et al.*, 2006; Dong *et al.*, 2007) in the general aspects described in the following paragraphs.

The protective effects of MT on diabetes-induced cardiac damages were confirmed at several levels: from subcellular and cellular to systemic levels (Fig. 4). These studies showed that cardiac MT overexpression significantly prevented diabetes-caused ultrastructural changes of cardiac myocytes [Fig. 4(A)], including mitochondrial and nuclear membrane damage, examined under light and electron microscopy (Liang *et al.*, 2002; Cai *et al.*, 2005). Diabetes induced a significant increase in cardiac lipid peroxidation and protein nitration in the WT diabetic mouse hearts, but not in MT-TG diabetic mice (Liang *et al.*, 2002; Ye *et al.*, 2003; Cai *et al.*, 2005; Fang *et al.*, 2005). Significant increases in serum levels of cardiac enzymes such as CPK [Fig. 4(B)] were observed only in the WT diabetic mice, but not in the MT-TG diabetic mice (Cai *et al.*, 2001; Cai *et al.*, 2005). Diabetes-induced cardiac fibrosis, shown by sirius red staining of collagen [Fig. 4(C)] and connective tissue growth factor (CTGF) protein expression, was significantly prevented in MT-TG diabetic mice or

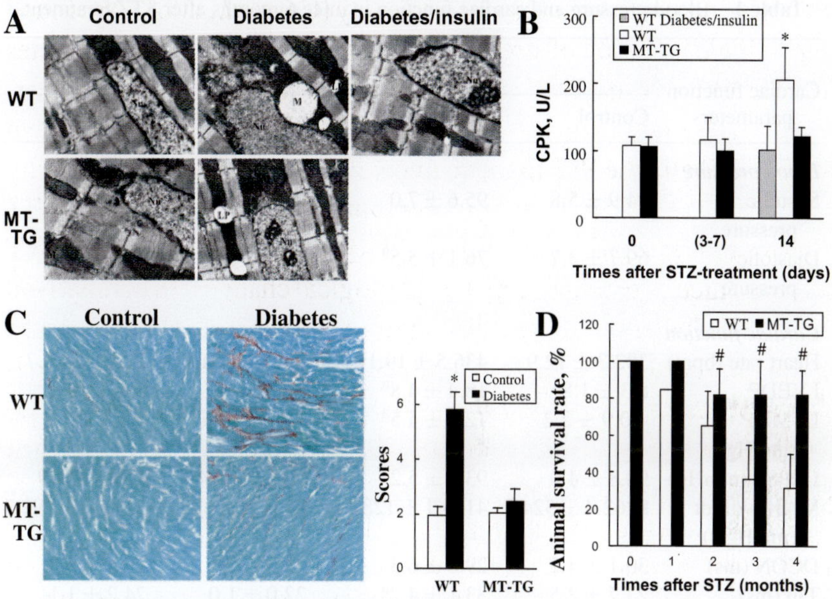

Fig. 4 MT prevention of diabetes-induced ultrastructural and cellular damages, cardiac remodeling, and animal mortality. (A) Heart tissues from both WT and MT-TG control and diabetic mice 2 weeks after the onset of diabetes were examined under electron microscope (×11000). LP and M indicate lipid droplets and mitochondria, respectively. (B) Serum was collected from WT and MT-TG control and diabetic mice at the indicated times to measure serum CPK. (C) Heart tissue from the WT and MT-TG diabetic mice 6 months after the onset of diabetes was examined for fibrosis using sirius red staining of collagen in combination with semiquantitative analysis. (D) The animal survival rate was also examined for both WT and MT-TG diabetic mice. These results were selected from the author's previous works (Cai *et al.*, 2005; Cai *et al.*, 2006). $^{*}P < 0.05$ vs. control in (B) and (C). $^{\#}P < 0.05$ vs. corresponding WT group in (D).

diabetic mice treated with Zn with significant induction of cardiac MT (Cai *et al.*, 2006; Wang *et al.*, 2006). Cardiac overexpression of MT significantly attenuated diabetes-induced cardiac dysfunction, examined by *in vivo* left ventricular hemodynamic performance as indicated in Table 1 (Cai *et al.*, 2005; Song *et al.*, 2005b), as well as cardiac and cellular contractility, examined by isolated Langendorff perfused heart model and isolated single cardiomyocytes from diabetic mice (Liang *et al.*, 2002; Ye *et al.*, 2003; Fang *et al.*, 2005; Ceylan-Isik *et al.*, 2006b; Wold *et al.*, 2006). In addition, because of

Table 1 Blood pressure and cardiac function in mice 6 months after STZ treatment.

Cardiac function parameters	WT		MT-TG	
	Control	Diabetes	Control	Diabetes
Blood pressure				
Systolic pressure	94.9 ± 5.8	95.6 ± 7.0	92.1 ± 2.7	94.8 ± 8.4
Diastolic pressure	69.7 ± 3.7	76.1 ± 5.5^a	68.9 ± 3.3	66.6 ± 8.5
Cardiac function				
Heart rate (bpm)	382.3 ± 12.9	436.5 ± 19.1^a	386.2 ± 10.4	399.2 ± 26.7
LVEDP	8.3 ± 1.8	18.4 ± 4.8^a	9.5 ± 2.1	8.4 ± 1.5
LVMDP (mmHg)	-0.9 ± 2.2	12.3 ± 5.5^a	-1.4 ± 3.6	-4.9 ± 1.8
LVPSP (mmHg)	98.8 ± 4.0	93.8 ± 5.2	103.2 ± 5.4	108.0 ± 4.9
MAX +dP/dt (mmHg/s)	8262.4 ± 524.7	4183.1 ± 1287.1^a	8132.0 ± 675.6	9113.6 ± 692.3
DCON (ms)	30.1 ± 4.2	29.0 ± 6.3	28.0 ± 2.1	29.6 ± 5.8
Tau (ms)	23.7 ± 2.5	33.1 ± 4.2^a	22.0 ± 1.0	24.2 ± 1.1
1/2 R (ms)	45.2 ± 3.6	40.8 ± 3.7	45.7 ± 3.7	47.2 ± 2.2

Note: LVEDP: left ventricular end-diastolic pressure; LVMDP: left ventricular minimum diastolic pressure; LVPSP: left ventricular peak systolic pressure; MAX +dP/dt: maximum rate of rise in intraventricular pressure during ventricular contraction; DCON: duration of contraction; Tau: time constant of the best-fit exponential pressure decay from the pressure at MAX +dP/dt to a positive pressure above the previous LVEDP by 10 mmHg; 1/2 R: time constant duration of 1/2 relaxation. Values are mean ± standard error (SE) ($n = 6$ or 7). [a] $p < 0.05$ vs. WT control.

the significant protection against diabetes-induced cardiac structural and functional changes, diabetes-induced animal mortality was also significantly attenuated in the MT-TG diabetic mice (Cai *et al.*, 2006).

The cardiac protection of MT from diabetic cardiomyopathy was not only observed in the STZ-induced type 1 diabetic mouse model (Cai *et al.*, 2005; Song *et al.*, 2005b; Ceylan-Isik *et al.*, 2006c; Wang *et al.*, 2006; Wold *et al.*, 2006), but was also noted in a genetic type 1 diabetic mouse model (Liang *et al.*, 2002; Ye *et al.*, 2003; Kralik *et al.*, 2005) and in a sucrose-feeding-induced insulin-resistant (prediabetic) mouse model (Fang *et al.*, 2005; Dong *et al.*, 2007). The spontaneously developed type 1 diabetic mouse model, OVE26 mice, was developed to carry a calmodulin minigene regulated by the rat insulin II promoter with a fivefold increase in the content of calmodulin in beta cells (Epstein *et al.*, 1989). These OVE26 mice normally develop severe diabetes within hours of birth. In the study by Liang *et al.* (2002), OVE26 mice were cross-bred with cardiac MT-TG mice to generate a transgenic mouse model (OVE26MT) with the features of both cardiac MT overexpression and spontaneously developing diabetes. They examined the potential for MT to prevent the occurrence of diabetic cardiomyopathy in OVE26MT diabetic mice, and found that OVE26 diabetic mice exhibited cardiomyopathy, characterized by morphological abnormalities and reduced contractility under ischemic conditions along with an increase in oxidative stress (measured by oxidized glutathione, GSSG); however, none of these abnormalities was observed in OVE26MT diabetic mice (Liang *et al.*, 2002). These results demonstrated that cardiac-specific overexpression of MT reduced diabetes-caused cardiac damage. Studies from Dr Ren's group (Fang *et al.*, 2005; Dong *et al.*, 2007) have shown that 3-month sucrose feeding caused an insulin resistance, shown by increased glucose intolerance, hyperinsulinemia, and hyperlipidemia, in both WT and MT-TG mice. The insulin resistance was accompanied by a significant increase in cardiac oxidative stress (increased GSSG) and reduced contractility of the cardiomyocytes only in the WT mice, but not in the MT-TG mice.

Enhanced cardiac MT provides cardiac protection from diabetes at a wide range of expression levels. Cardiac MT levels in the MT-TG mice were about 60-fold higher than those in WT mice (Cai *et al.*, 2005), while cardiac MT levels were 20-fold higher in OVE26MT mice (Liang *et al.*, 2002; Ye *et al.*, 2003; Kralik *et al.*, 2005) and 10-fold higher in the MT-TG mice used in Dr Ren's studies (Fang *et al.*, 2005; Ceylan-Isik *et al.*, 2006b; Wold *et al.*, 2006) than those in their corresponding control mice. Furthermore, whole-body MT-overexpressing mice that had twofold -to-threefold cardiac MT levels of the corresponding control mice were also resistant to diabetes-caused cardiac oxidative damage and fibrosis (Song *et al.*, 2005b).

Consistent with the finding from the whole-body MT transgenic mouse model (Song *et al.*, 2005b), induced cardiac MT synthesis which was also about twofold or threefold higher compared to control levels also significantly protected against diabetes-induced cardiac damage and dysfunction (Wang *et al.*, 2006). The fact that MT protection against diabetes-induced pathological changes was not affected by MT levels within the range of 3–60 folds of control levels was consistent with an early study regarding MT protection against DOX cardiotoxicity (Kang *et al.*, 1997). In this study, mice from transgenic lines expressing MT activity 10- or 130-fold higher than normal all showed significant protection against DOX-induced cardiac morphological changes and an increase in serum CPK activity without significant difference among MT-TG lines.

4. Possible Common Mechanisms for Metallothionein Prevention of Various Cardiomyopathies

The mechanisms by which MT prevents these cardiomyopathies remain unclear. Since MT functions as a potent antioxidant in the heart, and oxidative stress may also be one of the major mechanisms for these cardiomyopathies, the first possible common mechanism for MT prevention of these cardiomyopathies may be attributed to its antioxidant action, as indicated in Fig. 1. In addition, there may be other mechanisms involved in the cardiac protection from diabetes by MT. The following sections will discuss these possible common mechanisms.

4.1. *Antioxidant Action*

Glutathione (GSH) is one of the most prominent antioxidant defense components in the heart. During inactivation of various ROS such as hydrogen peroxide, GSH is converted to GSSG, i.e. there is an increase in the ratio of GSSG/GSH. This index is thus extensively used to indicate the existence of oxidative stress. The status of GSSG and GSH was measured for the first time in diabetic hearts, and a significantly increased level of GSSG without significant change in GSH was found; however, this did not occur in MT-TG diabetic hearts (Liang *et al.*, 2002). This study demonstrated that the elimination of diabetes-induced oxidative stress by MT in the heart is accompanied by a significant prevention of cardiac morphological and functional changes. In the subsequent studies using STZ-induced type 1 diabetic mice (Cai *et al.*, 2005; Song *et al.*, 2005b; Ceylan-Isik *et al.*, 2006a) and sucrose-feeding-induced insulin-resistant mice (Fang *et al.*, 2005), oxidative stress and damage were also evident in WT diabetic hearts, but not in MT-TG diabetic hearts (Table 2). More importantly, Fang *et al.* (2005) have confirmed that the increased GSSG/GSH ratio was seen in the heart, liver, and gastrocnemius muscle of sucrose-fed insulin-resistant WT mice, while in sucrose-fed MT-TG mice the increased GSSG/GSH ratio was attenuated only in the heart (not in the liver and gastrocnemius muscle), suggesting the direct effect of cardiac MT on oxidative stress (Fang *et al.*, 2005). Corresponding to the increased ratio of GSSG to GSH, lipid peroxide content was also demonstrated as another common index of oxidative damage (Cai *et al.*, 2005; Song *et al.*, 2005a).

Table 2 MT prevention of increased GSSG/GSH ratio and 3-NT accumulation in the heart.

Models		References
Diabetes	Increased GSSG	Liang *et al.* (2002)
	Increased GSSG/GSH ratio	Song *et al.* (2005b)
	Increased GSSG/GSH ratio	Fang *et al.* (2005), Ceylan-Isik *et al.* (2006a)
	Decreased mitochondrial GSH Increased mitochondrial GSSG Neither significant change of cytosolic GSH nor GSSG	Cai *et al.* (2006)
	Increased 3-NT	Cai *et al.* (2005), Cai *et al.* (2006)
DOX	Decreased total GSH	Sun and Kang (2002)
	Increased 3-NT	Shuai *et al.* (2007b)

Note: 3-NT, 3-nitrotyrosine-modified proteins as an index of nitrosative damage.

It is well known that superoxide interacts with nitric oxide to form peroxynitrite, which is the most active RNS to cause protein nitration. We have demonstrated the pathogenic role of nitrosative stress and MT prevention (Cai *et al.*, 2005; Cai, 2006; Cai *et al.*, 2006), which can be summarized as follows: (1) to directly show the existence of superoxide generation (by two methods, superoxide-specific probe and cytochrome c reduction assay) and peroxynitrite formation (by 3-nitrotyrosine (3-NT) generation, as summarized in Table 2); (2) both superoxide generation and 3-NT formation were abolished in the MT-TG mouse hearts, which was accompanied by a prevention of diabetic cardiomyopathy; and (3) using isolated cardiomyocytes, intracellular superoxide was induced by TNF-α/LPS, shown by a significant increase in 3-NT in TNF-α/LPS–treated cells. The cytotoxicity was significantly attenuated by urate (peroxynitrite scavenger) or MnTMPyP (superoxide scavenger) along with abolishment of 3-NT, suggesting the contribution of peroxynitrite-derived 3-NT to the cytotoxicity. More importantly, the significant increases in cytotoxicity and 3-NT in TNF-α/LPS–treated cells were not observed in the MT-TG cardiomyocytes, suggesting that MT prevention of TNF-α/LPS–induced cytotoxicity is mediated by suppression of peroxynitrite-induced 3-NT probably through prohibiting superoxide accumulation.

Consistent with the above finding regarding MT protection from diabetic hearts, DOX-induced decrease in total GSH in the heart was also attenuated in the MT-TG heart under the same condition of DOX treatment (Sun and Kang, 2002; see Table 1). DOX forms a complex with iron (Fe) that reacts spontaneously to generate hydrogen peroxide and hydroxyl radicals, which cause oxidative damage. (Kang *et al.*, 1997; Sun *et al.*, 2001; Sun and Kang, 2002). DOX also produces superoxide and peroxynitrite by different mechanisms (Vasquez-Vivar *et al.*, 1997; Pacher *et al.*, 2003). Wang *et al.* (2001a) have directly detected the formation of intracellular ROS by fluorescent confocal microscopy (probe: carboxy-H2-DCFDA, which mainly measures superoxide), and found that mitochondrial

ROS concentrations were dramatically elevated by DOX in non-transgenic cardiomyocytes while DOX-induced mitochondrial ROS formation was significantly prevented. Again, interaction of superoxide with nitric oxide will form peroxynitrite, which causes protein nitration. To support this notion, there was a significant increase in 3-NT formation in the DOX-treated hearts of WT mice, but not in the hearts of MT-TG mice (Shuai *et al.*, 2007b). As compared to WT mice, MT-KO mice were more sensitive to DOX-induced 3-NT in the heart along with the induction of cardiac damages, demonstrated by biochemical and histopathological alterations. These findings suggest that MT exhibits protective effects on the cardiac toxicology of DOX, which might be mediated through the prevention of superoxide generation and related nitrosative impairment (Shuai *et al.*, 2007b), as stated in a recent review (Takemura and Fujiwara, 2007).

Although there was no direct evidence for the increased GSSG/GSH ratio in the alcoholic heart, there was direct evidence showing that, in response to alcohol exposure, hepatic GSH concentrations were significantly decreased and GSSG concentrations were significantly increased in the WT mice, but not in the MT-TG mice (Zhou *et al.*, 2002). This study may indirectly indicate that cardiac protection of MT against alcohol-induced damage may also be mediated by its antioxidant action.

I/R-induced cardiac damage is well known through the induction of various ROS/RNS, and MT cardiac protection from I/R-induced damage is also predominantly mediated by its antioxidant action (Kang *et al.*, 2003). MDA, as a product of lipid peroxidation induced by a diversity of oxidative injuries, was found to be significantly increased in the hearts of both MT-TG and WT I/R-treated mice 4 hours after reperfusion; however, the concentration of MDA in the MT-TG mice was significantly lower than in the WT mice under the same I/R treatment. To demonstrate the role of oxidative stress in I/R, DMSO, a free-radical scavenger, was intravenously administrated 10 minutes before the animals were subjected to reperfusion. DMSO significantly reduced the I/R-induced infarct area. Analysis of all the slices indicated that the total volume of cardiac infarction was reduced by 49.3% in the DMSO-treated group. DMSO also efficiently decreased MDA concentrations (by 70.2%) in the heart (Kang *et al.*, 2003).

Although the exact mechanism by which MT prevents diabetes-caused oxidative stress remains largely unknown, indirect evidence indicates that MT prevention of NADPH oxidase (NOX) activation may be a critical mechanism involved in the prevention of cardiac oxidative damage. It was found that applying a NOX inhibitor (diphenyleneiodonium or apocynin) to cardiomyocytes can prevent high levels of glucose-induced reduction of cardiomyocyte contractility as compared to cells without NOX inhibitor treatment but exposed to the same conditions of high levels of glucose (Privratsky *et al.*, 2003; Ye *et al.*, 2003). Wold *et al.* (2006) further demonstrated the increased NOX ($p47^{phox}$, one isoform in the cytosol of cardiomyocytes) activation in the hearts of STZ-induced WT diabetic mice, but not MT-TG diabetic mice, suggesting that the cardiac protection by MT is probably through suppression of NOX activation, as indicated in Fig. 5. The inhibitory effect of MT against NOX activation, along with a significant prevention against aging effects, was also implicated in the hearts of MT-TG old mice as compared to age-matching mice (Yang *et al.*, 2006).

In summary, all of the above studies have strongly indicated that oxidative stress plays a critical role in the development of these cardiomyopathies, even though these cardiomyopathies have different etiologies. In support of this concept, Table 3 lists the studies that showed certain resistance of transgenic mice with various antioxidants to DOX-, I/R-, alcohol-, and diabetes-induced cardiomyopathies. As compared to these antioxidants, all 20 cysteine sulfur atoms in MT were involved in the radical quenching process, and the rate constant for the reaction of hydroxyl radical with MT is about 340-fold higher than that with glutathione (GSH) (Thornalley and Vašák, 1985; Abel and de Ruiter, 1989; Cai and Cherian, 2003). MT is thus about 800-fold more potent than GSH in preventing hydroxyl

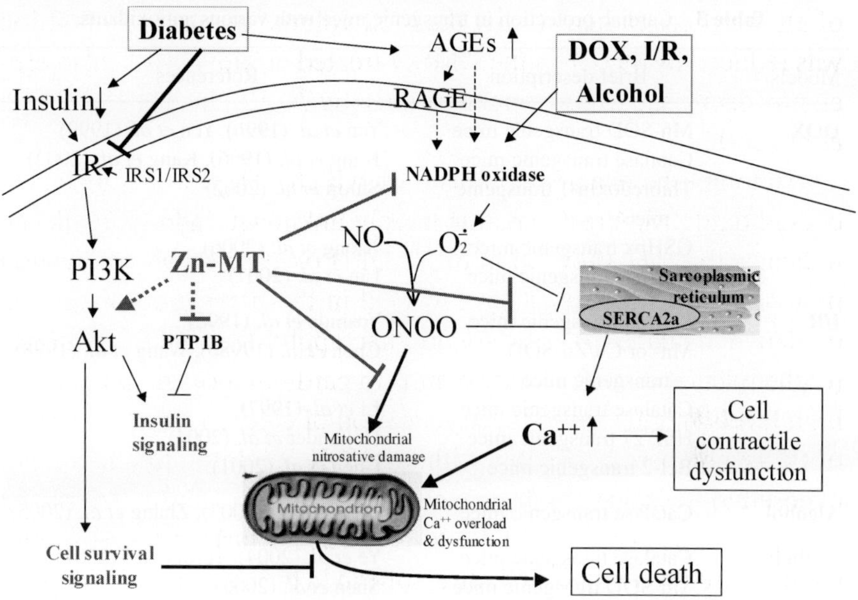

Fig. 5 Schematic outlines of the possible mechanisms responsible for cardiac protection by MT against DOX-, I/R-, alcohol-, and diabetes-induced cardiac cell death and remodeling, leading to cardiac dysfunction. DOX, I/R, alcohol, and diabetic pathogens — which include hyperglycemia, hyperglycemia, angiotensin II, inflammatory cytokines such as TNF-α, and AGEs — may all directly or indirectly induce NOX activation, leading to generation of ROS and RNS which in turn damage, to a certain extent, different organelles, resulting in organelle dysfunction and eventually cell death. MT somehow inhibits NOX activation and is also able to scavenge various ROS and RNS to prevent the organelles from ROS/RNS-induced damage. Akt activation is downregulated in type 1 diabetes or alcoholic heart, and insulin-stimulated Akt activation is blunted in insulin resistance; however, Zn-MT can, by unknown mechanisms, reserve the normal Akt activation in type 1 diabetes and insulin-stimulated Akt activation in the insulin-resistance model. Solid lines indicate the defined inhibitory or stimulatory action; and dashed lines indicate the uncertain, but possible, action.

radical-generated DNA damage *in vitro* (Abel and de Ruiter, 1989). MT was able to directly protect DNA and protein from peroxynitrite-induced damage in a cell-free system (Cai *et al.*, 2000). Unlike other antioxidants which protect against specific species of ROS (for instance, SOD protects against superoxide radicals, and catalase and GPx protect against hydrogen peroxide, as shown in Fig. 6), MT is a potent antioxidant to protect against a wide range of free radicals,

Table 3 Cardiac protection in transgenic mice with various antioxidants.

Models	Brief description	References
DOX	Mn-SOD transgenic mice	Yen *et al.* (1996), Yen *et al.* (1999)
	Catalase transgenic mice	Kang *et al.* (1996), Kang *et al.* (2002)
	Thioredoxin-1 transgenic mice	Shioji *et al.* (2002)
	GSHpx transgenic mice	Xiong *et al.* (2006)
	HSP 27 transgenic mice	Liu *et al.* (2007)
I/R	GSHpx transgenic mice	Yoshida *et al.* (1996)
	Mn- or Cu/Zn-SOD transgenic mice	Chen *et al.* (1998b), Wang *et al.* (1998)
	Catalase transgenic mice	Li *et al.* (1997)
	HSP 27 transgenic mice	Hollander *et al.* (2004)
	Bcl-2 transgenic mice	Chen *et al.* (2001)
Alcohol	Catalase transgenic mice	Zhang *et al.* (2003), Zhang *et al.* (2005)
Diabetes	Catalase transgenic mice	Ye *et al.* (2004), Turdi *et al.* (2007)
	Mn-SOD transgenic mice	Shen *et al.* (2006)
	GSHPx transgenic mice	Matsushima *et al.* (2006)

Note: HSP, heat shock protein.

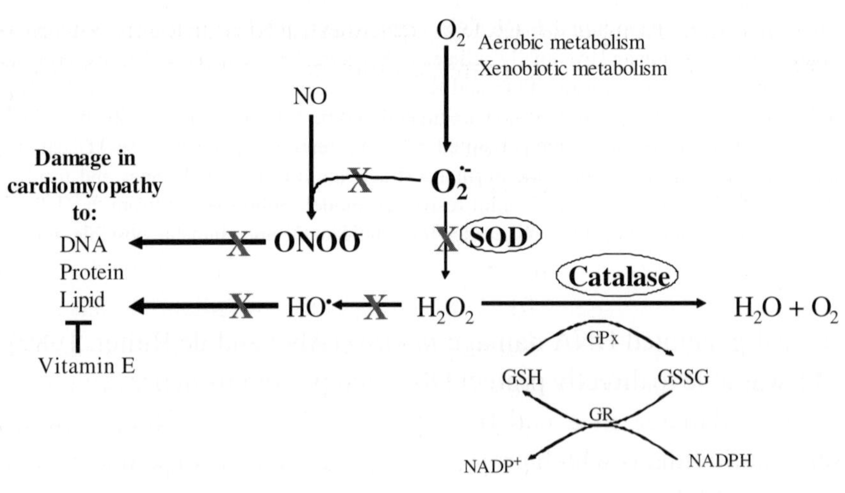

Fig. 6 Schematic illustration of various ROS and RNS generations and the sites of MT interaction with ROS and RNS (crosses). NO, nitric oxide; ONOO, peroxynitrite; GPx = GSHpx, glutathione peroxidase; GR, glutathione reductase.

as shown by the crosses in Fig. 6, including superoxide, hydrogen peroxide, hydroxyl radical, and peroxynitrite (Thornalley and Vašák, 1985; Abel and de Ruiter, 1989; Quesada *et al.*, 1996; Cai *et al.*, 2000; Cai and Cherian, 2003).

4.2. *Antiapoptotic Effect*

The apoptotic effect plays a critical role in the development of various cardiomyopathies (Garg *et al.*, 2005), and MT has been well documented for its potent antiapoptotic action under various conditions (Kang *et al.*, 2000b; Formigari *et al.*, 2007). Therefore, the second mechanism by which MT prevents these cardiomyopathies may be mediated by its antiapoptotic action, through suppression of oxidative stress or other cell death signalings.

Apoptotic cell death has been extensively documented in the hearts of animals or humans exposed chronically to DOX, I/R, alcohol, or diabetes (Kang *et al.*, 2000a; Cai *et al.*, 2002; Kang *et al.*, 2003; Fernandez-Sola *et al.*, 2006; Kuethe *et al.*, 2007). MT protects the heart against the apoptotic effects caused by these pathogenic factors. For instance, Kang *et al.* (2000a) demonstrated that when both MT-TG and their WT mice were treated intraperitoneally with DOX and sacrificed on the fourth day after treatment, apoptotic cells were significantly increased in the hearts of WT mice, but not MT-TG mice. Similarly, when both MT-TG and WT mice were subjected to an open-chest coronary artery occlusion and reperfusion procedure to produce I/R-induced left ventricle infarction, cardiac infarction produced after reperfusion for 4 hours was significantly reduced along with a significant inhibition of cardiac apoptotic cell death in the MT-TG mice as compared to the WT mice (Kang *et al.*, 2003).

Consistent with the above findings, we recently tested whether attenuation of early-phase cardiac cell death can prevent diabetic cardiomyopathy. For this purpose, diabetes was induced by a single dose of STZ in cardiac-specific MT-TG mice and WT controls. A significant reduction in diabetes-induced increases in TUNEL-positive cells, caspase-3 activation, and cytochrome c release from mitochondria was observed in the MT-TG mice as compared to WT mice. Double staining for cardiomyocytes with alpha-sarcomeric actin and caspase-3 confirmed the cardiomyocyte-specific apoptotic effects. A significant prevention of diabetic cardiomyopathy and enhanced animal survival rate were observed in the MT-TG diabetic mice as compared to WT diabetic mice. Thus, these results suggest that attenuation of early-phase cardiac cell death by MT results in a significant prevention in the development of diabetic cardiomyopathy (Cai *et al.*, 2006).

The antiapoptotic effect of MT against DOX, I/R, and diabetes was found to be mediated at least in part by its suppression of the mitochondrial cytochrome c release-dependent apoptotic pathway (Wang *et al.*, 2001a; Wang *et al.*, 2001b; Kang *et al.*, 2003; Cai *et al.*, 2006). As shown in Fig. 5, the mitochondrial protection of MT against the pathogenic effects on the membrane permeability caused by exposure to either DOX or diabetes was attributed to its suppression of mitochondrial ROS overproduction leading to oxidative stress (Wang *et al.*, 2001a; Cai *et al.*, 2002; Cai *et al.*, 2006). MT prevention of DOX-induced cardiac cell death is also mediated in part by suppression of DOX-activated p38 mitogen-activated protein kinase (MAPK), which is critically involved in the apoptotic process (Kang *et al.*, 2000a). These results showed that MT protects the heart from various cardiomyopathies likely through multiple mechanisms.

4.3. *Zinc Homeostasis and PI3/Akt-Dependent Insulin Signaling*

Besides the antioxidant and antiapoptotic actions of MT itself, Zn bound to MT may also mediate cardiac prevention of MT against these pathogenic factors. MT binds Zn under physiological conditions (Kang, 1999). Under oxidative stress conditions, Zn is released from MT (Malaiyandi *et al.*, 2004; Feng *et al.*, 2006; Maret and Krężel, 2007). As shown in Fig. 7, cardiac Zn is threefold higher in

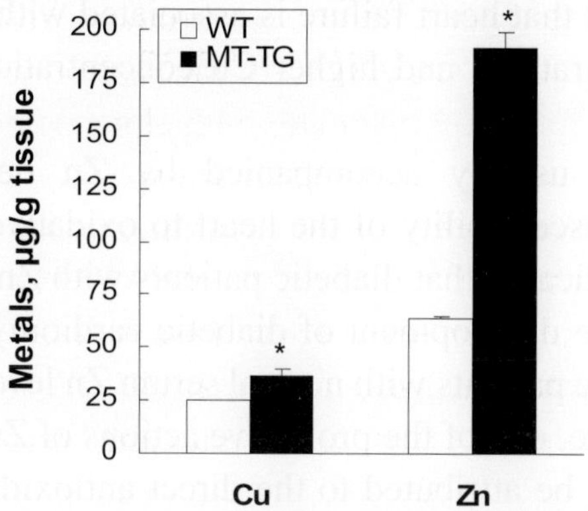

Fig. 7 Cardiac levels of Zn and Cu of the WT and MT-TG mice. Cardiac Zn and Cu contents were measured by atomic absorption spectrometry from 5 to 7 hearts of the WT and MT-TG mice (2 months old). *, $P < 0.05$ vs. WT group.

the MT-TG mice than in WT mice, while cardiac Cu is increased by about only 1.3 folds. Although several Zn transporters have been identified and characterized, these Zn transporters are all associated with plasma membranes to transport Zn across plasma membranes. Therefore, MT as an intracellular Zn-trafficking regulator delivers Zn to different targets either in mitochondria or in nucleus by redox signaling regulation.

Given that Zn plays a vital role in various cellular functions, Zn deficiency may be a risk factor for cardiac oxidative damage and supplementation with Zn may provide a significant prevention of oxidative damage to the heart. To support this notion, Kosar *et al.* (2006) performed a study to assess serum concentrations of selenium, Zn, and Cu in patients with heart failure and to compare idiopathic dilated cardiomyopathy and ischemic cardiomyopathy patients with healthy controls, which included 54 heart failure patients (26 idiopathic dilated cardiomyopathy patients and 28 ischemic cardiomyopathy patients) and 30 healthy subjects. They found that serum concentrations of selenium and Zn in heart failure patients were significantly lower than in healthy controls; however, serum Cu concentrations in these patients were significantly higher than in controls.

They concluded that heart failure is associated with lower selenium and Zn concentrations and higher Cu concentration (Kosar *et al.*, 2006).

Diabetes is usually accompanied by Zn deficiency, which increases the susceptibility of the heart to oxidative damage. It has been clearly indicated that diabetic patients with Zn deficiency have high risk for the development of diabetic cardiomyopathy as compared to diabetic patients with normal serum Zn levels (Soinio *et al.*, 2007). Therefore, one of the protective actions of Zn-MT in the diabetic heart may be attributed to the direct antioxidant action of Zn released from MT under diabetic oxidative stress and its subsequent uptake by plasma membranes to protect against lipid peroxidation and thereby stabilize membranes.

The protective effect of Zn administration to diabetic animals against diabetes-induced cardiomyopathy has been indicated in diabetic mice (Wang *et al.*, 2006). Karagulova *et al.* (2007) recently used fluorescent imaging to demonstrate decreased levels of labile Zn in I/R rat hearts. Phorbol 12-myristate 13-acetate, a known trigger of Zn release, liberated Zn ions in control hearts, but failed to produce any increase in Zn levels in I/R rat hearts. Adding the Zn ionophore pyrithione at reperfusion improved cardiac recovery up to 100% and reduced the incidence of arrhythmias more than twofold. This effect was dose-dependent, and high concentrations of Zn were toxic. Adding membrane-impermeable Zn chloride was ineffective. Hearts from rats receiving Zn pyrithione supplements in their diet fully recovered from I/R. The recovery was associated with the prevention of degradation of the two protein kinase C isoforms, delta and epsilon, during I/R. This study suggests a protective role of intracellular Zn in cardiac recovery from oxidative stress imposed by I/R (Karagulova *et al.*, 2007). Therefore, the level of intracellular labile Zn is changed in hearts subjected to I/R, and the maintenance of cardiac Zn status protects heart functions. Atahan *et al.* (2007) also provided a piece of confirming evidence to indicate the effect of Zn aspartate on I/R injury in skeletal muscle using the hind limb tourniquet operation approach. The viabilities of ischemic limbs (i.e. percentage of the contralateral control muscle) in the control, I/R, and I/R-Zn groups were 114% ± 12%, 20% ± 2%, and 95% ± 10%, respectively, indicating that I/R-induced muscle damage is significantly prevented by Zn aspartate. The decrease in SOD levels in the I/R group was also statistically reversed by the administration of Zn aspartate. Thus, Zn aspartate seems to be an effective treatment option against I/R-reperfusion injury (Atahan *et al.*, 2007).

In the above studies, Zn protection of the heart against heart failure might be related to the direct antioxidant action of Zn as discussed above. However, given that Zn is also a critical factor for many important proteins, enzymes, and transcription factors, its insulin-like function (Canesi *et al.*, 2001; Tang and Shay, 2001; Song *et al.*, 2005a) may also be one of the major responsible mechanisms (Fang *et al.*, 2005; Ceylan-Isik *et al.*, 2006a).

As shown in Fig. 5, binding of insulin to its receptor activates tyrosine kinase in the insulin receptor subunit and turns on either insulin receptor substrates (IRS-1 and IRS-2) or the Grb2/Sos/Ras/ERK1/ERK2 pathway. Several postreceptor components, including Akt, protein tyrosine phosphatase 1B (PTP1B), and the mammalian target of rapamycin, have been shown to participate in glucose metabolism (Huisamen, 2003; Fang *et al.*, 2005). Akt is positive and PTP1B is a negative regulator of insulin signaling. Studies have shown that Akt activation was downregulated in STZ-induced diabetic rats and mice (Huisamen, 2003; Ceylan-Isik *et al.*, 2006a). Fang *et al.* (2005) demonstrated that insulin resistance was induced by sucrose feeding in both WT and MT-TG mice, but that insulin-stimulated Akt activation was blunted only in the hearts of WT mice, not in MT-TG mice. Ceylan-Isik *et al.* (2006a) indicated that MT prevents the downregulation of Akt activation in STZ-induced diabetic mice. Fang *et al.* (2005) also provided evidence showing that sucrose-feeding-induced insulin resistance is accompanied with an increased expression of PTP1B in WT mice, but not in MT-TG mice. These pieces of information may suggest that cardiac protection by

MT from diabetes is probably, at least in part, via stimulating or preserving cardiac insulin signaling by preserving Akt activation in response to insulin and by depressing PTP1B upregulation, as indicated in Fig. 5.

Furthermore, reduced insulin sensitivity following alcohol intake also plays a role in alcohol-induced organ damage, although its precise mechanism is undefined. Alcohol intake dampened whole-body glucose tolerance in WT mice, but not in MT-TG mice. This difference was accompanied by reduced expression of total Akt, phosphorylated mTOR, and Akt1 kinase activity in alcohol-consuming WT mice. These data suggest that chronic alcohol intake interrupts cardiac contractile function and Akt/mTOR/p70s6k signaling. Akt, but unlikely mTOR and p70s6k, may contribute to the MT-elicited cardiac protective response (Li and Ren, 2006b). Consistent with this study, another study from the same group also indicated the importance of Akt activation in MT-mediated cardiac protection against age-induced damage (Fang *et al.*, 2005).

In short, these studies demonstrated that enhanced Akt-dependent insulin signaling is beneficial for diabetes-, alcohol-, and aging-induced cardiac dysfunction (Fig. 5). However, whether the stimulation of Akt and insulin signaling in the heart of MT-TG mice is mediated by the release of Zn from MT under oxidative stress requires further study.

4.4. *Other Possible Mechanisms*

4.4.1. *Calcium-Handling Proteins*

As mentioned above, calcium-handling abnormalities are one of the mechanisms for diabetic cardiac dysfunction. The intracellular Ca^{2+} clearing rate in the cardiomyocytes from WT diabetic mice (Ye *et al.*, 2003) and sucrose-fed insulin-resistant mice (Fang *et al.*, 2005) was significantly lower than that from control mice. However, these abnormal alterations were attenuated in MT-TG diabetic mice, suggesting the preventive effect of MT on diabetes-altered calcium handling. Nevertheless, the mechanisms by which MT affects this pathway remain unclear.

SERCA2a and NCX play crucial roles in cardiac Ca^{2+} regulation, including cytosolic Ca^{2+} removal, sarcoplasmic reticulum Ca^{2+} load, and therefore also the amount of Ca^{2+} available for cell contraction. Dysfunction of SERCA2a was extensively documented in diabetic hearts, and overexpression of SERCA2a can significantly prevent diabetes-caused calcium-handling dysfunction and cardiomyopathy (Trost *et al.*, 2002). Dr Ren and colleagues found that the abnormality of Ca^{2+} handling in the hearts of STZ-induced diabetic or spontaneously diabetic OVE26 mice was due to reduced expression and activation of SERCA2a and NCX (Duan *et al.*, 2003; Wold *et al.*, 2006). Diabetes-caused downregulation of SERCA2a (Kralik *et al.*, 2005) and NCX (Wold *et al.*, 2006) was abolished in the hearts of MT-TG mice, along with restoration of diabetes-induced reduction of cardiomyocyte contractility. The prevention of diabetes-reduced SERCA2a expression and activation was assumed to be due to the suppression of oxidative stress (Kralik *et al.*, 2005). To support their notion, Ayaz and Turan (2006) recently demonstrated that administration of selenium as an antioxidant to STZ-diabetic rats for 4 weeks significantly prevented diabetes-induced cardiomyocyte intracellular Ca^{2+} alteration, although SERCA2a expression and function were not measured (Ayaz and Turan, 2006). Indeed, SERCA2a was susceptible to oxidative or nitrosative damage, leading to its dysfunction (Ihara *et al.*, 2005; Lokuta *et al.*, 2005; French *et al.*, 2006). Therefore, the prevention of SERCA2a degradation and inactivation by MT under diabetic conditions may relate to MT suppression of NOX activation and associated peroxynitrite formation, leading to SERCA2a nitration, as illustrated in Fig. 5.

Li and Ren also demonstrated the critical role of MT in maintaining the calcium-handling prevention of NCX abnormality in the prevention of alcohol-induced cardiac dysfunction (Li and Ren, 2006a). Both WT and MT-TG mice were placed on a 4% alcohol or control diet for 12 weeks. Cardiac contractile function was evaluated in cardiomyocytes *in vitro*, including peak shortening, time to peak shortening, time to 90% relengthening, maximal velocity of shortening/relengthening (±dL/dt), intracellular Ca^{2+} rise (change in fura-2 fluorescent intensity [DeltaFFI]), and intracellular Ca^{2+} decay rate. Alcohol intake depressed the cell contractile function in WT mice, but not in MT-TG mice. Immunoblotting data showed reduced SERCA2a, NCX, and phospholamban expression in the WT mouse alcohol group, while all of these alcohol-induced changes in cardiac proteins were nullified in MT-TG mice (Li and Ren, 2006a).

4.4.2. *Advanced Glycation Endproducts*

The products of nonenzymatic glycation and oxidation of proteins and lipids, i.e. AGEs, accumulate when tissues or cells are exposed to chronic hyperglycemia. In turn, AGEs have multiple effects on vascular systems including the heart through a receptor (RAGE) to turn on several pathogenic pathways (Candido *et al.*, 2003; Kaneko *et al.*, 2005). However, ROS/RNS are the ones that act as both causative factors for AGE formation and effective mediators of AGEs under diabetic conditions (Kaneko *et al.*, 2005; Ceylan-Isik *et al.*, 2006c; Wu and Ren, 2006). Recently, Dr Ren and colleagues used antioxidant benfotiamine to treat STZ-induced diabetic rats or high levels of glucose-treated cardiac cells to show that benfotiamine significantly prevents diabetes-caused or high levels of glucose-caused oxidative damage along with cardiac function improvement without an effect on AGE contents, suggesting the critical role of AGE-derived oxidative stress in diabetes-caused cardiac dysfunction (Ceylan-Isik *et al.*, 2006c; Wu and Ren, 2006).

So far, there is no information regarding whether MT can prevent AGE formation and its pathogenic effect. However, since transition metals such as Cu and iron (Fe) play an important role in accelerating AGE formation (Qian *et al.*, 1998; Eaton and Qian, 2002a; Eaton and Qian, 2002b), MT may play a role in preventing its formation indirectly through binding transition metals to stop them from participating in Fenton reaction. We have demonstrated the preventive effect of MT directly on Cu- or Fe-induced DNA damage in a cell-free system due to a Fenton-type reaction to form hydroxyl radicals (Cai *et al.*, 1995; Cai *et al.*, 1998). Incubation of ceruloplasmin or Cu/Zn-SOD with high levels of glucose reportedly caused its fragmentation and a time-dependent release of its bound Cu, which then appeared to participate in a Fenton-type reaction (Ookawara *et al.*, 1992; Islam *et al.*, 1995). Under diabetic conditions, therefore, damage to Cu-regulating mechanisms through glycation or oxidization of ceruloplasmin and albumin could lead to elevated contents of catalytically active Cu in plasma, as assumed by Cooper *et al.* (2004 and 2005) and Shukla *et al.* (2006). Thus, the preventive effect of applying a Cu-selective chelator, trientine, on diabetic retinopathy, neuropathy, neurovascular dysfunction, and cardiomyopathy was indicated in STZ-induced diabetic rats (Cameron and Cotter, 1995; Cooper *et al.*, 2004; Cooper *et al.*, 2005; Hamada *et al.*, 2005) and even in diabetic patients (Cooper *et al.*, 2004).

Similarly, Fe chelation also suppressed angiotensin II-induced upregulation of fibrotic signaling in the heart (Saito *et al.*, 2005). Since MT has a strong affinity to bind Cu and Fe, it is reasonable to assume that it plays a role in removing Cu and Fe in the heart, either directly to prevent their participation in Fenton reaction to form free radicals or indirectly to prevent their acceleration of AGE formation. However, whether removal of cardiac Cu is the mechanism involved in the protective effect on diabetes-induced cardiomyopathy in the MT-TG diabetic mouse model remains unclear, since Cu content was not increased in the diabetic heart (Uriu-Adams *et al.*, 2005); therefore, further experiments in this area are urgently needed.

5. Possible Intervention for Various Cardiomyopathies Through Upregulated Cardiac Metallothionein Expression

As mentioned above, DOX-, I/R-, alcohol-, and diabetes-induced cardiac toxicity and eventually development of cardiomyopathy were significantly prevented not only in transgenic mice with 10- to 60-fold high expression of cardiac MT, but also in transgenic mice with only 2–3-fold high expression of cardiac MT, as compared to the WT mice (Kang *et al.*, 1997; Kang *et al.*, 2000a; Kang *et al.*, 2003; Cai *et al.*, 2005b; Song *et al.*, 2005b). This suggests that cardiac MT at the levels induced by traditionally used inducers may also prevent diabetic cardiomyopathy. Zn as an efficient MT inducer in multiple organs may be a candidate for clinical use to induce cardiac MT for protecting the heart against these pathogenic factors.

The protective effect of pre-induced cardiac MT by Zn has been indicated in early studies against DOX-induced cardiotoxicities (Naganuma *et al.*, 1988; Satoh *et al.*, 2000; Ali *et al.*, 2002). For instance, daunorubicin (DNR) is an anthracyline antibiotic used in the treatment of a variety of human cancers. Ali *et al.* (2002) investigated the effect of DNR administration on both cardiac and hepatic tissues, and the possible protective role of Zn on the cardiotoxicity and hepatotoxicity produced by DNR. Administration of DNR to Sprague–Dawley rats increased serum creatine kinase activity and blood troponin T levels (as cardiotoxicity indices), alanine aminotransferase activity (as a hepatotoxicity index), and cardiac and hepatic 2-thiobarbituric acid reactive substances (as an index of lipid peroxidation). Treatment with Zn prior to DNR dramatically induced MT mRNA and protein expression in both heart and liver; while DNR alone induced MT, but to a much lower degree than Zn. The increases in both MT protein and MT-1 mRNA can parallel with the reduction of cardiac and hepatic toxicities. These results indicate that MT induction by Zn is a highly effective approach in preventing cardiotoxicity and hepatotoxicity caused by DNR (Ali *et al.*, 2002). Shuai *et al.* (2007a and 2007b) also recently investigated the inhibitive effect of MT induced by Zn on DOX-induced cardiac nitrosative damage and apoptosis. Mice were treated with DOX with and without Zn pretreatment to induce cardiac MT. DOX administration decreased heart weight by 10% and caused remarkable cardiac apoptosis as demonstrated by DNA fragments as well as 3-NT accumulation, while Zn pretreatment significantly inhibited these effects of DOX. Elevated expression of Bax was obviously observed after DOX treatment, while this elevation was prevented by MT induction by Zn. These findings suggest that MT induced by Zn exhibits protective effects on the cardiac apoptosis of DOX, which might be mediated through the prevention of Bax protein upregulation by DOX and associated elevation of the Bax/Bcl-2 ratio (Shuai *et al.*, 2007a; Shuai *et al.*, 2007b).

In order to test the potential for the clinical use of Zn as a cardiac MT inducer, we recently examined the protective effects of Zn supplementation on diabetic cardiomyopathy in a mouse model (Wang *et al.*, 2006). STZ-induced diabetic mice were supplemented intraperitoneally with Zn sulfate every other day over a period of 3 months. At 6 months after the onset of diabetes, diabetic mice with and without 3-month Zn supplementation were examined for cardiomyopathy by functional and morphological analysis. Significant increases in cardiac fibril disarrangement, fibrosis, and dysfunction were observed in diabetic mice, but not in diabetic mice supplemented with Zn which induced a significant increase in cardiac MT expression. To determine the direct link of the protective effect of Zn supplementation against diabetic cardiomyopathy to cardiac MT synthesis, the cultured cardiac cells were exposed to high levels of glucose and free fatty acid (HG/FFA) treatment which mimics diabetic conditions. The cell survival rate was significantly decreased when cells were exposed to HG/FFA alone, but was unchanged when exposed to HG/FFA with a pretreatment of Zn which induces significant MT mRNA expression and MT protein synthesis. When MT expression was silenced using MT siRNA, the preventive effect of pretreatment with Zn was significantly abolished (Wang *et al.*, 2006). These results suggest that the prevention of diabetic cardiomyopathy by Zn supplementation is predominantly mediated by cardiac MT induction.

As mentioned above, Zn has an insulin-like function; therefore, Zn deficiency decreased the response of tissues to insulin (Hall *et al.*, 2005; Song *et al.*, 2005a) and Zn supplementation was able to improve hyperglycemia in STZ-induced and db/db diabetic mice

(Chen *et al.*, 1998a; Simon and Taylor, 2001). Since there is no free intracellular Zn, MT, via donating and storing Zn, plays an important role in its homeostasis and its insulin-like action in organs (Fang *et al.*, 2005), which is supported by the fact that MT-TG mice were protected against cardiac dysfunction caused by insulin resistance (Fang *et al.*, 2005; Dong *et al.*, 2007). Therefore, in addition to cardiac MT induction, Zn supplementation may partially improve glucose metabolism and reduce free acid β-oxidation via an insulin-like action.

These important findings highlight the great potential for a clinical application of Zn in the prevention of diabetic cardiomyopathy, given that Zn has been clinically used for other chronic diseases without significant toxicity (Gregorio *et al.*, 2007; Karaca *et al.*, 2007; Lin *et al.*, 2008; Moriarty-Craige *et al.*, 2007; Takahashi *et al.*, 2007). In addition, Zn supplementation may also provide other benefits for patients with diabetes (Roussel *et al.*, 2003; Al-Maroof and Al-Sharbatti, 2006).

In addition, although diabetic cardiomyopathy can occur without vascular disease, the coexistence of vascular disorders such as arteriosclerosis and hypertension also accelerates the progression of diabetic cardiomyopathy, as indicated in Fig. 8 (Cai and Kang, 2001). Zn deficiency or an increased Cu/Zn ratio, which is a common feature in diabetic individuals (Cooper *et al.*, 2005; Song *et al.*, 2005b), was associated with an increased risk of cardiovascular diseases (Canatan *et al.*, 2004; Reiterer *et al.*, 2005; Soinio *et al.*, 2007). Under diabetic conditions, Cu- and Fe-regulating proteins may be glycated, oxidized, or nitrated directly by chronic hyperglycemia or diabetes-derived ROS and RNS; as a result, Cu and Fe bound to these proteins will be released as free Cu and Fe into plasma (Cooper *et al.*, 2005; Shukla *et al.*, 2006; Van Campenhout *et al.*, 2006). The free Cu and Fe will be toxic to vascular endothelial cells, leading to arteriosclerosis and hypertension, which in turn accelerate the development of diabetic cardiomyopathy. Cooper and colleagues (2004) have provided direct evidence that, in STZ-induced diabetic animals with established heart failure, 7 weeks of oral trientine therapy could significantly alleviate

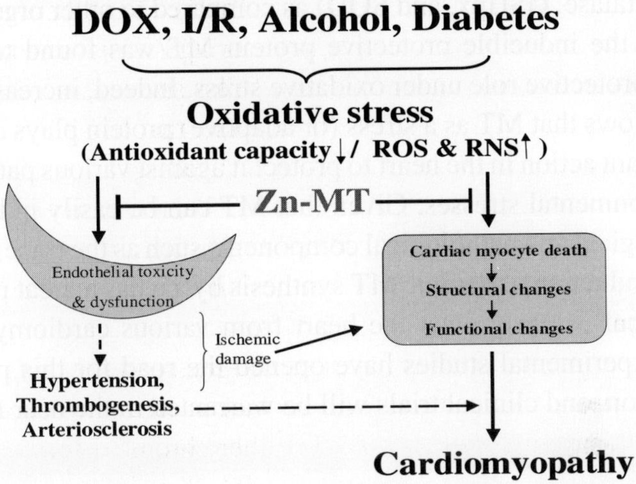

Fig. 8 Schematic illustration of the accelerating effects of vascular disorders on the development of cardiomyopathies caused by DOX, I/R, alcohol, or diabetes. Cardiac and vascular MT induced by Zn both play important roles in protecting the heart and blood vessels against DOX-, I/R-, alcohol-, or diabetes-induced cardiomyopathy.

heart failure without lowering blood glucose, substantially improve cardiomyocyte structure, and reverse elevations in left ventricular collagen. More importantly, they demonstrated similar evidence in humans with type 2 diabetes, where 6 months of treatment caused elevated left ventricular mass to decline significantly and reach a normal level. These data strongly suggest that removal of Cu may provide a new approach to the therapy of diabetic cardiomyopathy (Cooper *et al.*, 2004). Induced MT synthesis in vascular cells by systemic Zn supplementation may prevent the toxic effect of free Cu and Fe on endothelial cells, and indirectly protect the heart from vascular dysfunction, as indicated in Fig. 8.

6. Conclusion

Cardiomyopathy can be developed with a variety of etiologies; however, several kinds of these diseases, if not all, are somehow related to oxidative and/or nitrosative stress. Since the heart has a relatively low content of common known antioxidants (including

GSH, catalase, GSHpx, and SOD) as compared to other organs such as liver, the inducible protective protein MT was found to play a critical protective role under oxidative stress. Indeed, increasing evidence shows that MT as a stress (or adaptive) protein plays a critical antioxidant action in the heart to protect it against various pathogenic or environmental stresses. Given that MT can be easily induced by physiological and pathological components such as the trace element Zn, the induction of cardiac MT synthesis by Zn has a great potential for clinical use to protect the heart from various cardiomyopathic risks. Experimental studies have opened the road for this potential application, and clinical trials will be warranted in the near future.

Acknowledgments

The author thanks Dr Yi Tan for his excellent technical assistance in preparing this chapter. The work cited in this chapter was supported in part by grants from the American Diabetes Association (02-07-JF-02, 05-07-02) and the Juvenile Diabetes Research Foundation International (5-2006-382), and a start-up fund from the Chinese-American Research Institute for Diabetic Complications, Wenzhou Medical College, China.

References

Abel J, de Ruiter N. Inhibition of hydroxyl-radical-generated DNA degradation by metallothionein. *Toxicol Lett* 1989; **47**:191–196.

Al-Maroof RA, Al-Sharbatti SS. Serum zinc levels in diabetic patients and effect of zinc supplementation on glycemic control of type 2 diabetics. *Saudi Med J* 2006; **27**:344–350.

Ali MM, Frei E, Straub J, *et al.* Induction of metallothionein by zinc protects from daunorubicin toxicity in rats. *Toxicology* 2002; **179**:85–93.

Atahan E, Ergun Y, Belge Kurutas E, *et al.* Ischemia-reperfusion injury in rat skeletal muscle is attenuated by zinc aspartate. *J Surg Res* 2007; **137**:109–116.

Ayaz M, Turan B. Selenium prevents diabetes-induced alterations in $[Zn^{2+}]_i$ and metallothionein level of rat heart via restoration of cell redox cycle. *Am J Physiol Heart Circ Physiol* 2006; **290**:H1071–H1080.

Bauman JW, Liu J, Liu YP, Klaassen CD. Increase in metallothionein produced by chemicals that induce oxidative stress. *Toxicol Appl Pharmacol* 1991; **110**:347–354.

Cai L. Suppression of nitrative damage by metallothionein in diabetic heart contributes to the prevention of cardiomyopathy. *Free Radic Biol Med* 2006; **41**:851–861.

Cai L. Diabetic cardiomyopathy and its prevention by metallothionein: Experimental evidence, possible mechanisms and clinical implications. *Curr Med Chem* 2007; **14**:2193–2203.

Cai L, Cherian MG. Zinc-metallothionein protects from DNA damage induced by radiation better than glutathione and copper- or cadmium-metallothioneins. *Toxicol Lett* 2003; **136**:193–198.

Cai L, Kang YJ. Oxidative stress and diabetic cardiomyopathy: A brief review. *Cardiovasc Toxicol* 2001; **1**:181–193.

Cai L, Klein J, Kang YJ. Metallothionein prevents diabetic cardiomyopathy. *Toxicol Sci* 2001; **60**(Suppl 1):13.

Cai L, Klein JB, Kang YJ. Metallothionein inhibits peroxynitrite-induced DNA and lipoprotein damage. *J Biol Chem* 2000; **275**:38957–38960.

Cai L, Koropatnick J, Cherian MG. Metallothionein protects DNA from copper-induced but not iron-induced cleavage *in vitro*. *Chem Biol Interact* 1995; **96**:143–155.

Cai L, Li W, Wang G, *et al.* Hyperglycemia-induced apoptosis in mouse myocardium: Mitochondrial cytochrome c-mediated caspase-3 activation pathway. *Diabetes* 2002; **51**:1938–1948.

Cai L, Satoh M, Tohyama C, Cherian MG. Metallothionein in radiation exposure: Its induction and protective role. *Toxicology* 1999; **132**:85–98.

Cai L, Tsiapalis G, Cherian MG. Protective role of zinc-metallothionein on DNA damage *in vitro* by ferric nitrilotriacetate (Fe-NTA) and ferric salts. *Chem Biol Interact* 1998; **115**:141–151.

Cai L, Wang J, Li Y, *et al.* Inhibition of superoxide generation and associated nitrosative damage is involved in metallothionein prevention of diabetic cardiomyopathy. *Diabetes* 2005; **54**:1829–1837.

Cai L, Wang Y, Zhou G, *et al.* Attenuation by metallothionein of early cardiac cell death via suppression of mitochondrial oxidative stress results in a prevention of diabetic cardiomyopathy. *J Am Coll Cardiol* 2006; **48**:1688–1697.

Cameron NE, Cotter MA. Neurovascular dysfunction in diabetic rats. Potential contribution of autoxidation and free radicals examined using transition metal chelating agents. *J Clin Invest* 1995; **96**:1159–1163.

Canatan H, Bakan I, Akbulut M, *et al.* Relationship among levels of leptin and zinc, copper, and zinc/copper ratio in plasma of patients with essential hypertension and healthy normotensive subjects. *Biol Trace Elem Res* 2004; **100**:117–123.

Candido R, Forbes JM, Thomas MC, *et al.* A breaker of advanced glycation end products attenuates diabetes-induced myocardial structural changes. *Circ Res* 2003; **92**:785–792.

Canesi L, Betti M, Ciacci C, Gallo G. Insulin-like effect of zinc in Mytilus digestive gland cells: Modulation of tyrosine kinase-mediated cell signaling. *Gen Comp Endocrinol* 2001; **122**:60–66.

Ceylan-Isik AF, LaCour KH, Ren J. Gender disparity of streptozotocin-induced intrinsic contractile dysfunction in murine ventricular myocytes: Role of chronic activation of Akt. *Clin Exp Pharmacol Physiol* 2006a; **33**:102–108.

Ceylan-Isik AF, LaCour KH, Ren J. Sex difference in cardiomyocyte function in normal and metallothionein transgenic mice: The effect of diabetes mellitus. *J Appl Physiol* 2006b; **100**:1638–1646.

Ceylan-Isik AF, Wu S, Li Q, *et al.* High-dose benfotiamine rescues cardiomyocyte contractile dysfunction in streptozotocin-induced diabetes mellitus. *J Appl Physiol* 2006c; **100**:150–156.

Chen HS, Shan YX, Yang TL, *et al.* Insulin deficiency downregulated heat shock protein 60 and IGF-1 receptor signaling in diabetic myocardium. *Diabetes* 2005; **54**:175–181.

Chen K, Zhang J, Ye C. [Metallothionein involvement in the delayed protection after ischemic or anoxic preconditioning in myocardium or cultured cardiomyocytes]. *Zhonghua Yi Xue Za Zhi* 1997; **77**:106–110.

Chen MF, Chang CL, Liou SY. Increase in resting levels of superoxide anion in the whole blood of uremic patients on chronic hemodialysis. *Blood Purif* 1998a; **16**:290–300.

Chen Y, Saari JT, Kang YJ. Weak antioxidant defenses make the heart a target for damage in copper-deficient rats. *Free Radic Biol Med* 1994; **17**:529–536.

Chen Z, Chua CC, Ho YS, *et al.* Overexpression of Bcl-2 attenuates apoptosis and protects against myocardial I/R injury in transgenic mice. *Am J Physiol Heart Circ Physiol* 2001; **280**:H2313–H2320.

Chen Z, Siu B, Ho YS, *et al.* Overexpression of MnSOD protects against myocardial ischemia/reperfusion injury in transgenic mice. *J Mol Cell Cardiol* 1998b; **30**: 2281–2289.

Cooper GJ, Chan YK, Dissanayake AM, *et al.* Demonstration of a hyperglycemia-driven pathogenic abnormality of copper homeostasis in diabetes and its reversibility by selective chelation: Quantitative comparisons between the biology of copper and eight other nutritionally essential elements in normal and diabetic individuals. *Diabetes*, 2005; **54**:1468–1476.

Cooper GJ, Phillips AR, Choong SY, *et al.* Regeneration of the heart in diabetes by selective copper chelation. *Diabetes* 2004; **53**:2501–2508.

Davies MJ. The cardiomyopathies: An overview. *Heart* 2000; **83**:469–474.

Dong F, Li Q, Sreejayan N, *et al.* Metallothionein prevents high fat diet-induced cardiac contractile dysfunction: Role of peroxisome proliferator-activated receptor {gamma} coactivator-1{alpha} and mitochondrial biogenesis. *Diabetes* 2007; **56**:2201–2212.

Doroshow JH, Locker GY, Myers CE. Enzymatic defenses of the mouse heart against reactive oxygen metabolites: Alterations produced by doxorubicin. *J Clin Invest* 1980; **65**:128–135.

Duan J, Zhang HY, Adkins SD, *et al.* Impaired cardiac function and IGF-I response in myocytes from calmodulin-diabetic mice: Role of Akt and RhoA. *Am J Physiol Endocrinol Metab* 2003; **284**:E366–E376.

Eaton JW, Qian M. Interactions of copper with glycated proteins: Possible involvement in the etiology of diabetic neuropathy. *Mol Cell Biochem* 2002a; **234–235**:135–142.

Eaton JW, Qian M. Molecular bases of cellular iron toxicity. *Free Radic Biol Med* 2002b; **32**:833–840.

Epstein PN, Overbeek PA, Means AR. Calmodulin-induced early-onset diabetes in transgenic mice. *Cell* 1989; **58**:1067–1073.

Fang CX, Dong F, Ren BH, *et al.* Metallothionein alleviates cardiac contractile dysfunction induced by insulin resistance: Role of Akt phosphorylation, PTB1B, PPARgamma and c-Jun. *Diabetologia* 2005; **48**:2412–2421.

Feng W, Benz FW, Cai J, *et al.* Metallothionein disulfides are present in metallothionein-overexpressing transgenic mouse heart and increase under conditions of oxidative stress. *J Biol Chem* 2006; **281**:681–687.

Fernandez-Sola J, Fatjo F, Sacanella E, *et al.* Evidence of apoptosis in alcoholic cardiomyopathy. *Hum Pathol* 2006; **37**:1100–1110.

Formigari A, Irato P, Santon A. Zinc, antioxidant systems and metallothionein in metal mediated-apoptosis: Biochemical and cytochemical aspects. *Comp Biochem Physiol C Toxicol Pharmacol* 2007; **146**:443–459.

French JP, Quindry JC, Falk DJ, *et al.* Ischemia-reperfusion-induced calpain activation and SERCA2a degradation are attenuated by exercise training and calpain inhibition. *Am J Physiol Heart Circ Physiol* 2006; **290**:H128–H136.

Garg S, Narula J, Chandrashekhar Y. Apoptosis and heart failure: Clinical relevance and therapeutic target. *J Mol Cell Cardiol* 2005; **38**:73–79.

Gregorio GV, Dans LF, Cordero CP, Panelo CA. Zinc supplementation reduced cost and duration of acute diarrhea in children. *J Clin Epidemiol* 2007; **60**:560–566.

Gross ER, Hsu AK, Gross GJ. Diabetes abolishes morphine-induced cardioprotection via multiple pathways upstream of glycogen synthase kinase-3beta. *Diabetes* 2007; **56**:127–136.

Hall AG, Kelleher SL, Lonnerdal B, Philipps AF. A graded model of dietary zinc deficiency: Effects on growth, insulin-like growth factor-I, and the glucose/insulin axis in weanling rats. *J Pediatr Gastroenterol Nutr* 2005; **41**:72–80.

Hamada Y, Nakashima E, Naruse K, *et al.* A copper chelating agent suppresses carbonyl stress in diabetic rat lenses. *J Diabetes Complications* 2005; **19**:328–334.

Hollander JM, Martin JL, Belke DD, *et al.* Overexpression of wild-type heat shock protein 27 and a nonphosphorylatable heat shock protein 27 mutant protects against ischemia/reperfusion injury in a transgenic mouse model. *Circulation* 2004; **110**: 3544–3552.

Huisamen B. Protein kinase B in the diabetic heart. *Mol Cell Biochem* 2003; **249**:31–38.

Ihara Y, Kageyama K, Kondo T. Overexpression of calreticulin sensitizes SERCA2a to oxidative stress. *Biochem Biophys Res Commun* 2005; **329**:1343–1349.

Islam KN, Takahashi M, Higashiyama S, *et al.* Fragmentation of ceruloplasmin following non-enzymatic glycation reaction. *J Biochem* 1995; **118**:1054–1060.

Kaneko M, Bucciarelli L, Hwang YC, *et al.* Aldose reductase and AGE-RAGE pathways: Key players in myocardial ischemic injury. *Ann NY Acad Sci* 2005; **1043**:702–709.

Kang YJ. The antioxidant function of metallothionein in the heart. *Proc Soc Exp Biol Med* 1999; **222**:263–273.

Kang YJ, Chen Y, Epstein PN. Suppression of doxorubicin cardiotoxicity by overexpression of catalase in the heart of transgenic mice. *J Biol Chem* 1996; **271**:12610–12616.

Kang YJ, Chen Y, Yu A, *et al.* Overexpression of metallothionein in the heart of transgenic mice suppresses doxorubicin cardiotoxicity. *J Clin Invest* 1997; **100**:1501–1506.

Kang YJ, Li G, Saari JT. Metallothionein inhibits ischemia-reperfusion injury in mouse heart. *Am J Physiol* 1999; **276**:H993–H997.

Kang YJ, Li Y, Sun X, Sun X. Antiapoptotic effect and inhibition of ischemia/reperfusion-induced myocardial injury in metallothionein-overexpressing transgenic mice. *Am J Pathol* 2003; **163**:1579–1586.

Kang YJ, Sun X, Chen Y, Zhou Z. Inhibition of doxorubicin chronic toxicity in catalase-overexpressing transgenic mouse hearts. *Chem Res Toxicol* 2002; **15**:1–6.

Kang YJ, Zhou ZX, Wang GW, *et al.* Suppression by metallothionein of doxorubicin-induced cardiomyocyte apoptosis through inhibition of p38 mitogen-activated protein kinases. *J Biol Chem* 2000a; **275**:13690–13698.

Kang YJ, Zhou ZX, Wu H, *et al.* Metallothionein inhibits myocardial apoptosis in copper-deficient mice: Role of atrial natriuretic peptide. *Lab Invest* 2000b; **80**:745–757.

Karaca Z, Tanriverdi F, Kurtoglu S, *et al.* Pubertal arrest due to Zn deficiency: The effect of zinc supplementation. *Hormones (Athens)* 2007; **6**:71–74.

Karagulova G, Yue Y, Moreyra A, *et al.* Protective role of intracellular zinc in myocardial ischemia/reperfusion is associated with preservation of protein kinase C isoforms. *J Pharmacol Exp Ther* 2007; **321**:517–525.

Kosar F, Nisanoglu V, Aksoy Y, *et al.* Effects of coronary revascularization and concomitant aneurysmectomy on QT interval duration and dispersion. *J Electrocardiol* 2006; **39**:194–198.

Kralik PM, Ye G, Metreveli NS, *et al.* Cardiomyocyte dysfunction in models of type 1 and type 2 diabetes. *Cardiovasc Toxicol* 2005; **5**:285–292.

Kuethe F, Sigusch HH, Bornstein SR, *et al.* Apoptosis in patients with dilated cardiomyopathy and diabetes: A feature of diabetic cardiomyopathy? *Horm Metab Res* 2007; **39**:672–676.

Lang CH, Frost RA, Summer AD, Vary TC. Molecular mechanisms responsible for alcohol-induced myopathy in skeletal muscle and heart. *Int J Biochem Cell Biol* 2005; **37**: 2180–2195.

Li G, Chen Y, Saari JT, Kang YJ. Catalase-overexpressing transgenic mouse heart is resistant to ischemia-reperfusion injury. *Am J Physiol* 1997; **273**:H1090–H1095.

Li Q, Ren J. Cardiac overexpression of metallothionein attenuates chronic alcohol intake-induced cardiomyocyte contractile dysfunction. *Cardiovasc Toxicol* 2006a; **6**:173–182.

Li Q, Ren J. Cardiac overexpression of metallothionein rescues chronic alcohol intake-induced cardiomyocyte dysfunction: Role of Akt, mammalian target of rapamycin and ribosomal p70s6 kinase. *Alcohol Alcohol* 2006b; **41**:585–592.

Liang Q, Carlson EC, Donthi RV, *et al.* Overexpression of metallothionein reduces diabetic cardiomyopathy. *Diabetes* 2002; **51**:174–181.

Lin LC, Que J, Lin KL, *et al.* Effects of zinc supplementation on clinical outcomes in patients receiving radiotherapy for head and neck cancers: A double-blinded randomized study. *Int J Radiat Oncol Biol Phys* 2008; **70**:368–373.

Liu L, Zhang X, Qian B, *et al.* Over-expression of heat shock protein 27 attenuates doxorubicin-induced cardiac dysfunction in mice. *Eur J Heart Fail* 2007; **9**: 762–769.

Lokuta AJ, Maertz NA, Meethal SV, *et al.* Increased nitration of sarcoplasmic reticulum Ca^{2+}-ATPase in human heart failure. *Circulation* 2005; **111**:988–995.

Malaiyandi LM, Dineley KE, Reynolds IJ. Divergent consequences arise from metallothionein overexpression in astrocytes: Zinc buffering and oxidant-induced zinc release. *Glia* 2004; **45**:346–353.

Maret W, Krężel A. Cellular zinc and redox buffering capacity of metallothionein/thionein in health and disease. *Mol Med* 2007; **13**:371–375.

Matsushima S, Kinugawa S, Ide T, *et al.* Overexpression of glutathione peroxidase attenuates myocardial remodeling and preserves diastolic function in diabetic heart. *Am J Physiol Heart Circ Physiol* 2006; **291**:H2237–H2245.

Moriarty-Craige SE, Ha KN, Sternberg P Jr, *et al.* Effects of long-term zinc supplementation on plasma thiol metabolites and redox status in patients with age-related macular degeneration. *Am J Ophthalmol* 2007; **143**:206–211.

Naganuma A, Satoh M, Imura N. Specific reduction of toxic side effects of adriamycin by induction of metallothionein in mice. *Jpn J Cancer Res* 1988; **79**:406–411.

Nath R, Kumar D, Li T, Singal PK. Metallothioneins, oxidative stress and the cardiovascular system. *Toxicology* 2000; **155**:17–26.

Onody A, Zvara A, Hackler L Jr, *et al.* Effect of classic preconditioning on the gene expression pattern of rat hearts: A DNA microarray study. *FEBS Lett* 2003; **536**:35–40.

Ookawara T, Kawamura N, Kitagawa Y, Taniguchi N. Site-specific and random fragmentation of Cu,Zn-superoxide dismutase by glycation reaction. Implication of reactive oxygen species. *J Biol Chem* 1992; **267**:18505–18510.

Oshima Y, Fujio Y, Nakanishi T, *et al.* STAT3 mediates cardioprotection against ischemia/reperfusion injury through metallothionein induction in the heart. *Cardiovasc Res* 2005; **65**:428–435.

Pacher P, Liaudet L, Bai P, *et al.* Potent metalloporphyrin peroxynitrite decomposition catalyst protects against the development of doxorubicin-induced cardiac dysfunction. *Circulation* 2003; **107**:896–904.

Privratsky JR, Wold LE, Sowers JR, *et al.* AT1 blockade prevents glucose-induced cardiac dysfunction in ventricular myocytes: Role of the AT1 receptor and NADPH oxidase. *Hypertension* 2003; **42**:206–212.

Qian M, Liu M, Eaton JW. Transition metals bind to glycated proteins forming redox active "glycochelates": Implications for the pathogenesis of certain diabetic complications. *Biochem Biophys Res Commun* 1998; **250**:385–389.

Quesada AR, Byrnes RW, Krezoski SO, Petering DH. Direct reaction of H_2O_2 with sulfhydryl groups in HL-60 cells: Zinc-metallothionein and other sites. *Arch Biochem Biophys* 1996; **334**:241–250.

Reiterer G, MacDonald R, Browning JD, *et al.* Zinc deficiency increases plasma lipids and atherosclerotic markers in LDL-receptor-deficient mice. *J Nutr* 2005; **135**:2114–2118.

Roussel AM, Kerkeni A, Zouari N, *et al.* Antioxidant effects of zinc supplementation in Tunisians with type 2 diabetes mellitus. *J Am Coll Nutr* 2003; **22**:316–321.

Saito K, Ishizaka N, Aizawa T, *et al.* Iron chelation and a free radical scavenger suppress angiotensin II-induced upregulation of TGF-beta1 in the heart. *Am J Physiol Heart Circ Physiol* 2005; **288**:H1836–H1843.

Satoh M, Naganuma A, Imura N. Involvement of cardiac metallothionein in prevention of adriamycin induced lipid peroxidation in the heart. *Toxicology* 1988; **53**:231–237.

Satoh M, Naganuma A, Imura N. Modulation of adriamycin toxicity by tissue-specific induction of metallothionein synthesis in mice. *Life Sci* 2000; **67**:627–634.

Shen X, Zheng S, Metreveli NS, Epstein PN. Protection of cardiac mitochondria by overexpression of MnSOD reduces diabetic cardiomyopathy. *Diabetes* 2006; **55**: 798–805.

Shioji K, Kishimoto C, Nakamura H, *et al.* Overexpression of thioredoxin-1 in transgenic mice attenuates adriamycin-induced cardiotoxicity. *Circulation* 2002; **106**:1403–1409.

Shuai Y, Guo J, Peng SQ, Zhang LS. [Metallothionein inhibited DOX-induced cardiac apoptosis in mice]. *Sichuan Da Xue Xue Bao Yi Xue Ban* 2007a; **38**:620–623.

Shuai Y, Guo JB, Peng SQ, *et al.* Metallothionein protects against doxorubicin-induced cardiomyopathy through inhibition of superoxide generation and related nitrosative impairment. *Toxicol Lett* 2007b; **170**:66–74.

Shukla N, Maher J, Masters J, *et al.* Does oxidative stress change ceruloplasmin from a protective to a vasculopathic factor? *Atherosclerosis* 2006; **187**:238–250.

Simon SF, Taylor CG. Dietary zinc supplementation attenuates hyperglycemia in db/db mice. *Exp Biol Med (Maywood)* 2001; **226**:43–51.

Soinio M, Marniemi J, Laakso M, *et al.* Serum zinc level and coronary heart disease events in patients with type 2 diabetes. *Diabetes Care* 2007; **30**:523–528.

Song Y, Wang J, Li XK, Cai L. Zinc and the diabetic heart. *Biometals* 2005a; **18**:325–332.

Song Y, Wang J, Li Y, *et al.* Cardiac metallothionein synthesis in streptozotocin-induced diabetic mice, and its protection against diabetes-induced cardiac injury. *Am J Pathol* 2005b; **167**:17–26.

Sun X, Kang YJ. Prior increase in metallothionein levels is required to prevent doxorubicin cardiotoxicity. *Exp Biol Med (Maywood)* 2002; **227**:652–657.

Sun X, Zhou Z, Kang YJ. Attenuation of doxorubicin chronic toxicity in metallothionein-overexpressing transgenic mouse heart. *Cancer Res* 2001; **61**:3382–3387.

Takahashi M, Saito H, Higashimoto M, Hibi T. Possible inhibitory effect of oral zinc supplementation on hepatic fibrosis through downregulation of TIMP-1: A pilot study. *Hepatol Res* 2007; **37**:405–409.

Takemura G, Fujiwara H. Doxorubicin-induced cardiomyopathy from the cardiotoxic mechanisms to management. *Prog Cardiovasc Dis* 2007; **49**:330–352.

Tang X, Shay NF. Zinc has an insulin-like effect on glucose transport mediated by phosphoinositol-3-kinase and Akt in 3T3-L1 fibroblasts and adipocytes. *J Nutr* 2001; **131**:1414–1420.

Thiene G, Angelini A, Basso C, *et al.* The new definition and classification of cardiomyopathies. *Adv Clin Path* 2000; **4**:53–57.

Thornalley PJ, Vašák M. Possible role for metallothionein in protection against radiation-induced oxidative stress. Kinetics and mechanism of its reaction with superoxide and hydroxyl radicals. *Biochim Biophys Acta* 1985; **827**:36–44.

Trost SU, Belke DD, Bluhm WF, *et al.* Overexpression of the sarcoplasmic reticulum Ca(2+)-ATPase improves myocardial contractility in diabetic cardiomyopathy. *Diabetes* 2002; **51**:1166–1171.

Turdi S, Li Q, Lopez FL, Ren J. Catalase alleviates cardiomyocyte dysfunction in diabetes: Role of Akt, Forkhead transcriptional factor and silent information regulator 2. *Life Sci* 2007; **81**:895–905.

Turko IV, Marcondes S, Murad F. Diabetes-associated nitration of tyrosine and inactivation of succinyl-CoA:3-oxoacid CoA-transferase. *Am J Physiol Heart Circ Physiol* 2001; **281**:H2289–H2294.

Uriu-Adams JY, Rucker RB, Commisso JF, Keen CL. Diabetes and dietary copper alter ^{67}Cu metabolism and oxidant defense in the rat. *J Nutr Biochem* 2005; **16**:312–320.

Van Campenhout A, Van Campenhout C, Lagrou AR, *et al.* Iron-binding antioxidant capacity is impaired in diabetes mellitus. *Free Radic Biol Med* 2006; **40**:1749–1755.

Vasquez-Vivar J, Martasek P, Hogg N, *et al.* Endothelial nitric oxide synthase-dependent superoxide generation from adriamycin. *Biochemistry* 1997; **36**:11293–11297.

Wang GW, Klein JB, Kang YJ. Metallothionein inhibits doxorubicin-induced mitochondrial cytochrome c release and caspase-3 activation in cardiomyocytes. *J Pharmacol Exp Ther* 2001a; **298**:461–468.

Wang GW, Zhou Z, Klein JB, Kang YJ. Inhibition of hypoxia/reoxygenation-induced apoptosis in metallothionein-overexpressing cardiomyocytes. *Am J Physiol Heart Circ Physiol* 2001b; **280**:H2292–H2299.

Wang J, Song Y, Elsherif L, *et al.* Cardiac metallothionein induction plays the major role in the prevention of diabetic cardiomyopathy by zinc supplementation. *Circulation* 2006; **113**:544–554.

Wang L, Zhou Z, Saari JT, Kang YJ. Alcohol-induced myocardial fibrosis in metallothionein-null mice: Prevention by zinc supplementation. *Am J Pathol* 2005; **167**:337–344.

Wang P, Chen H, Qin H, *et al.* Overexpression of human copper, zinc-superoxide dismutase (SOD1) prevents postischemic injury. *Proc Natl Acad Sci USA* 1998; **95**:4556–4560.

Wold LE, Ceylan-Isik AF, Fang CX, *et al.* Metallothionein alleviates cardiac dysfunction in streptozotocin-induced diabetes: Role of Ca^{2+} cycling proteins, NADPH oxidase, poly(ADP-ribose) polymerase and myosin heavy chain isozyme. *Free Radic Biol Med* 2006; **40**:1419–1429.

Wu S, Ren J. Benfotiamine alleviates diabetes-induced cerebral oxidative damage independent of advanced glycation end-product, tissue factor and TNF-alpha. *Neurosci Lett* 2006; **394**:158–162.

Xiong Y, Liu X, Lee CP, *et al.* Attenuation of doxorubicin-induced contractile and mitochondrial dysfunction in mouse heart by cellular glutathione peroxidase. *Free Radic Biol Med* 2006; **41**:46–55.

Yang X, Doser TA, Fang CX, *et al.* Metallothionein prolongs survival and antagonizes senescence-associated cardiomyocyte diastolic dysfunction: Role of oxidative stress. *FASEB J* 2006; **20**:1024–1026.

Ye G, Metreveli NS, Donthi RV, *et al.* Catalase protects cardiomyocyte function in models of type 1 and type 2 diabetes. *Diabetes* 2004; **53**:1336–1343.

Ye G, Metreveli NS, Ren J, Epstein PN. Metallothionein prevents diabetes-induced deficits in cardiomyocytes by inhibiting reactive oxygen species production. *Diabetes* 2003; **52**:777–783.

Yen HC, Oberley TD, Gairola CG, *et al.* Manganese superoxide dismutase protects mitochondrial complex I against adriamycin-induced cardiomyopathy in transgenic mice. *Arch Biochem Biophys* 1999; **362**:59–66.

Yen HC, Oberley TD, Vichitbandha S, *et al.* The protective role of manganese superoxide dismutase against adriamycin-induced acute cardiac toxicity in transgenic mice. *J Clin Invest* 1996; **98**:1253–1260.

Yin X, Wu H, Chen Y, Kang YJ. Induction of antioxidants by adriamycin in mouse heart. *Biochem Pharmacol* 1998; **56**:87–93.

Yoshida T, Watanabe M, Engelman DT, *et al.* Transgenic mice overexpressing glutathione peroxidase are resistant to myocardial ischemia reperfusion injury. *J Mol Cell Cardiol* 1996; **28**:1759–1767.

Zhang X, Dong F, Li Q, *et al.* Cardiac overexpression of catalase antagonizes ADH-associated contractile depression and stress signaling after acute ethanol exposure in murine myocytes. *J Appl Physiol* 2005; **99**:2246–2254.

Zhang X, Klein AL, Alberle NS, *et al.* Cardiac-specific overexpression of catalase rescues ventricular myocytes from ethanol-induced cardiac contractile defect. *J Mol Cell Cardiol* 2003; **35**:645–652.

Zhou Z, Sun X, Kang YJ. Metallothionein protection against alcoholic liver injury through inhibition of oxidative stress. *Exp Biol Med (Maywood)* 2002; **227**:214–222.

Chapter 13

ZINC STATUS, METALLOTHIONEINS AND ATHEROSCLEROSIS IN THE ELDERLY

Eugenio Mocchegiani, Robertina Giacconi and Marco Malavolta

Zinc status, inflammation, and genetic determinants are prominent mechanisms in the pathogenesis of atherosclerosis (AT) and its compliances (cardiovascular diseases). In this review, we report the possible impact of zinc on AT development as well as the role played by a significant genetic determinant involved in inflammation, such as interleukin-6 (IL-6). Genetic polymorphism of IL-6 may affect a different inflammatory response as well as zinc turnover, predisposing to AT. Indeed, zinc deficiency is suggested as a risk factor for AT with advancing aging. The increment of dysfunctional proteins involved in zinc homeostasis, i.e. metallothioneins (MT), caused by persistent inflammation and oxidative stress may further contribute to zinc deficiency and consequently to the development of AT. A zinc supplementation may be useful to achieve healthy aging and, as such, to prevent AT, but it is necessary to consider the individual genetic background (especially when referred to IL-6 and MT polymorphisms) for the success of zinc intervention. Therefore, a zinc genomic approach may offer a reasonable hope for understanding the impact of zinc on molecular processes that maintain health and prevent the development of AT.

Keywords: Zinc; metallothioneins; inflammation; oxidative stress; genetic background; polymorphism; atherosclerosis; cardiovascular diseases; elderly.

1. Introduction

The process of life for individuals is to guarantee and preserve its integrity, as postulated by the evolutionistic theories (Kirkwood and Austad, 2000). On the other hand, the preservation of an organism's integrity has "a price to pay," represented by a chronic low-grade inflammatory status, as documented in aged populations

(Krabbe *et al.*, 2004; De Martinis *et al.*, 2005). This peculiar inflammatory status, leading to long-term tissue damage, is detrimental for longevity and has been found to be related to mortality risk in old people. According to this point of view, it seems potentially involved in the onset and progression of some age-related inflammatory diseases such as atherosclerosis (AT) and its compliances (Bruunsgaard *et al.*, 2001).

These observations suggest the key role played by inflammatory molecules in the complex pathogenesis of age-related diseases, as shown by different studies. Evidence is also accumulating on the crucial role of genetic background to modulate the susceptibility to age-related diseases (Franceschi *et al.*, 2000). In particular, it has been shown that single nucleotide polymorphisms (SNPs) or other genetic variations, located within the promoter regions of proinflammatory cytokines, influence the susceptibility to age-related diseases by increasing gene transcription and therefore cytokine production (Bennermo *et al.*, 2004). Conversely, SNPs or other genetic variations, determining the increased or decreased production of anti-inflammatory and proinflammatory cytokines, seem to be associated with successful aging (Lio *et al.*, 2003). In this context, intracellular zinc homeostasis plays a peculiar role because zinc is implicated in the inflammatory/immune response, in particular in keeping under control the inflammatory status with the possible achievement of health longevity (Mocchegiani *et al.*, 2006). In other words, the presence of proinflammatory/anti-inflammatory genotypes or a specific genetic background affecting zinc status seems to be associated with inflammation, influencing the susceptibility to age-related diseases or promoting the health longevity, respectively (Mocchegiani *et al.*, 2006; Cipriano *et al.*, 2006; Vasto *et al.*, 2006).

Zinc exerts an important role in modulating the inflammatory/immune response during aging due to its influence on the correct balance between anti-inflammatory and proinflammatory cytokines produced by Th1 and Th2 cells, respectively (Prasad, 2000). The intracellular zinc availability is regulated by metallothioneins (MTs), which are specific cysteine-rich proteins that bind and transfer zinc to other apoenzymes or apoproteins during a transient stress or inflammatory condition (Maret and Krężel, 2007). In this way, zinc activates some zinc-dependent antioxidant enzymes or zinc-related signaling pathways to fight the inflammation and oxidative stress (Mocchegiani *et al.*, 2000). During aging, the inflammatory status is chronic, leading to a limited zinc release by MT with subsequent low intracellular free zinc ion availability. Therefore, a reduced intracellular zinc availability, via an altered MT homeostasis, occurs in aging strictly related to higher inflammatory status (Mocchegiani *et al.*, 2002). The loss of zinc and the subsequent low zinc intracellular availability may promote atherosclerosis (AT) for two fundamental reasons. Firstly, the chronic inflammation by high circulating levels of proinflammatory cytokines (IL-6 and TNF-alpha) is the main background for the development of AT and its compliances (Hansson *et al.*, 2006); secondly, zinc is a fundamental trace element in counterbalancing atherogenesis, protecting the cells from oxidative stress and preserving endothelial cell integrity (Hennig *et al.*, 1999). A reduced zinc ion availability has been recently reported in peripheral blood mononuclear cells (PBMCs) from patients with carotid atherosclerosis (Giacconi *et al.*, 2007). Moreover, it has been shown that a higher intake of zinc might be beneficial in relation to cardiovascular disease (CVD) mortality (Lee *et al.*, 2005). Taking into account these clinical evidences, it is clear that zinc is a key trace element for the prevention of AT and its compliances (especially CVD). In the present short review, we report the role played by zinc and MT in AT in relationship to some molecules involved in inflammation and oxidative stress in order to suggest a possible nutritional approach in preventing AT in the elderly.

2. Zinc Deficiency and AT

The contribution of zinc deficiency in AT development and progression is still debated, despite the fact that several studies have implicated perturbations of zinc metabolism in the etiology of CVD (Beattie and Kwun, 2004). Some cross-sectional studies in patients with various forms of CVD reported decreased zinc concentrations in serum (Martin-Lagos *et al.*, 1997) or plasma (Oster *et al.*, 1989); other authors found no differences between patients with a history of CVD or lower limb atherosclerosis and controls when looking at zinc concentrations in toenails (Martin-Moreno *et al.*, 2003) or plasma (Tiber *et al.*, 1986). Conversely, some authors reported increased zinc concentrations in the serum (Iskra *et al.*, 1997) of patients with lower limb atherosclerosis or CVD compared to control groups.

However, longitudinal or nested case-control studies within a prospective population describe a general involvement of zinc deficiency in increased CVD mortality (Reunanen *et al.*, 1996), especially when a concomitant dismetabolism of other trace elements (copper and magnesium) is found (Leone *et al.*, 2006). When no change in the risk of death from CVD was observed, the protective effect of a satisfactory zinc status was found to match the data, suggesting analysis of larger prospective data sets (Kok *et al.*, 1988). The rationale behind these investigations could be attributed to the antiatherogenic role played by zinc exerted through the inhibition of oxidative stress and apoptosis of endothelial cells during inflammatory conditions (Hennig *et al.*, 1999; Meerarani *et al.*, 2000). In fact, the importance of zinc for optimal antioxidant stress response is well documented by the recent literature supporting its role in enhancing the activity and expression of stress-related and antioxidant proteins including metallothioneins (MT) (Davis and Cousins, 2000), chaperones (Klosterhalfen *et al.*, 1996), ApoJ (Mocchegiani *et al.*, 2007), poly(ADP-ribose) polymerase-1 (PARP-1) (Kunzmann *et al.*, 2008), methionione sulfoxide reductase (Msr) (Cabriero *et al.*, 2008), and superoxide dismutase (SOD) (Mariani *et al.*, 2008). These results could explain why LDL fractions of rats fed with a zinc-deficient diet were more susceptible to oxidative damage (Yousef *et al.*, 2002), and why zinc ions inhibited LDL modification in isolated macrophages (Wilkins and Leake, 1994).

It has been demonstrated that oxidative stress, induced by zinc deficiency, activates the NF-κB cascade and an inflammatory cytokine response in different cellular lines, including endothelial cells (Hennig *et al.*, 1999; Herbein *et al.*, 2006). These results may appear paradoxical in view of the fact that zinc is known to activate NF-κB signaling, enhancing its translocation from the cytosol to the nucleus (Mackenzie *et al.*, 2002) and its binding activity (Ho and Ames, 2002). However, zinc induction of the NF-κB activation pathway appears to be cell-specific and counterbalanced by the concomitant activation of NF-κB activation inhibitors such as A20, which is a zinc finger-transactivating factor involved in reducing IL-1β and TNF-α–induced NF-κB activation (Prasad *et al.*, 2004). Zinc is also required for the PPAR-α and PPAR-γ DNA-binding activity, which in turn downregulates the TNF-α–mediated induction of inflammatory transcription factors NF-κB and AP-1, reducing the expression of their target genes, VCAM-1 and IL-6 (Reiterer *et al.*, 2004). It has been recently reported that *in vitro* zinc treatment (50 μM) increases PPAR-α gene expression from the PBMCs of elderly people (Mazzatti *et al.*, 2007). The modulatory effect of zinc on PPAR signaling suggests a possible involvement of zinc in modulating the lipid profile, but some results from association studies between serum zinc and cholesterol are inconclusive (Hughes and Samman, 2006). Differences in the length and dosage of zinc supplementation may in part explain the uncertain results related to this phenomenon, but age-specific related changes might also be implicated. In fact, opposite results were found between elderly and adult donors on PPAR-α regulation as well as on many other genes involved in the lipid and cholesterol metabolism (e.g. PRKAR1A, PPP1R3C, ACSL4, ALPI, FABP4, GPLD1, LPL, IL-6, PPAR-γ) of PBMCs following zinc treatment (Mazzatti *et al.*, 2008). In any case, age-related alterations in several of these lipid-regulatory genes may have an impact on disease risk, including cardiovascular diseases (for a review, see Bruunsgaard (2006)).

In contrast, the majority of trials with zinc supplementation described unchanged levels of plasma cholesterol; whereas in an elderly population selected on the basis of low zinc intake or serum zinc, zinc supplementation decreased plasma cholesterol concentrations (Hughes and Samman, 2006). Similar results showing benefits in the elderly and adverse or lack of effects of zinc supplementation in adults have been reported for HDL cholesterol and other lipid markers (Hughes and Samman, 2006). Although these findings may seem in part contradictory to the real benefit of zinc on parameters linked to the appearance of cardiovascular diseases, an antiatherogenic effect of zinc is strictly related to the age of subjects and age-related conditions such as the inflammatory status and lipid profile. Such an assumption may be supported by data on animal models in which zinc may have an antiatherogenic effect, including inhibition of iron-catalyzed free radical reactions in the lesion (Ren *et al.*, 2006; Jenner *et al.*, 2007). On the other hand, high-dose zinc has also been suggested among strategies of prophylaxis and therapy of the AT process to terminate angina pectoris (Eby and Halcomb, 2006), but most of the studies in humans suggest that fibrinolytic factors and platelet zinc concentrations are unaffected by zinc supplementation (Hughes and Samman, 2006).

Taking into account that inflammatory processes associated with atherogenesis and AT progression are regulated by several polymorphisms of cytokine network genes (for a review, see Vasto *et al.* (2007)), it is not excluded that the individual genetic background might influence the response to zinc supplementation or the negative impact of zinc deficiency on health. In this context, a recent trial of zinc supplementation in selected old people with specific polymorphisms for IL-6 (particularly in old subjects carrying GG genotypes named C−) has demonstrated beneficial effects of zinc supplementation in innate immune response and improvements in the gene expression of a stress-related protein ApoJ. This finding suggests that not all old people may be prone to zinc supplementation, but rather selected people with a specific genetic background;

for the other people, a correct diet containing food rich in zinc may be sufficient to maintain the healthy status (Mocchegiani *et al.*, 2007). However, further research is required, taking into consideration factors such as differences that may exist in the extent of AT, diet, lifestyle, and the possible involvement of genetic and age-related changes affecting the individual zinc homeostasis. On this aspect, there is considerable evidence of the possible impact of aging on MT expression and function that deserves particular consideration.

3. Zinc Dyshomeostasis in AT: Role of Metallothioneins (MTs)

MTs are highly conserved, low-molecular-weight, thiol-rich proteins. The mammalian MTs have 61 amino acids, including 20 cysteine residues, which confer to this protein a very high affinity for zinc (Kd $= 5 \times 10^{-13}$ M) (Vašák and Kägi, 1983). MTs are genetically polymorphous protein families with subfamilies, subgroups, and various isoforms. Humans possess genes for four subfamilies (encoded by at least 10 functional MT genes), all located in chromosome 16: the brain-specific MT-3, the squamous epithelium-specific MT-4, and the ubiquitous MT-1 and MT-2 (West *et al.*, 1990). MTs act as a metal chaperone for the regulation of gene expression and for the synthesis and functional activity of proteins such as metalloproteins, metal-dependent transcription factors, and antioxidant enzymes (Maret, 1995). MTs, through their antioxidant role, confer cytoprotection from injury by reactive oxygen and nitrogen species. For instance, zinc-MT has been shown to scavenge hydroxyl radical *in vitro* and to be more effective than glutathione in preventing hydroxyl radical-induced DNA degradation (Abel and de Ruiter, 1989). Under transient oxidative stress, zinc is released from MT and transferred to zinc fingers and other transcription factors in order to modulate their DNA-binding efficiency and the expression of several genes involved in the intracellular antioxidant response (Mocchegiani *et al.*, 2000). Functions of zinc-MT are based on reversible dissociation of its zinc ions and oxidoreduction of the cysteine sulphur donors in zinc/thiolate clusters (Maret and Krężel, 2007). This property is unusual for zinc proteins and provides a mechanism whereby zinc availability and redox metabolism are linked (Maret and Krężel, 2007). When oxidants react with MT, zinc ions are released and the oxidized protein is formed. This occurs in the case of superoxide/peroxide signaling (Fliss and Ménard, 1992) as well as other redox-active mediators. MT interaction with glutathione disulphide (GSSG) determines the release of zinc and its transfer to other apoenzymes with lower zinc-binding affinity during oxidative stress (Maret *et al.*, 1999). Nitric oxide (NO) also releases zinc from MT by transnitrosation, i.e. the transfer of NO from *S*-nitrosothiols to the sulfur donor of zinc ligands (Chen *et al.*, 2002).

The basal level of MT in biological systems is very low, although it may vary with the age and type of tissue. However, this protein is strongly upregulated in response to zinc and other metal ions, as well as in response to agents which cause oxidative stress and/or inflammation (Andrews, 2000). The six-zinc-finger metal-responsive transcription factor MTF-1 plays a central role in transcriptional activation of the MT genes in response to metals and oxidative stress (Heuchel *et al.*, 1994). Induction of MTs by proinflammatory cytokines (IL-1β, TNF-α, IL-6) has been recently linked to mechanisms related to modulation of MTF-1 (Cousins *et al.*, 2006). During inflammation, cells can respond to cytokines by upregulating inducible nitric oxide synthase (iNOS), which generates large amounts of NO from arginine (Spahl *et al.*, 2003). Release of labile zinc from pre-existing MT or other thiolate ligands by NO may provide the means, via zinc binding to MTF-1 and other zinc finger proteins, for the regulation of zinc-dependent genes including MT itself (Cousins *et al.*, 2006). Moreover, proinflammatory cytokines (IL-1β, TNF-α, IL-6) stimulate cellular zinc influx, via STAT-mediated signaling, through upregulation of the zinc transporters Zip14 and Zip6 (Cousins *et al.*, 2006), which in turn directly activate MTF-1 transcription. Similar mechanisms, mediated by proinflammatory cytokines, could be involved in MT induction following lipopolysaccharide (LPS) stimulation (Leibbrandt and Koropatnick, 1994).

During aging and in AT, the persistence of a chronic low-grade inflammation mediated by proinflammatory cytokines has been associated with persistent high MT levels (Mocchegiani *et al.*, 2002; Giacconi *et al.*, 2007). Although MT response has been proven to be beneficial against stressor agents, the constant upregulation of MT may be involved in a proinflammatory response, as suggested by the lower TNF-α secretion observed in LPS-stimulated MT-null mice macrophages compared to wild-type controls (Kanekiyo *et al.*, 2002). This phenomenon may be related to the decreased zinc ion availability as feedback response to MT induction (Mocchegiani *et al.*, 2000). But, emerging data suggest that more complex mechanisms leading to the formation of dysfunctional MT with consequent loss of zinc homeostasis could occur during aging (Mocchegiani *et al.*, 2006). In this case, irreversibly or persistently oxidized MT may no longer bind or release zinc ions with consequent impaired response to oxidative stress (Maret and Krężel, 2007). It has been noticed that AT is a condition associated with hyperhomocysteinemia (Papatheodorou and Weiss, 2007). Interestingly, L-homocysteine can target intracellular MT by forming dysfunctional mixed-disulfide conjugates implicated in the disruption of zinc and redox homeostasis (Barbato *et al.*, 2007), leading to endothelial dysfunction and AT.

Taking into account the existence of a strict relationship between AT and type 2 diabetes mellitus (DM2), another interesting point is the suggestion that the reduced defenses in DM2 against oxidative stress are due to low levels of MT-I+II in muscle (Scheede-Bergdahl *et al.*, 2005). Consistent with this interpretation, different investigations show that a model of MT transgenic mice developing diabetes is protected from cardiomiopathy as compared to controls (Ceylan-Isik *et al.*, 2006; Liang *et al.*, 2002). Overexpression of MT in pancreatic beta cells provides broad resistance to oxidative stress by scavenging reactive oxygen species (Li *et al.*, 2004). Moreover, MTs protect against thromboembolic disorders that characterize DM2, inhibiting platelet aggregation (Sheu *et al.*, 2003).

However, during aging and in AT, the high MT levels are associated with lower zinc ion availability, increased inflammation, and reduced capacity to fight oxidative stress (Mocchegiani *et al.*, 2002). Old patients affected by DM2 show increased oxidative damage, chronic inflammation, and zinc deficiency (DiSilvestro, 2000). As a result, the zinc release from MT is reduced with an impaired intracellular zinc availability to counteract oxidative damage and to prevent diabetes complications (Giacconi *et al.*, 2005).

To better clarify the role of MT in aging and age-related diseases (such as DM2 and AT), novel SNPs of MT-1A and MT-2A genes have been investigated. The polymorphism at the –209 A/G locus of the MT-2A gene was found to be associated with hyperglycemia and enhanced glycosylated hemoglobin (HbA1c) in DM2-atherosclerotic patients (Giacconi *et al.*, 2005). Patients with AA genotype displayed higher fasting glucose levels, HbA1c, and lower plasma zinc than those with AG genotype. Moreover, AA carriers showed a major prevalence of ischemic cardiomyopathy, suggesting a decreased capability to counteract oxidative stress through zinc signaling via MT. Another recent investigation showed an association between the +838 C/G MT-2A polymorphism and the susceptibility of carotid artery disease. The +838 C− carriers displayed increased inflammatory markers, impaired zinc availability, and NK cell cytotoxicity as compared to those with C+ genotype (Giacconi *et al.*, 2007). In addition, the +647 A/C MT-1A polymorphism modulated zinc availability and inflammatory status, and was associated with longevity in Italian elderly females (Cipriano *et al.*, 2006). All of these findings suggest that aging and the individual genetic background affect the regulatory role of MT in zinc homeostasis, and that this phenomenon may play a crucial role in the progression of AT.

4. Concluding Remarks

From the data herein presented, the peculiar role played by zinc in the progression of AT through MT homeostasis emerges. This fact is strictly related to the individual inflammatory status, in particular by high levels of proinflammatory cytokines (especially IL-6) and by the individual genetic background linked to specific polymorphisms for IL-6 and MT. The genetic background is also pivotal for a possible beneficial effect of zinc supplementation in order to reach healthy aging and consequently to delay/prevent AT or to limit the damage of endothelial cells in the presence of high inflammatory status and altered lipid profile. This effect is also dependent on the age of the subjects. Therefore, in considering zinc interventions in the elderly, populations must be stratified by their age, especially by age-related conditions such as inflammatory status and lipid profile. This selection should reduce interindividual variance in response and increase the rate of success for zinc intervention studies. Although the complexities of AT and its compliances are significant, a zinc genomic approach offers reasonable hope for understanding the impact of zinc on molecular processes that maintain health and prevent the development of AT.

Acknowledgments

This work was supported by INRCA and by the European Commission (ZINCAGE project, no. FOOD-CT-2004-506850; coordinated by Dr Eugenio Mocchegiani) (www.zincage.org).

References

Abel J, de Ruiter N. Inhibition of hydroxyl-radical-generated DNA degradation by metallothionein. *Toxicol Lett* 1989; **47**:191–196.

Andrews GK. Regulation of metallothionein gene expression by oxidative stress and metal ions. *Biochem Pharmacol* 2000; **59**:95–104.

Barbato JC, Catanescu O, Murray K, *et al.* Targeting of metallothionein by L-homocysteine: A novel mechanism for disruption of zinc and redox homeostasis. *Arterioscler Thromb Vasc Biol* 2007; **27**:49–54.

Beattie JH, Kwun IS. Is zinc deficiency a risk factor for atherosclerosis? *Br J Nutr* 2004; **91**:177–181.

Bennermo M, Held C, Green F, *et al.* Prognostic value of plasma interleukin-6 concentrations and the −174 G>C and −572 G>C promoter polymorphisms of the interleukin-6 gene in patients with acute myocardial infarction treated with thrombolysis. *Atherosclerosis* 2004; **174**:157–163.

Bruunsgaard H. The clinical impact of systemic low-level inflammation in elderly populations. With special reference to cardiovascular disease, dementia and mortality. *Dan Med Bull* 2006; **53**:285–309.

Bruunsgaard H, Pederson M, Pederson BK. Aging and proinflammatory cytokines. *Curr Opin Hematol* 2001; **8**:131–136.

Cabreiro F, Perichon M, Jatje J, *et al.* Zinc supplementation in the elderly subjects: Effect on oxidized protein degradation and repair systems in peripheral blood lymphocytes. *Exp Gerontol* 2008; **43**:483–487.

Ceylan-Isik AF, LaCour KH, Ren J. Sex difference in cardiomyocyte function in normal and metallothionein transgenic mice: The effect of diabetes mellitus. *J Appl Physiol* 2006; **100**:1638–1646.

Chen Y, Irie Y, Keung WM, Maret W. *S*-nitrosothiols react preferentially with zinc thiolate clusters of metallothionein III through transnitrosation. *Biochemistry* 2002; **41**:8360–8367.

Cipriano C, Malavolta M, Costarelli L, *et al.* Polymorphisms in MT1a gene coding region are associated with longevity in Italian Central female population. *Biogerontology* 2006; **7**:357–365.

Cousins RJ, Liuzzi JP, Lichten LA. Mammalian zinc transport, trafficking, and signals. *J Biol Chem* 2006; **281**:24085–24089.

Davis SR, Cousins RJ. Metallothionein expression in animals: A physiological perspective on function. *J Nutr* 2000; **130**:1085–1088.

De Martinis M, Franceschi C, Monti D, Ginaldi L. Inflamm-aging and lifelong antigenic load as major determinants of aging rate and longevity. *FEBS Lett* 2005; **579**:2035–2039.

DiSilvestro RA. Zinc in relation to diabetes and oxidative stress. *J Nutr* 2000; **130**: 1509S–1511S.

Eby GA, Halcomb WW. High-dose zinc to terminate angina pectoris: A review and hypothesis for action by ICAM inhibition. *Med Hypotheses* 2006; **66**:169–172.

Fliss H, Ménard M. Oxidant-induced mobilization of zinc from metallothionein. *Arch Biochem Biophys* 1992; **293**:195–199.

Franceschi C, Bonafè M, Valensin S, *et al.* Inflamm-aging. An evolutionary perspective on immunosenescence. *Ann NY Acad Sci* 2000; **908**:244–254.

Giacconi R, Cipriano C, Muti E, *et al.* Novel −209A/G MT2A polymorphism in old patients with type 2 diabetes and atherosclerosis: Relationship with inflammation (IL-6) and zinc. *Biogerontology* 2005; **6**:407–413

Giacconi R, Muti E, Malavolta M, *et al.* The +838 C/G MT2A polymorphism, metals, and the inflammatory/immune response in carotid artery stenosis in elderly people. *Mol Med* 2007; **13**:388–395.

Hansson GK, Robertson AK, Söderberg-Nauclér C. Inflammation and atherosclerosis. *Annu Rev Pathol* 2006; **1**:297–329.

Hennig B, Meerarani P, Toborek M, McClain CJ. Antioxidant-like properties of zinc in activated endothelial cells. *J Am Coll Nutr* 1999; **18**:152–158.

Herbein G, Varin A, Fulop T. NF-kappaB, AP-1, zinc-deficiency and aging. *Biogerontology* 2006; **7**:409–419.

Heuchel R, Radtke F, Georgiev O, *et al.* The transcription factor MTF-1 is essential for basal and heavy metal-induced metallothionein gene expression. *EMBO J* 1994; **13**: 2870–2875.

Ho E, Ames BN. Low intracellular zinc induces oxidative DNA damage, disrupts p53, NFkappa B, and AP1 DNA binding, and affects DNA repair in a rat glioma cell line. *Proc Natl Acad Sci USA* 2002; **99**:16770–16775.

Hughes S, Samman S. The effect of zinc supplementation in humans on plasma lipids, antioxidant status and thrombogenesis. *J Am Coll Nutr* 2006; **25**:285–291.

Iskra M, Patelski J, Majewski W. Relationship of calcium, magnesium, zinc and copper concentrations in the arterial wall and serum in atherosclerosis obliterans and aneurysm. *J Trace Elem Med Biol* 1997; **11**:248–252.

Jenner A, Ren M, Rajendran R, *et al.* Zinc supplementation inhibits lipid peroxidation and the development of atherosclerosis in rabbits fed a high cholesterol diet. *Free Radic Biol Med* 2007; **42**:559–566.

Kanekiyo M, Itoh N, Kawasaki A, *et al.* Metallothionein modulates lipopolysaccharide-stimulated tumour necrosis factor expression in mouse peritoneal macrophages. *Biochem J* 2002; **361**:363–369.

Kirkwood TB, Austad SN. Why do we age? *Nature* 2000; **408**:233–238.

Klosterhalfen B, Töns C, Hauptmann S, *et al.* Influence of heat shock protein 70 and metallothionein induction by zinc-bis-(DL-hydrogenaspartate) on the release of inflammatory mediators in a porcine model of recurrent endotoxemia. *Biochem Pharmacol* 1996; **52**:1201–1210.

Kok FJ, Van Duijn CM, Hofman A, *et al.* Serum copper and zinc and the risk of death from cancer and cardiovascular disease. *Am J Epidemiol* 1988; **128**:352–359.

Krabbe KS, Pederesen M, Bruunsgaard H. Inflammatory mediators in the elderly. *Exp Gerontol* 2004; **39**:687–699.

Kunzmann A, Dedoussis G, Jajte J, *et al.* Effect of zinc on cellular poly(ADP-ribosyl)ation capacity. *Exp Gerontol* 2008; **43**:409–414.

Lee DH, Folsom AR, Jacobs DR Jr. Iron, zinc, and alcohol consumption and mortality from cardiovascular diseases: The Iowa Women's Health Study. *Am J Clin Nutr* 2005; **81**:787–791.

Leibbrandt ME, Koropatnick J. Activation of human monocytes with lipopolysaccharide induces metallothionein expression and is diminished by zinc. *Toxicol Appl Pharmacol* 1994; **124**:72–81.

Leone N, Courbon D, Ducimetiere P, Zureik M. Zinc, copper, and magnesium and risks for all-cause, cancer, and cardiovascular mortality. *Epidemiology* 2006; **17**:308–314.

Li X, Chen H, Epstein PN. Metallothionein protects islets from hypoxia and extends islet graft survival by scavenging most kinds of reactive oxygen species. *J Biol Chem* 2004; **279**:765–771.

Liang Q, Carlson EC, Donthi RV, *et al.* Overexpression of metallothionein reduces diabetic cardiomyopathy. *Diabetes* 2002; **51**:174–181.

Lio D, Scola L, Crivello A, *et al.* Inflammation, genetics, and longevity: Further studies on the protective effects in men of IL-10 −1082 promoter SNP and its interaction with TNF-alpha −308 promoter SNP. *J Med Genet* 2003; **40**:296–299.

Mackenzie GG, Zago MP, Keen CL, Oteiza PI. Low intracellular zinc impairs the translocation of activated NF-kappa B to the nuclei in human neuroblastoma IMR-32 cells. *J Biol Chem* 2002; **277**:34610–346177.

Maret W. Metallothionein/disulfide interactions, oxidative stress, and the mobilization of cellular zinc. *Neurochem Int* 1995; **27**:111–117.

Maret W, Jacob C, Vallee BL, Fischer EH. Inhibitory sites in enzymes: Zinc removal and reactivation by thionein. *Proc Natl Acad Sci USA* 1999; **96**:1936–1940.

Maret W, Krężel A. Cellular zinc and redox buffering capacity of metallothionein/thionein in health and disease. *Mol Med* 2007; **13**:371–375.

Mariani E, Mangialasche F, Feliziani FT, *et al.* Effects of zinc supplementation on antioxidant enzyme activities in healthy old subjects. *Exp Gerontol* 2008; **43**:445–451.

Martin-Lagos F, Navarro-Alarcon M, Terres-Martos C, *et al.* Serum copper and zinc concentrations in serum from patients with cancer and cardiovascular disease. *Sci Total Environ* 1997; **204**:27–35.

Martin-Moreno JM, Gorgojo L, Riemersma RA, *et al.* Heavy Metals and Myocardial Infarction Study Group. Myocardial infarction risk in relation to zinc concentration in toenails. *Br J Nutr* 2003; **89**:673–678.

Mazzatti DJ, Malavolta M, White AJ, *et al.* Differential effects of *in vitro* zinc treatment on gene expression in peripheral blood mononuclear cells derived from young and elderly individuals. *Rejuvenation Res* 2007; **10**:603–620.

Mazzatti DJ, Mocchegiani E, Powell JR. Age-specific modulation of genes involved in lipid and cholesterol homeostasis by dietary zinc. *Rejuvenation Res* 2008; **11**:281–285.

Meerarani P, Ramadass P, Toborek M, *et al.* Zinc protects against apoptosis of endothelial cells induced by linoleic acid and tumor necrosis factor alpha. *Am J Clin Nutr* 2000; **71**:81–87.

Mocchegiani E, Costarelli L, Giacconi R, *et al.* Nutrient–gene interaction in aging and successful aging. A single nutrient (zinc) and some target genes related to inflammatory/immune response. *Mech Ageing Dev* 2006; **127**:517–525.

Mocchegiani E, Giacconi R, Cipriano C, *et al.* MtmRNA gene expression, via IL-6 and glucocorticoids, as potential genetic marker of immunosenescence: Lessons from very old mice and humans. *Exp Gerontol* 2002; **37**:349–357.

Mocchegiani E, Giacconi R, Cipriano C, *et al.* Zinc, metallothioneins, and longevity: Effect of zinc supplementation: Zincage study. *Ann NY Acad Sci* 2007; **1119**:129–146.

Mocchegiani E, Muzzioli M, Giacconi R. Zinc and immunoresistance to infection in aging: New biological tools. *Trends Pharmacol Sci* 2000; **21**:205–208.

Oster O, Dahm M, Oelert H, Prellwitz W. Concentrations of some trace elements (Se, Zn, Cu, Fe, Mg, K) in blood and heart tissue of patients with coronary heart disease. *Clin Chem* 1989; **35**:851–856.

Papatheodorou L, Weiss N. Vascular oxidant stress and inflammation in hyperhomocysteinemia. *Antioxid Redox Signal* 2007; **9**:1941–1958.

Prasad AS. Effects of zinc deficiency on Th1 and Th2 cytokine shifts. *J Infect Dis* 2000; **182**:S62–S68.

Prasad AS, Bao B, Beck FW, *et al.* Antioxidant effect of zinc in humans. *Free Radic Biol Med* 2004; **37**:1182–1190.

Reiterer G, Toborek M, Hennig B. Peroxisome proliferator activated receptors alpha and gamma require zinc for their anti-inflammatory properties in porcine vascular endothelial cells. *J Nutr* 2004; **134**:1711–1715.

Ren M, Rajendran R, Ning P, *et al.* Zinc supplementation decreases the development of atherosclerosis in rabbits. *Free Radic Biol Med* 2006; **41**:222–225.

Reunanen A, Knekt P, Marniemi J, *et al.* Serum calcium, magnesium, copper and zinc and risk of cardiovascular death. *Eur J Clin Nutr* 1996; **50**:431–437.

Scheede-Bergdahl C, Penkowa M, Hidalgo J, *et al.* Metallothionein-mediated antioxidant defense system and its response to exercise training are impaired in human type 2 diabetes. *Diabetes* 2005; **54**:3089–3094.

Sheu JR, Hsiao G, Shen MY, *et al.* Inhibitory mechanisms of metallothionein on platelet aggregation in *in vitro* and platelet plug formation in *in vivo* experiments. *Exp Biol Med* 2003; **228**:1321–1324.

Spahl DU, Berendji-Grun D, Suschek CV, *et al.* Regulation of zinc homeostasis by inducible NO synthase-derived NO: Nuclear metallothionein translocation and intranuclear Zn^{2+} release. *Proc Natl Acad Sci USA* 2003; **100**:13952–13957.

Tiber AM, Sakhaii M, Joffe CD, Ratnaparkhi MV. Relative value of plasma copper, zinc, lipids and lipoproteins as markers for coronary artery disease. *Atherosclerosis* 1986; **62**:105–110.

Vašák M, Kägi JHR. Spectroscopic properties of metallothionein. In: Siegel H (ed.), *Metal Ions in Biological Systems*, Marcel Dekker, New York, 1983, pp. 213–273.

Vasto S, Mocchegiani E, Candore G, *et al.* Inflammation, genes and zinc in aging and age-related diseases. *Biogerontology* 2006; **7**:315–327.

Vasto S, Mocchegiani E, Malavolta M, *et al.* Zinc and inflammatory/immune response in aging. *Ann NY Acad Sci* 2007; **1100**:111–122.

West AK, Stallings R, Hildebrand CE, *et al.* Human metallothionein genes: Structure of the functional locus at 16q13. *Genomics* 1990; **8**:513–518.

Wilkins GM, Leake DS. The oxidation of low density lipoprotein by cells or iron is inhibited by zinc. *FEBS Lett* 1994; **341**:259–262.

Yousef MI, El-Hendy HA, El-Demerdash FM, Elagamy EI. Dietary zinc deficiency induced-changes in the activity of enzymes and the levels of free radicals, lipids and protein electrophoretic behavior in growing rats. *Toxicology* 2002; **175**:223–234.

PART IV

METALLOTHIONEIN IN HEPATOLOGY

Chapter 14

METALLOTHIONEINS AND LIVER DISEASES

Lydia Oliva, Renata D'Incà, Valentina Medici and Giacomo Carlo Sturniolo

Metallothioneins (MTs) are cysteine-rich proteins capable of scavenging free radicals and sequestering metal ions. In the liver, these proteins are involved in copper and zinc metabolism, in the chelation of heavy metals, and in protection against oxidative damage. Because of their properties, MTs are involved in many liver diseases, which can be sorted into the following:

1. *Metal storage liver diseases*. Zinc, which is an important anticopper agent for Wilson's disease, acts by increasing the concentration of MTs in the enterocytes, thereby reducing metal absorption. Copper also accumulates in the liver in cholestatic diseases, in which MTs are reportedly overexpressed and induced by ursodeoxycholic acid (UDCA), the main drug used to treat cholestasis. The role of MTs in hemochromatosis, an iron-accumulating disease, has yet to be established; but in animal models, it has been suggested that zinc, by increasing MT concentration, could exert a beneficial effect.
2. *Toxic liver diseases*. By sequestering metal ions and scavenging free radicals, MTs protect against damage caused by exogenous toxic substances, such as cadmium and arsenic, and by the toxic effects on hepatocytes of ethanol and fat in alcoholic and nonalcoholic liver diseases.
3. *Chronic viral hepatitis*. By lowering the inflammatory injury, MTs have a protective action against chronic liver damage; a relationship has also been described between MTs and the severity of liver disease and the response to therapy.
4. *Hepatocellular carcinoma.* MTs are downexpressed and inversely correlated with tumor stage; an inverse correlation has also been reported between MT concentrations and response to platinum chemotherapy.

Given the ability of zinc to strongly induce MT synthesis, zinc supplementation could be useful not only in Wilson's disease, but also in other liver diseases in which MTs exert a protective effect.

Keywords: Metallothionein; liver disease; copper; zinc; hepatitis; toxic; hepatocellular carcinoma.

1. Introduction

Metallothioneins (MTs) are proteins occurring in virtually all living organisms with a highly conserved structure among species, and are mostly ubiquitous in eukaryotes (Coyle *et al.*, 2002). Due to their high cysteine content, MTs can scavenge free radicals and sequester metal ions. In the liver, which is one of the organs with the highest MT content, these proteins are involved in the metabolism of trace elements like zinc and copper, in the chelation of heavy metals (especially cadmium), and in the protection against oxidative damage. Liver MT production is enhanced by a number of metals, cytokines, and stress hormones (Coyle *et al.*, 2002). Zinc is the primary physiological inducer, although Cu, Cd, Hg, Au, and Bi also bind to the metal transcription factor MTF-1, which then binds to metal response elements (MREs) in the promoter region of the MT gene. MT synthesis is enhanced in inflammatory conditions as a consequence of the release of cytokines like IL-6, IL-1, TNF-α, and IFN-γ, which probably take effect in MT-gene nucleotide sequences other than MREs. Catecholamines, glucocorticoids, and glucagon have also been found to induce MT production.

The aim of this chapter is to describe the role of MTs in liver diseases: from metal storage-driven diseases like Wilson's disease and hemochromatosis to diseases in which the crucial pathogenic event is oxidative stress (e.g. viral, alcoholic or nonalcoholic, and toxic hepatitis). Finally, some attention will be paid to the role of MTs in liver tumors.

2. Metal Storage Diseases

2.1. *Wilson's Disease*

Wilson's disease is an inherited, autosomal recessive condition that affects about 30 individuals per million population. It is caused by mutations in the ATP7B gene, which encodes a P-type ATPase important for copper excretion in the bile (Fig. 1). The disease is thus characterized by copper accumulation, particularly in the liver, but also in the brain, cornea, kidney, and joints. The main clinical presentations of Wilson's disease in humans are (1) hepatic disease varying from asymptomatic hepatomegaly to fatty liver, acute hepatitis,

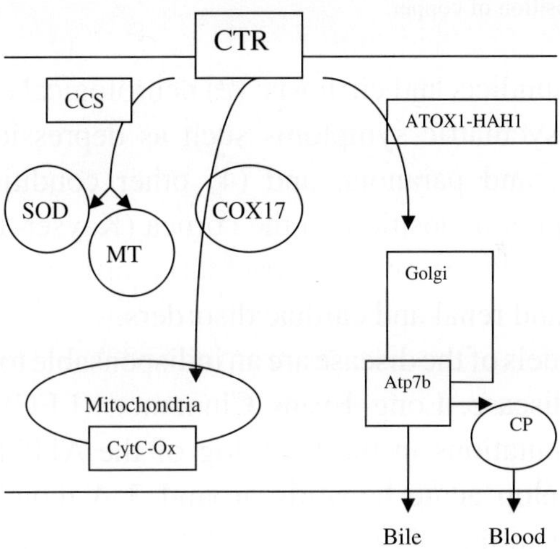

Fig. 1 Metabolism of copper into the hepatocyte. Three main pathways are involved in copper transporter (CTR) activity: (1) Golgi network through the chaperones ATOX1-HAH1. Here, Cu is delivered to the ATP7B protein, which brings copper into the bile or into ceruloplasmin (CP), the blood-transporter protein. (2) Mitochondria through the chaperone COX17. Here, Cu is required for cytochrome c oxidase (CytC-Ox), the final enzyme of the respiratory chain. (3) Storage and protection through the chaperone CCS, which delivers Cu to superoxide dismutase (SOD) or MT.

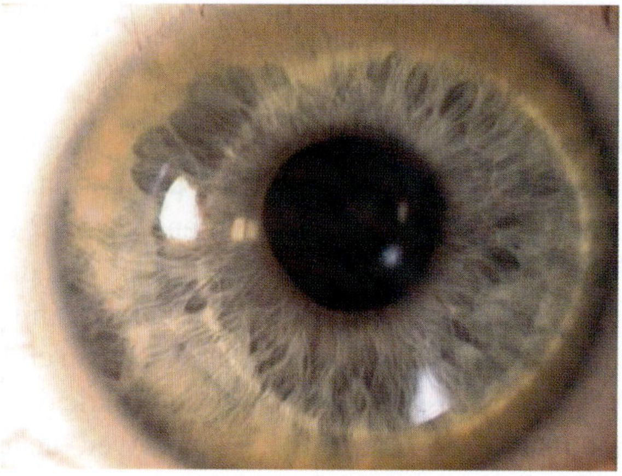

Fig. 2 Kayser–Fleischer ring. A gold-brownish ring appears on the external border of the cornea due to deposition of copper.

liver failure, jaundice, and cirrhosis; (2) neurological extrapyramidal disease; (3) psychiatric symptoms such as depression, irritability, hallucinations, and paranoia; and (4) other conditions including hemolysis, copper deposition on the cornea (Kayser–Fleischer ring; Fig. 2), nephrolithiasis and aminoaciduria, arthritis and premature osteoporosis, and renal and cardiac disorders.

Animal models of the disease are an indispensable tool for research on Wilson's disease. Long–Evans Cinnamon (LEC) rats (Fig. 3), which carry mutations in the homolog of the ATP7B gene, spontaneously develop acute hepatitis around 3–4 months after birth, and 50% of them die of fulminant hepatitis (Komatsu *et al.*, 2002; Cuthbert, 1995). Almost all of them develop cirrhosis and liver cancer within a year. LEC rats differ from Wilson's disease patients in that they do not have high brain copper levels or any apparent neurological dysfunctions.

Copper is an essential trace metal needed for survival, and serves as an important catalytic cofactor in redox reactions required for many biological functions involved in growth and development (Tapiero *et al.*, 2003). In fact, copper deficiency, as in Menkes disease, for instance, leads to neurological impairment and death in

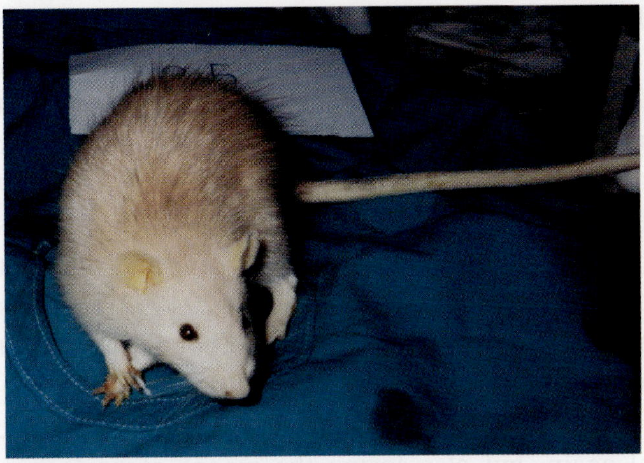

Fig. 3 A Long–Evans Cinnamon (LEC) rat, with the characteristic color of the coat.

infancy. On the other hand, when in excess of cellular needs, copper can be toxic because, like iron, it participates in reactions which produce highly reactive oxygen species (ROS) that are responsible for lipid membrane peroxidation, direct protein oxidation, and DNA breakdown and base modifications, resulting in tissue damage and DNA mutagenesis. Most cellular copper is bound to cytoplasmic proteins such as MTs (which have a marked affinity for copper), and binding makes it nontoxic. In one Wilson's disease patient, an increased liver concentration of MTs was reported, but only 36% of copper was found to bind to MTs in these patients (Nartey *et al.*, 1987; Hunziker and Sternlieb, 1991).

There are three therapeutic strategies for preventing Cu-related damage in Wilson's disease (Medici *et al.*, 2007): (1) copper-chelating drugs like penicillamine and trientine; (2) zinc; and (3) tetrathiomolybdate. MTs are involved in the mode of action of most drugs used in Wilson's disease.

2.1.1. *Copper-Chelating Drugs*

D-penicillamine was the first drug to be used in Wilson's disease, as the sulfhydryl group enables it to chelate divalent cations such as copper. The logic for its use is based on its ability to remove copper accumulation in the liver and to form copper–penicillamine complexes which are then excreted in the urine. It has also been suggested that *D*-penicillamine can induce MT production and increase copper excretion in urine. This would explain why patients who withdraw from chelating therapy die of liver disease which progresses more rapidly than one might expect, judging from the time it takes for copper to accumulate again (McQuaid *et al.*, 1992). The assumption is that this situation is rather like a time bomb, in which there are still high levels of copper in the liver, but it is incorporated in MTs; stopping penicillamine therapy means not only starting to accumulate the metal again, but also losing the means of protection against this toxic metal. This mechanism finds confirmation in the fact that penicillamine-treated patients often still have high copper concentrations in the liver, in spite of any clinical improvement.

McQuaid *et al.* (1992) found that penicillamine alone does not increase MT concentrations in the liver, but only when in combination with copper. This may be because *D*-penicillamine can remove copper from intracellular ligands, making it available for binding MREs, which are promoters of MT synthesis. However, copper is a weaker MT inducer, while zinc is more powerful. In the same study, they found that intraperitoneal injection of *D*-penicillamine and copper in rats induced a rise not only in hepatic MT levels, but also in intracellular zinc distribution. Hence, the hypothesis is that *D*-penicillamine facilitates the access of copper to zinc binding sites, leaving zinc free to bind MRE sites in the promoter sequence of the MT gene.

No studies have been published to date on the relationships between trientine therapy and MTs.

2.1.2. *Zinc*

Zinc is a trace element essential to the structure and function of a large number of macromolecules and for over 300 enzymatic reactions.

It has antioxidant properties due to several factors (Tapiero and Tew, 2003): it induces MT synthesis; it is a component of superoxide dismutase; it is a protective agent for thiols; and it prevents interaction between chemical groups and iron to form free radicals.

Zinc was developed as a treatment for Wilson's disease in the late 1980s. It is the strongest inducer of MTs, so its effect in Wilson's disease consists of increasing MT concentrations in the intestine and liver. Orally administered zinc causes an increase in MT concentrations in the epithelial cells of the duodenum, where MTs have a much greater affinity for copper than for zinc, so copper is trapped in these cells and then eliminated in the stool by normal cell exfoliation. This mechanism not only reduces food copper absorption, but also inhibits the reabsorption of large amounts of this endogenously secreted metal. Zinc therefore serves to remove copper from the body as well as prevent its accumulation. The induction of intestinal MTs by zinc seems to be a rapid process: high levels of MTs have been found in Wilson's disease patients 5–6 days after beginning oral administration (Yuzbasiyan-Gurkan *et al.*, 1992). On the other hand, zinc enhances MT concentrations in the liver (Santon *et al.*, 2003a), so even the copper remaining in the tissue is neutralized by binding with these proteins, which makes it innocuous.

Other benefits of zinc include the antioxidant activity of MTs. These proteins have free-radical scavenging properties, even independently of any metals, because of their cysteine residues (glutathione-like functions). The toxicity of copper is due to pro-oxidative actions; thus, by increasing MTs, zinc functions in two ways, i.e. it reduces copper accumulation and absorption as well as protects against oxidative damage.

We measured MT concentrations and trace metals in duodenal biopsies from Wilson's disease patients (Sturniolo *et al.*, 1999) and found that zinc therapy increased duodenal MT concentrations by 1500%, as opposed to an increase of only 150% achieved with penicillamine (Fig. 4). An increase in iron cell concentration was also described in those patients (Sturniolo *et al.*, 1999). Moreover,

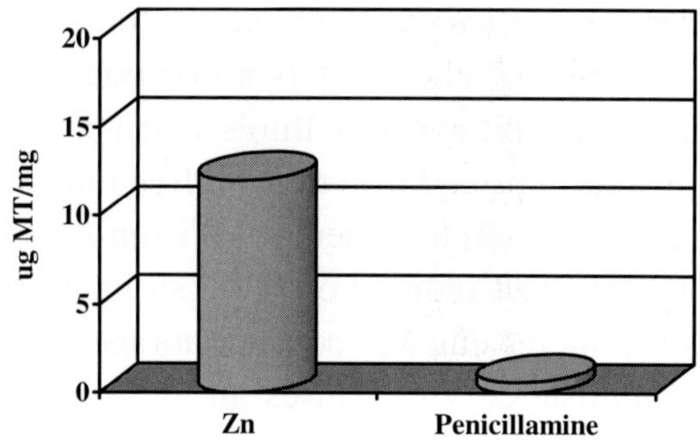

Fig. 4 Duodenal concentration of MTs in Wilson's disease patients treated with zinc or penicillamine.

experimental studies on LEC rats (Kato *et al.*, 1993) demonstrated higher liver concentrations not only of copper, but also of iron. It is well known that there is a close link between copper and iron metabolism; so even if iron metabolism is not altered in Wilson's disease patients, it is likely that a zinc-mediated increase in duodenal MTs also improves the sequestering of iron in the gut lumen, thereby lowering its absorption. Like copper, iron is involved in oxidative reactions, so the lower iron absorption induced by zinc therapy may be another reason for its therapeutic effect in Wilson's disease.

Santon *et al.* (2003b) provided a further demonstration of the beneficial effects of zinc therapy. Apoptotic liver cells of LEC rats, some of them treated with zinc, were assessed using the immunofluorescence technique, and the zinc-treated LEC rats had significantly fewer apoptotic and necrotic hepatoparenchymal cells (Fig. 1). The concentration of apoptotic cells also correlated inversely with the hepatic concentration of MTs (Figs. 5 and 6).

The discovery of the effectiveness of zinc therapy in Wilson's disease has brought many benefits in clinical management. Unlike penicillamine, zinc has very limited side effects and is a very safe drug. The only side effect observed is gastric discomfort, recorded in 10%–15% of cases with zinc sulfate, but rapidly overcome by switching to zinc acetate (Brewer *et al.*, 1998; Medici *et al.*, 2006).

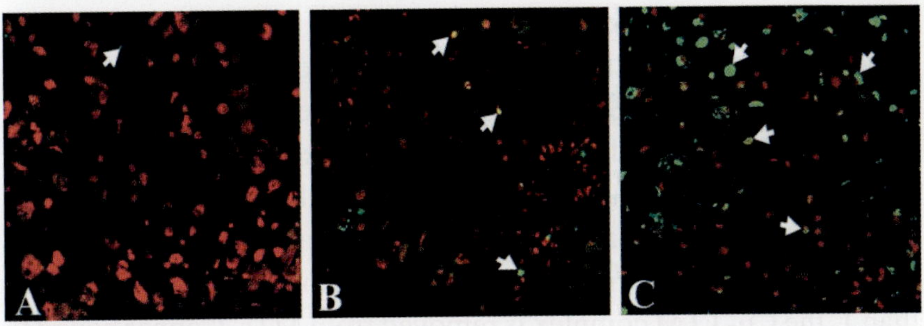

Fig. 5 Localization of apoptotic and necrotic cells (arrows) by TUNEL (TdT-mediated fluorescein dUTP nick end labeling) assay in hepatic tissues of (A) basal, (B) Zn-treated, and (C) untreated groups. Reproduced from Santon *et al.* (2003b).

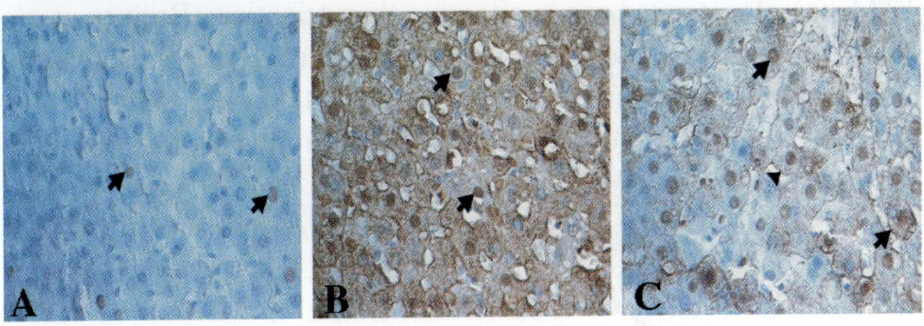

Fig. 6 Immunodetection of MTs (arrows) in hepatic tissues of (A) basal, (B) Zn-treated, and (C) untreated groups. Reproduced from Santon *et al.* (2003b).

2.1.3. *Tetrathiomolybdate*

Treating Wilson's disease patients with neurological symptoms can be challenging. A deterioration in neurological status may be seen during initial treatment with chelating drugs and zinc. Penicillamine and trientine are effective in binding copper, but this causes an initial increase in serum copper levels because large amounts of the metal are mobilized, and the neurons may suffer from this process. On the other hand, zinc takes 4–6 months to control the toxic effects of copper, during which time the disease may progress to some degree (Brewer *et al.*, 2006).

A new drug has consequently been developed to meet this need: tetrathiomolybdate (TTM), comprising one molybdenum atom and four sulfur atoms. TTM reacts with inorganic copper to form hetero-bimetallic complexes through Mo-S-Cu clusters. It has been demonstrated (Ogra *et al.*, 1996) that TTM can also remove copper bound to MTs, unlike the other drugs that only act on free copper. The selective copper-removal mechanism is explained by a dose-dependent affinity of TTM for copper: at a TTM/Cu molecular ratio lower than 1, TTM binds to copper to form a TTM-Cu-MT complex; at a ratio higher than 1, the Cu-TTM complex is removed from MTs. Copper bound to TTM is subsequently released into the bloodstream, where a complex is formed with albumin; this explains why copper removed by TTM is unavailable for cell uptake, so any neurological damage is avoided. TTM has also proved more effective than trientine (Brewer *et al.*, 2006) in Wilson's disease patients with neurological presentation, and it has recently emerged (Klein *et al.*, 2004) that TTM is also effective in Wilson's disease with a hepatic presentation, proving useful in treating acute hepatitis in LEC rats.

2.2. *Hemochromatosis*

Hemochromatosis is an inherited, autosomal recessive disease caused by mutations in the HFE gene, which regulates iron metabolism. HFE impairment coincides with an excessive intestinal absorption of iron and its consequent accumulation. Like copper, iron can be toxic in excessive amounts because of its power to promote the Fenton reaction and the production of ROS. The organs most involved in iron-induced damage are the liver, heart, pancreas, skin, joints, and hypophysis. The main clinical manifestations are hepatic, ranging from hepatomegaly to cirrhosis and hepatocarcinoma. The only therapeutic approach available at present is regular phlebotomy, which removes the excess iron from the bloodstream. Even liver transplantation is not a definitive cure because the site of the mutated protein is the bowel, so iron would also accumulate in the graft. Chronic iron overload can also occur in hematological diseases such as thalassemia, requiring several blood transfusions; in this case, the therapeutic strategy is administration of chelating agents, which lower iron storage.

We could not find any work specifically assessing the potential role of MTs in hemochromatosis in the available literature, but a study by Formigari *et al.* (2007) mentioned several discoveries that could pave the way to future developments. They studied the effect of zinc supplementation on rat hepatoma cells treated with iron, finding that zinc supplementation reduced iron content in the cells and induced MT synthesis. It has consequently been argued that zinc could have beneficial effects in two ways: by exercising a competitive inhibitory effect on iron uptake by the divalent metal transporter, and by increasing MT concentration. The latter would protect cells against iron-driven injury by scavenging free radicals produced in Fenton reactions. Further studies, either in animal models or in humans, will better elucidate the role of MTs in iron toxicity and could provide the basis for new treatment strategies in hemochromatosis.

2.3. *Cholestatic Liver Diseases*

Wilson's disease is the perfect example of impaired copper excretion in the bile, so it is hardly surprising that there is copper overload in all diseases entailing a bile excretion defect. Several studies have reported high liver copper concentrations in cholestatic diseases such as primary biliary cirrhosis (PBC) and primary sclerosing cholangitis (PSC), both immune-mediated diseases, and in congenital biliary abnormalities such as biliary atresia (Mulder *et al.*, 1992; Hunziker and Sternlieb, 1991; Mulder *et al.*, 1991).

It has consequently been argued that MTs could have an important role in protecting against metal accumulation by binding the excessive amounts of copper. There is reportedly a higher concentration of MTs in livers with copper overload (Mulder *et al.*, 1992), even if this is not as prominent as in Wilson's disease. The same study (Mulder *et al.*, 1992) also determined plasma MT concentrations in cholestatic disease, finding them increased in the same manner.

In addition, MT concentrations were found to be correlated with the severity of PBC, so MTs may represent a tool for assessing disease progression.

The main therapeutic approach to cholestatic liver disease is to administer ursodeoxycholic acid (UDCA), which facilitates the progression of bile into the biliary tracts. It has also been suggested that UDCA can have other beneficial effects, such as protecting cell membranes from damage induced by hydrophobic bile acids, and modulating immune system functions. A Japanese group (Mitsuyoshi *et al.*, 1999) assessed whether UDCA could also protect against oxidative stress, and found that this was another of the drug's mechanisms of action via the induction of antioxidant substances. They evaluated the vulnerability of hepatocytes to H_2O_2 and cadmium with or without UDCA treatment; in the case of UDCA treatment, the lactate dehydrogenase (LDH) activity released from the cytosol of hepatocytes — a sensitive marker of cell damage — was significantly lower than in control cells, and this was dependent on increased GSH and MT synthesis. This explains why UDCA also has a beneficial effect in noncholestatic liver diseases like hepatitis B. Therefore, UDCA may arguably be useful in all diseases in which oxidative stress has a central role. Further studies are needed to explain which mechanisms are involved in UDCA-related MT induction.

3. Toxic Liver Diseases

The fact that MTs have a highly conserved structure within the species, are present in virtually all living organisms, and are ubiquitous in eukaryotes suggests that they have an important role in survival. An example of their indispensable function lies in their ability to protect against exogenous toxic agents. Because of its particular structure, with no aromatic amino acids and many cysteines, MT can bind and store metal ions and scavenge free radicals (the means by which many toxic substances cause damage). The best-known protective effect of MTs is against cadmium, but they have also proved capable of reducing the damage induced by organic chemicals such as CCl_4 (carbon tetrachloride), drugs such as acetaminophen, and metalloids such as arsenic.

3.1. *Cadmium*

Cadmium (Cd) (The University of Edinburgh, 2007) is a heavy metal with no essential biological function, but is extremely toxic to humans. It is widely used in industry, and may thus contaminate the soil and consequently also the vegetables that absorb it readily. It is a constituent of pigments, batteries, and metal coatings, and is found in cigarette smoke and vehicle fumes. Humans have a daily intake of cadmium by ingestion and inhalation. After intestinal or pulmonary absorption, Cd is carried to the liver, where it is complexed with MTs; the complex then reaches the kidney, being filtered at the glomerulus and reabsorbed in the proximal tubule, where it is stored. Cadmium toxicity in human cells depends on its ability to inhibit various enzyme systems and induce the uncoupling of oxidative phosphorylation in mitochondria. Cd poisoning has several clinical effects and the organs most involved are the kidney, lung, liver, and bone (osteomalacia). A carcinogenic effect of Cd has also been described in the lung and prostate.

Cadmium is responsible for hepatotoxicity in cases of both acute and chronic exposure. After acute exposure, Cd produces hepatocyte swelling, fatty changes, necrosis, and apoptosis with increased transaminases. After chronic Cd exposure, the necrosis is mild with little or no increase in transaminase levels, but connective tissue proliferation results in interstitial fibrosis (Habeebu *et al.*, 2000). The mechanisms behind the cause of Cd hepatotoxicity are still being investigated. Cd is mainly bound to MTs, and each MT molecule can trap seven Cd atoms.

3.1.1. *Role of MT in Acute Damage*

It has been demonstrated that MTs protect against Cd toxicity. Constitutive MT overexpression in transgenic animals is related to a decrease in the lethal and hepatotoxic effects of Cd (Liu *et al.*, 1995), whereas MT-null mice have been found very susceptible to such damage (Liu *et al.*, 1996). There is also evidence (Goering and Klaassen, 1984) that animals pretreated with low doses of Cd (which induces MT synthesis) suffer less from subsequent toxic doses of the metal.

3.1.2. *Role of MT in Chronic Damage*

In the study by Habeebu *et al.* (2000), wild-type and knock-out mice for MTs were treated with Cd for 10 weeks. MT-null mice had lower hepatic concentrations of Cd, but much more severe liver injury. The knock-out mice had the same lesions after about 1/10 of the dose given to wild-type mice, and their maximum tolerable dose was 1/8 of the one given to the wild-type animals. The higher hepatic concentrations of Cd in wild-type animals is thought to be due to MT binding and neutralizing of Cd. This mechanism protects the hepatocytes because MTs can sequester Cd in the cytosol, preventing its passage into critical organelles. The higher the hepatic concentration, the lower the systemic concentrations of Cd; thus, MTs can protect not only against hepatic toxicity, but also from other parenchymal toxicities, such as nephrotoxicity.

The fibrogenic effect of Cd was studied by del Carmen *et al.* (2002). Cd was found capable of inducing the synthesis of α_1 collagen and MT-II genes in hepatic stellate cells, in combination with an increase in oxidative stress. It was consequently postulated that the profibrogenic effect of Cd is related to oxidative damage, and the increase in MT concentration is probably an intrinsic defense mechanism against this exogenous insult.

3.2. *Arsenic*

Arsenic is a metalloid naturally present in the earth, but is also a byproduct of ore production and coal consumption. The main mechanism of arsenic-induced damage depends on its chemical form and is due to its ability to interact with cellular sulfhydryl groups in proteins, leading to functional alterations and GSH depletion.

In a mouse model, it was demonstrated (Liu *et al.*, 2000) that MT-null mice are more sensitive to arsenic-induced liver damage, revealing more frequent and more severe lesions. This does not seem to depend on apoptosis because wild-type and MT-null mice revealed a similar caspase-3 activity after arsenic administration. The protective effect of MT against arsenic toxicity is consequently thought to be mediated by the antioxidant properties of MTs. Similarly, MTs have been found to protect against acetaminophen- and CCl_4-induced liver damage (Liu *et al.*, 1999; Klaassen and Liu, 1998). Both substances take effect by inducing oxidative stress, and MT-null mice had more severe liver damage than wild-type mice after the administration of the two toxic substances.

3.3. *Alcoholic Liver Disease*

3.3.1. *Alcohol Toxicity*

Alcoholic liver disease (ALD) (Gramenzi *et al.*, 2006) develops in people who consume excessive amounts of alcohol. The extent of alcohol abuse and liver injury varies considerably, but the severity of disease does not always correlate with the quantitative alcohol intake: there is an interplay of other factors, such as genetics and the environment. The alcohol intake threshold for defining abuse is 40–80 g/day for males and 20–40 g/day for females.

ALD is a frequent cause of liver disease. In 1997, the age-adjusted death rate from alcohol-induced liver disease in the United States was 3.8 per 100 000, which corresponded to 28% of all deaths from liver diseases (Sass and Shaikh, 2006). Alcohol-induced liver injury can take a wide range of clinical forms, from fatty liver to cirrhosis, via alcoholic steatohepatitis (ASH). It is estimated that the majority of heavy drinkers have a fatty liver, though only 10%–35% of them develop hepatitis and 8%–20% of the latter progress to cirrhosis.

The mechanisms of alcoholic liver injury are numerous and well known (Gramenzi *et al.*, 2006):

- a reduced mitochondrial respiratory chain activity with an impaired liver cell energy metabolism;
- oxidative damage: oxidation of ethanol to water and carbon dioxide occurs through the production of toxic metabolites such as acetaldehyde, which is a reactive substance responsible for the formation of protein adducts, for lipid peroxidation and nucleic acid oxidation. Further acetaldehyde metabolism is accompanied by an imbalance in the NADH/NAD ratio: the reduced state of NAD facilitates the transfer of electrons to oxygen, generating reactive species such as superoxide anions. Finally, alcohol induces the CYP2E1 pathway, which is associated with an excessive production of superoxide radicals via interaction with cytochrome reductase;
- immune response: acetaldehyde can interact with proteins, leading to the formation of adducts that act as neoantigens. Alcohol intake has also been associated with an impaired intestinal barrier against numerous substances, e.g. bacterial lipopolysaccharide (LPS), as demonstrated by an altered intestinal permeability to sugar probes. LPS is a nonspecific T-cell activator, increasing the production of cytokines like TNF-α, IL-6, and TGF-β. These mediators have been found involved in the mechanisms of hepatic fibrosis, inflammation, necrosis, and apoptosis.

3.3.2. *Role of MT in ALD*

Mice overexpressing MT display a significant resistance to acute alcoholic liver damage, with lower levels of lipid and protein peroxidation as well as oxidized glutathione (Kang and Zhou, 2005). On the other hand, MT-null mice reveal exactly the opposite picture, with greater oxidative damage and early degenerative morphological changes and hepatocytes that look necrotic (Zhou *et al.*, 2002a). These negative effects were suppressed by Zn supplementation; since we know that MT is required to maintain high zinc levels in the liver, we can therefore assume that the protective role of MT in ALD is mediated by zinc. Studies on animal models have shown that zinc supplementation prevents acute and chronic alcohol-induced liver injury (Zhou *et al.*, 2002b). This is mediated by many hepatic and extrahepatic effects of zinc. Ethanol abuse has been found to increase Zn excretion in urine and reduce Zn absorption from intestine, leading to Zn deficiency.

Zn deficiency contributes to hepatic alcoholic damage by the following mechanisms:

- causing a shift in the metabolic pathway of alcohol: Zn deficiency coincides with a loss of ALDH (aldehyde-dehydrogenase) activity, with an increase in the CYP2E1 pathway — the system with the highest output of ROS;
- increasing the amount of apoptosis and TNFα production; and
- altering intestinal permeability, with a greater risk of endotoxemia, which is a well-recognized factor in alcohol-induced liver injury.

In clinical practice, studies are needed to examine the potential therapeutic use of zinc in the prevention and treatment of ALD.

3.4. *Nonalcoholic Fatty Liver Disease*

It is estimated that 17%–33% of the population in the USA have nonalcoholic fatty liver disease (NAFLD) (McCullough, 2006). Although fatty liver can be caused by drugs and surgical procedures, the main risk factors associated with this disease are obesity and insulin resistance. Though it is a common and benign condition, NAFLD has the potential to evolve into chronic hepatitis (nonalcoholic steatohepatitis or NASH, 30% of cases) and cirrhosis (15%–20% of NASH cases). Insulin resistance gives rise to hyperinsulinemia and an increased flux of free fatty acids (FFAs) into the liver. Insulin resistance has a paradoxical effect in the liver: while there is a systemic resistance to this hormone, it retains its effects in the liver, promoting lipogenesis.

Thus, NAFLD is a result of the combination of an increased FFA flux and lipogenesis.

The evolution of NAFLD into NASH is mediated by oxidative stress, which has three potential etiologies (McCullough, 2006):

- hyperinsulinemia: insulin has many effects that are often overlooked, such as stimulating ROS production (a profibrogenic effect due to its ability to induce connective tissue growth factors) and causing endoplasmic reticulum stress with unfolding protein response and apoptosis;
- lipid peroxidation: FFAs can be oxidized in mitochondria, peroxisomes, and microsomes, each of the three pathways being able to generate ROS and subsequently induce lipid peroxidation; and
- hepatic iron accumulation: the role of iron in NAFLD is still unclear, but there are findings to suggest its importance, such as the fact that many NAFLD patients have hyperferritinemia and that phlebotomy has improved liver histology in NAFLD cases. Iron is a well-recognized pro-oxidant, after all, so it is hardly surprising that it should have a role in the onset of NASH.

3.4.1. *Role of MT in NASH*

The importance of oxidative stress as the causative factor in NASH suggests a potential role for MTs in these processes. A possible link between NASH and ASH in the role of MTs also emerges from a recent finding that leptin resistance relates to the suppression of MTs, which leads a more severe alcoholic liver damage (Tomita *et al.*, 2004).

It is well established that obesity is a risk factor for the progression of ALD in humans, and leptin deficiency/resistance is thought to be an important dysregulatory mechanism involved in the pathogenesis of steatosis and fat accumulation. Leptin-deficient ob/ob mice and leptin-receptor-deficient Zucker fa/fa rats are known to be obese and insulin-resistant, and they are often used to study the mechanisms behind the progression from steatosis to steatohepatitis. The study by Tomita *et al.* (2004) showed that leptin-receptor-deficient rats were more susceptible to alcoholic liver damage, and this was associated with a lower concentration and synthesis of MTs. One explanation could be that alcohol-induced oxidative stress in a liver already damaged by fat accumulation is even more toxic if the organism is short of natural protectors like MTs. Why the leptin system impairment leads to a lower MT concentration remains to be seen, but the findings of this study could provide the basis for future developments in our understanding of NAFLD and in new therapies such as zinc, which — by increasing the MTs — could have beneficial effects not only in alcoholic, but also in nonalcoholic, liver disease.

4. Chronic Viral Hepatitis

4.1. *Chronic Hepatitis C*

The hepatitis C virus (HCV), first discovered in 1989, infects approximately 179 million individuals worldwide, or 3% of the world's population (Batey, 2007). It only infects humans and is transmitted essentially by blood exchange. Chronic liver infection has been estimated to be responsible for approximately 250 000–350 000 deaths/year, mostly due to cirrhosis, end-stage liver disease, and hepatocellular carcinoma. HCV infection frequently becomes a chronic disease, with 50%–80% of the people infected developing chronic hepatitis. There are many host and viral factors responsible for the persistence of the infection without clearance of the virus. One is the virus's ability to modulate innate and adaptive immune responses, and another is the great variability of viral antigens, so the virus can always escape the immune system. The natural history of chronic hepatitis C involves progression to cirrhosis over 20–30 years in 20% of cases.

Current standard treatment (Chevaliez and Pawlotsky, 2007) for chronic hepatitis C is a combination of pegylated interferon-alpha (IFN-α) and ribavirin, which both inhibit virus replication. Response to therapy depends largely on the virus genotype: it is approximately 40% for genotype 1; 80% for genotype 2 or 3; and still unknown for genotypes 4, 5, and 6.

The mechanisms of liver damage (Pawlotsky, 2004) after HCV infection are not well known, but there is evidence to suggest that the virus itself does not cause any damage. In fact, there is no relationship between viral load and the severity and progression of liver disease; the only lesion directly due to the virus is steatosis, found especially in cases of HCV genotype 3. Almost all of the parenchymal damage is due to the immune response to the virus, in which central pathogenic roles are played by oxidative stress and apoptosis.

It has been demonstrated that chronic inflammation and several viral proteins trigger an oxidative burst in human cells, and that apoptosis of liver cells infected by HCV is an important host defense mechanism against the infection. MTs have a central role in these two pathogenic mechanisms (Carrera *et al.*, 2003).

4.1.1. *MT and Oxidative Stress Induced by HCV*

MT is one of the most powerful hydroxyl radical scavengers, given its high content of cysteine residues and its role in zinc metabolism — a trace element itself responsible for protection against oxidative stress. In chronic HCV infection, ROS are produced in several ways: first, chronic inflammation leads to a chronic release of inflammatory mediators by cells of the immune system, including cytokines, lysosomal enzymes, and ROS; next, the virus itself contains proteins like NS3, which can trigger oxidative bursts, and proteins that can alter mitochondrial function, whose impairment leads to increased ROS production. MTs can block these mechanisms, neutralizing the free radicals produced.

4.1.2. *MT and Apoptosis in HCV Infection*

Apoptosis is a host's defense mechanism against viral infection that eliminates the infected cells without promoting viral diffusion. It has been demonstrated that NF-κB has a crucial role in inhibiting liver cell apoptosis mediated by HCV, and MTs have been found to decrease NF-κB activation; thus, MTs indirectly promote apoptotic pathways.

4.1.3. *MT in Chronic Hepatitis C*

A Spanish study (Carrera *et al.*, 2003) analyzed MT concentration and expression in patients with chronic hepatitis C. It was found that HCV patients had a significantly lower liver MT concentration and expression than controls; and, more to the point, MT concentrations correlated inversely with the disease's histological activity index, necroinflammatory score, and fibrosis score. Therefore, it seems clear that MT could play a part in the progression and severity of HCV-induced liver disease. The same study also showed that response to interferon therapy depended to some degree on MT; in fact, patients responding better to this treatment had higher hepatic concentrations of MTs.

A Japanese group (Takagi *et al.*, 2001) performed a trial in which they added zinc supplementation to the usual therapy with interferon (IFN). The result was that patients treated with IFN + Zn had a lower viral load and lower aminotransferase levels than patients treated with interferon alone. Zinc may have various beneficial effects, from immunomodulation to its role in the dimerization of IFN. Zinc is also an undeniably strong inducer of MTs, however, and — given the above-described protective role of these proteins — it can be argued that the value of zinc in HCV hepatitis is partly thanks to MTs. Further studies are needed, in mouse models for example, to verify just how important MTs are in this process.

4.2. *Chronic Hepatitis B*

HBV infection is a very common disease affecting 5% of the world's population, especially in Asia and Africa (Pramoolsinsup, 2002). It is transmitted via sexual, parenteral, and vertical pathways. Exposure to HBV causes an acute infection, often without any symptoms. The chronicization rate varies considerably, depending on the subject's age: from 90% in the newborns of infected mothers to 10% in immunocompetent adults. A vaccine protecting against HBV has been available in Italy since 1991, and has led to a decline in the rate of new infections. Chronic HBV infection is associated in 50%–75% of cases with active virus replication and chronic hepatitis, and chronic hepatitis is a strong risk factor for cirrhosis and hepatocellular carcinoma.

HBV is considered (Baumert *et al.*, 2007) an essentially non-cytopathic virus, though some variants have been associated with apoptosis induced by the intracellular accumulation of viral DNA or proteins, which can have a toxic effect. Nevertheless, there is ample consensus that the host's immune response determines the onset and progression of liver disease. The main cells involved in immune response against HBV are $CD8^+$ T cells, which recognize HBV-infected cells and can react in two ways: a cytopathic reaction, leading to cell death and inflammation (i.e. hepatitis); and a reaction driven by IFNγ and TNFα, in which viral replication is inhibited without cell damage. HBV chronicization is believed to depend on host factors (mainly T-cell dysfunction) and viral factors, especially the ability to escape the immune system by producing viral variants.

4.2.1. *Role of MT in Chronic Hepatitis B*

Chronic liver disease, cirrhosis, and hepatocellular carcinoma are a consequence of chronic inflammation, which drives cell damage by producing toxic endogenous substances like free radicals. Given the importance of oxidative stress in chronic inflammation, it is of interest to study the role of MTs in chronic hepatitis B. The available literature contains only one published study on this topic (Quaife *et al.*, 1999), in which two mouse models were considered: one transgenic, constitutively expressing HbsAg (the model of chronic hepatitis B) and characterized by the development of hepatocellular carcinoma beyond 18 months of age; and one double transgenic, in which the

HbsAg expression was accompanied by an overexpression of MT-I. The double transgenic mice were found to have lower levels of aneuploidy and hyperplasia than the mice without overexpression of MTs. It is therefore argued that MTs protecting against free radicals could also have beneficial effects on liver damage caused by chronic inflammation in hepatitis B, and may protect against the onset of hepatocellular carcinoma.

5. Hepatocellular Carcinoma

Hepatocellular carcinoma (HCC) (Llovet *et al.*, 2004) is the fifth most common tumor in the world. It develops mainly in cirrhosis and is the leading cause of death in cirrhotic patients. In 80% of cases, HCC arises from a liver that is cirrhotic due to HCV infection or alcohol abuse. Current therapeutic strategies for HCC are as follows:

- resection: for patients with single asymptomatic HCC and preserved liver function;
- liver transplantation: for single HCC masses with a diameter <5 cm, or three HCC masses each <3 cm in diameter, in the absence of extrahepatic or vascular spread;
- percutaneous alcohol injection: for unresectable and early-stage HCC. This achieves a 70% response rate in solitary tumors with a diameter <3 cm;
- chemoembolization: used mainly for unresectable intermediate- or advanced-stage HCC. This technique consists of administering embolizing agents, e.g. gelatin, together with intra-arterial chemotherapy (doxorubicin, mitomicin, cisplatin) mixed with lipiodol. It achieves a partial response in 15%–55% of cases, and significantly delays tumor progression and vascular invasion.

5.1. *Role of MT in HCC*

Recent evidence shows that MTs play an important part in human tumors, being involved in carcinogenic and apoptotic processes and cell differentiation. MT concentrations can vary widely with specific tumor cell types, however, some having high and others low levels of this protein (Huang and Yang, 2002).

Several studies have attempted to assess MT concentrations in HCC (Huang and Yang, 2002; Waalkes *et al.*, 1996; Endo *et al.*, 2004; Kawata *et al.*, 2006). In all cases, the result was that MT concentrations were significantly lower in HCC livers than in control livers. It also emerged (Endo *et al.*, 2004) that MT levels were inversely correlated with tumor stage and differentiation, being the lowest in poorly differentiated and stage III–IV HCCs. More specifically, Kawata *et al.* (2006) reported that MT concentrations were not only lower in HCC livers than in control livers, but also lower in juxtatumoral tissue than in portions of the tumor. On differentiating by isoforms, they found that MT-I was especially responsible for the difference in concentration. The same authors (Nakamura *et al.*, 2004) tried to find specific MT isoform functions and found some evidence to suggest that MT-I relates to metabolism and detoxification of toxic metals, while MT-II is responsible for homeostasis of essential metals. High levels of copper have been found in HCC tissue (Ebara *et al.*, 2000); it may be that one of the mechanisms behind carcinogenesis in the liver relies on the downregulation of MT-I, giving rise to a greater amount of free copper, which is a promoter of oxidative stress. Given the role of MT in protecting against oxidative stress, a deficiency of those proteins might promote the activation of oncogenes or the inactivation of oncosuppressor genes. The exact role of copper and MTs in the activation and progression of carcinogenic events is not yet clear, however.

MTs also have a role in response to chemotherapy: because of their metal ion-chelating property, MTs can influence the effectiveness of metal-containing chemotherapeutic compounds, inhibiting their ability to reach intracellular DNA targets within tumor cells. Platinum agents can be used in HCC. It has been reported (Endo *et al.*, 2004) that response to platinum chemotherapy depended heavily on MT concentration, being very low in tumor cells expressing high MT levels (Fig. 7).

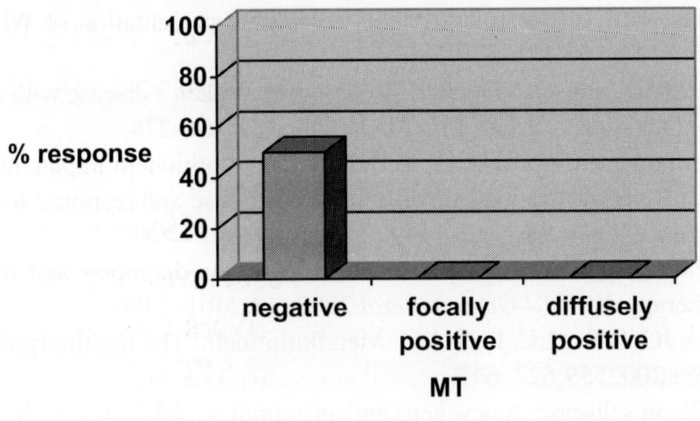

Fig. 7 Response to platinum chemotherapy and degree of MT staining in HCC. Study on 22 patients. Reproduced from Endo *et al.* (2004).

Another interesting study (Waalkes *et al.*, 1996) analyzed the action of Cd, a potential carcinogenic metal, on a mouse model of HCC (*N*-nitrosodiethylamine–induced tumor in B6C3F1 mice). It surprisingly found that Cd had a remarkable tumor-suppressive action because the incidence and growth of liver tumors were much reduced by Cd administration. Further analyzing hepatocytes undergoing Cd-induced necrosis showed that these cells were negative for MT immunoreaction, whereas the surviving surrounding tissue was positive. It was consequently assumed that the absence of MTs led to an extreme sensitivity to cadmium toxicity, thus selecting which neoplastic cells would be damaged. MT quantification in HCC tissue may thus help us to choose the right chemotherapeutic drug, avoiding platinum being administered in cases strongly expressing MTs. Further studies are needed to establish whether cadmium can be used as chemotherapy for HCC.

References

Batey RG. Controversies in and challenges to our understanding of hepatitis C. *World J Gastroenterol* 2007; **13**:4168–4176.

Baumert TF, Thimme R, von Weizsäcker F. Pathogenesis of hepatitis B virus infection. *World J Gastroenterol* 2007; **13**:82–90.

Brewer GJ, Askari F, Lorincz MT, *et al.* Treatment of Wilson's disease with ammonium tetrathiomolybdate: IV. Comparison of tetrathiomolybdate and trientine in a double-blind study of treatment of the neurologic presentation of Wilson's disease. *Arch Neurol* 2006; **63**:521–527.

Brewer GJ, Dick RD, Johnson VD, *et al.* Treatment of Wilson's disease with zinc: XV long-term follow-up studies. *J Lab Clin Med* 1998; **132**:264–278.

Carrera G, Paternain JL, Carrere N, *et al.* Hepatic metallothionein in patients with chronic hepatitis C: Relationship with severity of liver disease and response to treatment. *Am J Gastroenterol* 2003; **98**:1142–1149.

Chevaliez S, Pawlotsky JM. Hepatitis C virus: Virology, diagnosis and management of antiviral therapy. *World J Gastroenterol* 2007; **13**:2461–2466.

Coyle P, Philcox JC, Carey LC, Rofe AM. Metallothionein: The multipurpose protein. *Cell Mol Life Sci* 2002; **59**:627–647.

Cuthbert JA. Wilson's disease: A new gene and an animal model for an old disease. *J Investig Med* 1995; **43**:323–336.

del Carmen EM, Souza V, Bucio L, *et al.* Cadmium induces alpha(1)collagen (I) and metallothionein II gene and alters the antioxidant system in rat hepatic stellate cells. *Toxicology* 2002; **170**:63–73.

Ebara M, Fukuda H, Hatano R, *et al.* Relationship between copper, zinc and metallothionein in hepatocellular carcinoma and its surrounding liver parenchyma. *J Hepatol* 2000; **33**:415–422.

Endo T, Yoshikawa M, Ebara M, *et al.* Immunohistochemical metallothionein expression in hepatocellular carcinoma: Relation to tumor progression and chemoresistance to platinum agents. *J Gastroenterol* 2004; **39**:1196–1201.

Formigari A, Santon A, Irato P. Efficacy of zinc treatment against iron-induced toxicity in rat hepatoma cell line H4-II-E-C3. *Liver Int* 2007; **27**:120–127.

Goering PL, Klaassen CD. Tolerance to cadmium-induced hepatotoxicity following cadmium pretreatment. *Toxicol Appl Pharmacol* 1984; **74**:308–313.

Gramenzi A, Caputo F, Biselli M, *et al.* Review article: Alcoholic liver disease — Pathophysiological aspects and risk factors. *Aliment Pharmacol Ther* 2006; **24**:1151–1161.

Habeebu SS, Liu J, Liu Y, Klaassen CD. Metallothionein-null mice are more sensitive than wild-type mice to liver injury induced by repeated exposure to cadmium. *Toxicol Sci* 2000; **55**:223–232.

Huang GW, Yang LY. Metallothionein expression in hepatocellular carcinoma. *World J Gastroenterol* 2002; **8**:650–653.

Hunziker PE, Sternlieb I. Copper metallothionein in patients with hepatic copper overload. *Eur J Clin Invest* 1991; **21**:466–471.

Kang YJ, Zhou Z. Zinc prevention and treatment of alcoholic liver disease. *Mol Aspects Med* 2005; **26**:391–404.

Kato J, Kohgo Y, Sugawara N, *et al.* Abnormal hepatic iron accumulation in LEC rats. *Jpn J Cancer Res* 1993; **84**:219–222.

Kawata T, Nakamura S, Nakayama A, *et al.* An improved diagnostic method for chronic hepatic disorder: Analyses of metallothionein isoforms and trace metals in the liver of patients with hepatocellular carcinoma as determined by capillary zone electrophoresis and inductively coupled plasma-mass spectrometry. *Biol Pharm Bull* 2006; **29**:403–409.

Klaassen CD, Liu J. Induction of metallothionein as an adaptive mechanism affecting the magnitude and progression of toxicological injury. *Environ Health Perspect* 1998; **106**:S297–S300.

Klein D, Arora U, Lichtmannegger J, *et al.* Tetrathiomolybdate in the treatment of acute hepatitis in an animal model for Wilson's disease. *J Hepatol* 2004; **40**:409–416.

Komatsu Y, Ogra Y, Suzuki KT. Copper balance and ceruloplasmin in chronic hepatitis in a Wilson's disease animal model, LEC rats. *Arch Toxicol* 2002; **76**:502–508.

Liu J, Liu Y, Goyer RA, *et al.* Metallothionein-I/II null mice are more sensitive than wild-type mice to the hepatotoxic and nephrotoxic effects of chronic oral or injected inorganic arsenicals. *Toxicol Sci* 2000; **55**:460–467.

Liu J, Liu Y, Hartley D, *et al.* Metallothionein-I/II knockout mice are sensitive to acetaminophen-induced hepatotoxicity. *J Pharmacol Exp Ther* 1999; **289**:580–586.

Liu J, Liu Y, Michalska AE, *et al.* Metallothionein plays less of a protective role in cadmium-metallothionein-induced nephrotoxicity than in cadmium-chloride-induced hepatotoxicity. *J Pharmacol Exp Ther* 1996; **276**:1216–1223.

Liu Y, Liu J, Iszard MB, *et al.* Transgenic mice that overexpress metallothionein-I are protected from cadmium lethality and hepatotoxicity. *Toxicol Appl Pharmacol* 1995; **135**:222–228.

Llovet JM, Fuster J, Bruix J; Barcelona-Clinic Liver Cancer Group. The Barcelona approach: Diagnosis, staging, and treatment of hepatocellular carcinoma. *Liver Transpl* 2004; **10**:S115–S120.

McCullough AJ. Pathophysiology of nonalcoholic steatohepatitis. *J Clin Gastroenterol* 2006; **40**:S17–S29.

McQuaid A, Lamand M, Mason J. The interactions of penicillamine with copper *in vivo* and the effect on hepatic metallothionein levels and copper/zinc distribution: The implications for Wilson's disease and arthritis therapy. *J Lab Clin Med* 1992; **119**: 744–750.

Medici V, Rossaro L, Sturniolo GC. Wilson's disease — A practical approach to diagnosis, treatment and follow-up. *Dig Liver Dis* 2007; **39**:601–609.

Medici V, Trevisan CP, D'Incà R, *et al.* Diagnosis and management of Wilson's disease: Results of a single center experience. *J Clin Gastroenterol* 2006; **40**:936–941.

Mitsuyoshi H, Nakashima T, Sumida Y, *et al.* Ursodeoxycholic acid protects hepatocytes against oxidative injury via induction of antioxidants. *Biochem Biophys Res Commun* 1999; **263**:537–542.

Mulder TP, Janssens AR, Verspaget HW, Lamers CB. Plasma metallothionein concentration in patients with liver disorders: Special emphasis on the relation with primary biliary cirrhosis. *Hepatology* 1991; **14**:1008–1012.

Mulder TP, Janssens AR, Verspaget HW, *et al.* Metallothionein concentration in the liver of patients with Wilson's disease, primary biliary cirrhosis, and liver metastasis of colorectal cancer. *J Hepatol* 1992; **16**:346–350.

Nakamura S, Kawata T, Nakayama A, *et al.* Implication of the differential roles of metallothionein 1 and 2 isoforms in the liver of rats as determined by polyacrylamide-coated capillary zone electrophoresis. *Biochem Biophys Res Commun* 2004; **320**: 1193–1198.

Nartey NO, Frei JV, Cherian MG. Hepatic copper and metallothionein distribution in Wilson's disease (hepatolenticular degeneration). *Lab Invest* 1987; **57**:397–401.

Ogra Y, Ohmichi M, Suzuki KT. Mechanisms of selective copper removal by tetrathiomolybdate from metallothionein in LEC rats. *Toxicology* 1996; **106**:75–83.

Pawlotsky JM. Pathophysiology of hepatitis C virus infection and related liver disease. *Trends Microbiol* 2004; **12**:96–102.

Pramoolsinsup C. Management of viral hepatitis B. *J Gastroenterol Hepatol* 2002; **17**: S125–S145.

Quaife CJ, Cherne RL, Newcomb TG, *et al.* Metallothionein overexpression suppresses hepatic hyperplasia induced by hepatitis B surface antigen. *Toxicol Appl Pharmacol* 1999; **155**:107–116.

Santon A, Irato P, Medici V, *et al.* Effect and possible role of Zn treatment in LEC rats, an animal model of Wilson's disease. *Biochim Biophys Acta* 2003a; **1637**:91–97.

Santon A, Sturniolo GC, Albergoni V, Irato P. Metallothionein-1 and metallothionein-2 gene expression and localisation of apoptotic cells in Zn-treated LEC rat liver. *Histochem Cell Biol* 2003b; **119**:301–308.

Sass DA, Shaikh OS. Alcoholic hepatitis. *Clin Liver Dis* 2006; **10**:219–237.

Sturniolo GC, Mestriner C, Irato P, *et al.* Zinc therapy increases duodenal concentrations of metallothionein and iron in Wilson's disease patients. *Am J Gastroenterol* 1999; **4**:334–338.

Takagi H, Nagamine T, Abe T, *et al.* Zinc supplementation enhances the response to interferon therapy in patients with chronic hepatitis C. *J Viral Hepat* 2001; **8**:367–371.

Tapiero H, Tew KD. Trace elements in human physiology and pathology: Zinc and metallothioneins. *Biomed Pharmacother* 2003; **57**:399–411.

Tapiero H, Townsend DM, Tew KD. Trace elements in human physiology and pathology. Copper. *Biomed Pharmacother* 2003; **57**:386–398.

The University of Edinburgh. Faculty of Medicine. Electronic Medical Curriculum Portfolio. Available at www.portfolio.mvm.ed.ac.uk/studentwebs/session2/group29/cadtox.htm/. Downloaded on Nov. 29, 2007.

Tomita K, Azuma T, Kitamura N, *et al.* Leptin deficiency enhances sensitivity of rats to alcoholic steatohepatitis through suppression of metallothionein. *Am J Physiol Gastrointest Liver Physiol* 2004; **287**:G1078–G1085.

Waalkes MP, Diwan BA, Rehm S, *et al.* Down-regulation of metallothionein expression in human and murine hepatocellular tumors: Association with the tumor-necrotizing and antineoplastic effects of cadmium in mice. *J Pharmacol Exp Ther* 1996; **277**: 1026–1033.

Yuzbasiyan-Gurkan V, Grider A, Nostrant T, *et al.* Treatment of Wilson's disease with zinc: X. Intestinal metallothionein induction. *J Lab Clin Med* 1992; **120**:380–386.

Zhou Z, Sun X, Kang YJ. Metallothionein protection against alcoholic liver injury through inhibition of oxidative stress. *Exp Biol Med (Maywood)* 2002a; **227**:214–222.

Zhou Z, Sun X, Lambert JC, *et al.* Metallothionein-independent zinc protection from alcoholic liver injury. *Am J Pathol* 2002b; **160**:2267–2274.

Index